MIXING

THEORY AND PRACTICE

Volume II

Contributors to This Volume

P. H. CALDERBANK

CURTIS W. CLUMP

H. F. IRVING

RICHARD P. LEEDOM

EMERSON J. LYONS

JON H. OLSON

NORMAN H. PARKER

R. L. SAXTON

LAWRENCE E. STOUT, Jr.

MIXING

THEORY AND PRACTICE

EDITED BY

VINCENT W. UHL

DEPARTMENT OF CHEMICAL ENGINEERING
UNIVERSITY OF VIRGINIA
CHARLOTTESVILLE, VIRGINIA

JOSEPH B. GRAY

ENGINEERING DEPARTMENT
E. I. DU PONT DE NEMOURS & COMPANY, INC.
WILMINGTON, DELAWARE

VOLUME II

1967

ACADEMIC PRESS   New York and London

CHEMISTRY

ACADEMIC PRESS INC.
111 Fifth Avenue, New York, New York 10003

United Kingdom Edition published by
ACADEMIC PRESS INC. (LONDON) LTD.
Berkeley Square House, London W.1

LIBRARY OF CONGRESS CATALOG CARD NUMBER: 65-26039

PRINTED IN THE UNITED STATES OF AMERICA

List of Contributors

Numbers in parentheses indicate the pages on which the authors' contributions begin.

P. H. CALDERBANK, University of Edinburgh, Edinburgh, Scotland (1)

CURTIS W. CLUMP, Department of Chemical Engineering, Lehigh University, Bethlehem, Pennsylvania (263)

H. F. IRVING, Baker Perkins Inc., Saginaw, Michigan (169)

RICHARD P. LEEDOM, E.I. du Pont de Nemours & Company, Wilmington, Delaware (287)

EMERSON J. LYONS, General American Transportation Corporation, Chicago, Illinois (225)

JON H. OLSON, University of Delaware, Newark, Delaware (115)

NORMAN H. PARKER, Croton-on-Hudson, New York (287)

R. L. SAXTON, E. I. du Pont de Nemours & Company, Wilmington, Delaware (169)

LAWRENCE E. STOUT, JR., Monsanto Company, St. Louis, Missouri (115)

Preface

Mixing is a widely practiced operation; it occurs whenever fluids are moved in the conduits and vessels of laboratory and industrial-processing equipment. Mixing is of interest not only when it results in the dispersion of one component in another but also when it is an agency for the promotion of heat transfer, mass transfer, solid suspension, and reaction.

Although much has been published on the theory and practice of mixing, these writings are spread throughout the literature. This situation calls for a work devoted to organizing, summarizing, and interpreting this substantial mine of source material. A book which provides a complete and practical summary of mixing knowledge should save hours of literature searching and review by research workers and students. Such a book should also markedly improve the application of mixing knowledge by development, design, and operating engineers.

A dozen authors have cooperated in the writing of this treatise to meet these needs. Their interests cover most of the theoretical and practical aspects of various mixing operations, which range from the statistical theory of turbulence to construction details of various types of equipment. Each chapter has its special flavor and emphasis which reflect the kinds of problems involved in the various mixing operations and the viewpoints and insights of the authors. The extent to which a fundamental approach is used to relate the process variables differ among the mixing operations discussed. The theoretical relationships are inherently more complex in some cases than others. For example, the mathematics of tensors is required for understanding the turbulent behavior of fluids, while mathematically simple rate equations are adequate for some heat and mass transfer operations. In most cases, correlations of dimensionless groups provide practical relationships among the process variables.

The subject matter of this book has been divided into two volumes because a single volume would be too bulky and awkward for the reader to handle. In the first volume, the chapters deal with mixing in turbulent flow, the power consumption of rotating impellers, the mixing process in vessels, and mechanically aided heat transfer. In the second volume, the subject areas for the chapters are mass transfer in two-phase systems, the effects of mixing on chemical reactions, the mixing of highly viscous materials, the suspension of particles in liquids, the mixing of dry solid particles, and the mechanical design of impeller-type mixers.

The editors wish to acknowledge the efforts of those people who have contributed to the successful completion of this book—foremost, of course, are the authors of the many chapters and their patient, understanding wives. The assistance of others who helped shape the work by their comments, criticisms, and suggestions is also acknowledged. Those who helped especially in this way are L. C. Eagleton, S. I. Atallah, E. D. Grossmann, and H. E. Grethlein.

VINCENT W. UHL
JOSEPH B. GRAY

April, 1967

Contents

Chapter 6. MASS TRANSFER

P. H. Calderbank

Chapter 7. MIXING AND CHEMICAL REACTIONS

Jon H. Olson and Lawrence E. Stout, Jr.

Chapter 8. MIXING OF HIGH VISCOSITY MATERIALS

H. F. Irving and R. L. Saxton

Chapter 9. SUSPENSION OF SOLIDS

EMERSON J. LYONS

Chapter 10. MIXING OF SOLIDS

CURTIS W. CLUMP

Chapter 11. MECHANICAL DESIGN OF IMPELLER-TYPE LIQUID MIXING EQUIPMENT

RICHARD P. LEEDOM AND NORMAN H. PARKER

Contents of Volume I

CHAPTER 6

Mass Transfer

P. H. Calderbank

University of Edinburgh
Edinburgh, Scotland

I. The Role of Dispersion in Mass Transfer

The mixing or dispersion of one phase in another phase with which the first is immiscible is important in many chemical engineering operations such as heat and mass transfer. Dispersion not only brings about a large increase in the interfacial area available for material or heat transfer but also places the fluids in a state of motion which serves to increase the specific rates of both of the above transfer processes.

In the design of two-phase contacting equipment, operating conditions are first chosen to secure favorable mass-transfer driving forces so that batch, cocurrent, or countercurrent operation is selected and the times of exposure of the two phases are determined. Having evaluated the minimum number of transfer units required for the proposed duty, from considerations of maximum driving force, attention must next be directed to the dispersion process whose effectiveness, in continuous flow systems, determines the height of a transfer unit. The latter, which should be minimal for economic design, is inversely dependent on the magnitudes of the mass-transfer coefficient and interfacial area and directly dependent on the flow velocity. Thus, if the flow velocity and output is to be high, the mass-transfer coefficient and interfacial area must be as large as possible. In multistage processes, efficiency in dispersion must be accompanied by a corresponding efficiency in phase separation since the effects of entrainment are often deleterious.

Dispersions of one fluid in another are produced by injecting one phase into the other as a jet or sheet when surface tension forces cause the latter to collapse into a dispersion of drops or bubbles. Fluid dispersions produced in this way may be further reduced in particle size if they are then subjected to shear or turbulent forces as a result of the subsequent flow of the dispersion. Very high rates of fluid flow through orifices can induce dispersion by shear as the particles of the dispersed phase are propelled at high velocity through the continuous phase. High velocities can also be developed for the same purpose by centrifugal force in spinning disk atomizers. A dispersion may be mechanically agitated or sheared by high velocity flow in a conduit to reduce its particle size through the action of turbulence.

These techniques are used to produce high interfacial areas in contacting immiscible fluids for mass-transfer operations such as distillation, gas absorption, spray drying, and liquid-liquid extraction. The dissolution and crystallization of solids in liquids is facilitated by suspending solid particles in a fluid stream as a dispersion of high interfacial area while solids in gas dispersions are used in several catalytic gas reactions and in the combustion of solid fuels. The mixing or interdispersion of immiscible phases is thus an important and common operation designed to produce high interfacial areas for mass transfer, chemical reaction, or both.

Dispersion by mechanical agitation is resorted to when very high rates of transfer, unattainable by other means, are demanded. Even when the process requires many stacked mixing stages it has sometimes proved economical to employ mechanical agitation as in rotary and centrifugal extraction equipment. In gas-liquid contacting, plate columns employ the principle of dispersion through orifices and the vapor flow rate is generally high enough to secure a thorough agitation of the liquid phase at the same time as a large interfacial area is presented. In fermentation processes, the gas flow rate is relatively low and mechanical agitation is used to bring about better contacting.

Some principles of efficient contacting in continuous dispersed-phase flow may be seen by consideration of the relationship

$$a = \frac{6H}{D_p}$$

where a = interfacial area per unit volume of mixed phases,
H = fractional volumetric hold-up of dispersed phase, and
D_p = equivalent, mean spherical diameter of one drop or bubble of the dispersed phase

The interfacial area may be increased by increasing the hold-up or by reducing the bubble or drop diameter. For a fine-grained dispersion in which the bubbles or drops obey Stoke's law (S9), $v_t \propto D_p^2$, where v_t is the velocity of rise or fall of the sphere. Moreover $H \propto 1/v_t$, for a fixed dispersed-phase flow rate, so that $a \propto 1/D_p^3$. These arguments, which favor the use of fine-grained dispersions, have to be scrutinized in the light of the energy necessary to produce them. Also it is known that the fluid comprising small drops or bubbles is essentially static, while that comprising larger drops or bubbles is in a state of internal circulation and external oscillation which serves to increase the mass-transfer coefficients. In contacting liquids with gases, the hold-up may be increased by increasing the gas flow rate through multiple orifices until the foam density reaches a value of about $0.35\rho_L$ where ρ_L is the liquid density. Under normal conditions, little further increase in hold-up can be realized, this critical condition corresponding fairly closely with what would be expected from the closest packing of equal spheres. However, with liquids of low surface tension, for example, detergent solutions, fire-fighting foam, and ore flotation fluids, much lower foam densities may be realized. It is clear in these cases that the dispersed phase no longer exists as spheres but as stacked straight-sided bodies, probably rhombic dodecahedrons (H7). A similarly low density may be produced with liquids of high surface tension by aeration on specially designed sieve trays, as described later. All the evidence available shows that variation of the orifice diameter, of itself, does not solely determine the bubble size in gas-liquid contacting. The stable bubble size also depends on the intensity of turbulence existing in the continuous phase, and this level will of course be greater as the orifice diameter is decreased with consequent increase in energy dissipation, observed as pressure drop. With increasing viscosity of the continuous phase, increasing amounts of energy are absorbed by viscous dissipation and smaller amounts appear as turbulent fluctuations. Under these conditions, small bubbles, which may be initially formed from small orifices, may coalesce into larger ones as they move away from the orifices, their ultimate size being a function of the local turbulence intensity. The point which is emphasized is that the bubble size is not solely determined by conditions obtaining at its birth but also by the hydrodynamic field in which it subsequently finds itself. In conventional plate columns, the fraction of the plate area occupied by holes and slots and the size of aperture is found by experience and is a practical balance to secure a high level of turbulence in the continuous phase with an economic energy loss and a reasonably low entrainment. In fermentation equipment, mechanical agitation is used to create a turbulent field in which only small bubbles survive. Here it would be expected that the type of gas sparger used would be unimportant, and this seems form practical experience to be the case.

Liquid-liquid extraction columns employing orifice dispersers are operated at close to flooding to obtain a maximum practical value of the hold-up. Such equipment may be filled with baffles or packing or agitated or pulsed to create increased turbulence in the continuous phase. For packed towers it has been shown (P3) that a stable drop size is rapidly reached as the dispersed phase passes into the packing, this size being independent of that obtaining at the injection nozzles. This observation is in accord with the principle that the hydrodynamic field determines the stable drop or bubble size.

For rigorous design of contacting equipment, the foregoing qualitative observations, although a useful guide, will not by themselves suffice. Although far from complete, recent work has put such design methods on a more quantitative basis. A proper approach to these problems must take account of the physical properties of the dispersion and of the mass-transfer coefficients in the dispersed and continuous phases. In evaluating mass-transfer coefficients it is necessary to consider the internal circulation and oscillation of the particles of the dispersed phase and to recognize the effects of the presence of surface active materials. If complete information on the above factors were available, it would be theoretically possible to design equipment using the operating and equilibrium relationships for the system of interest. Even if operating and equilibrium lines were linear, such an analysis would be extremely laborious as the summation of resistances to diffusion within dispersed particles and through the continuous phase is not straightforward and involves much labor. If this difficulty were ignored, one would still have the problem of estimating "end effects" caused by the usually enhanced rates of mass transfer at the dispersing and coalescing sections of the equipment. These difficulties arise in contemplating the possibilities of a generalized solution although, in particular cases, a simpler situation capable of analytical treatment may exist. At the present juncture, however, one must be content to identify and estimate the effect of the important parameters which determine contacting efficiency. From such knowledge, some simplification of these apparently complex problems may be realized.

II. Measurement of the Physical Properties of Fluid Dispersions

It is necessary to have some knowledge of the physical properties of dispersions in order to understand mass transfer in dispersed systems. The pertinent properties are the interfacial area per unit volume of dispersion, the mean particle size, and the volume fraction of dispersed phase. Any two of these quantities are sufficient to define all three since they are interrelated as, $a = 6H/D_{S.M.}$ where $D_{S.M.}$ is the Sauter mean, or surface volume mean diameter defined as

$$D_{S.M.} = \frac{\Sigma_1^n n D^3}{\Sigma_1^n n D^2} = \frac{6H}{a} \tag{1}$$

The following discussion of these properties is confined to dispersions of relatively coarse particles ($> 50 \mu$ mean diameter). Finer grained dispersions than these are undesirable in continuous multistage mass-transfer processes because of their slow speed of phase separation, since the high mass-transfer rates obtained in dispersions can only be taken advantage of if they are followed by a comparably high rate of phase separation. The properties of fine-grained dispersion (mists, colloids, emulsions) present special problems which are outside the scope of this chapter.

A. INTERFACIAL AREAS

1. Optical Methods

Sauter (S1) first determined the interfacial areas of carburetor sprays by the use of a light-obscuration method in 1928. Vermeulen (V4) used light transmission to determine the interfacial areas of gas-liquid and liquid-liquid dispersions produced in mixing vessels and Rogers, *et al.* (R1), by using the same technique, made similar observations. Calderbank (C2) adapted Sauter's technique to measure interfacial areas in gas-liquid and liquid-liquid dispersions in mixing vessels and sieve-plate columns and in a later paper (C4) described a method for measuring the interfacial area of dense dispersions from their optical reflectivity.

All of these techniques make use of the light-scattering properties of dispersions. The simplest in application is the obscuration method used by Sauter (S1) and Calderbank (C2) in which, as described in the latter paper,

P. H. Calderbank

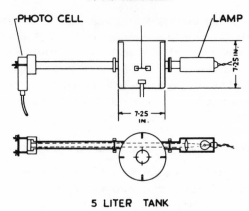

FIG. 1. Light-transmission probe in a small mixing vessel.

a parallel beam of light is passed through the dispersion onto a photocell placed some distance away from the scattering region, at the extremity of an internally blackened tube (Figs. 1 and 2). In this way, only the light which meets no

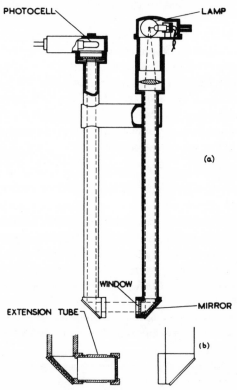

FIG. 2. Portable light-transmission probe for large mixing vessels.

obstruction in its path is received by the photocell and scattered light which is not coincident with the parallel beam is absorbed on the blackened tube. The light attenuation is evidently dependent on the projected area of the particles of the dispersion and if the particles are randomly dispersed, the fact that some particles lie in the shadow of others is accountable for as described below.

If A is the cross-sectional area of the light beam, this is progressively reduced by obscuration in its passage through the dispersion so that for an incremental path length, dl, containing particles of projected area, a_p, per unit volume of dispersion,

$$dA = Aa_p \, dl$$

or

$$d(\ln A) = a_p \, dl$$

and integrating,

$$\ln(A_0/A) = a_p l$$

and since the intensity of light reaching the photocell is proportional to the area illuminated,

$$\ln(I_0/I) = a_p l$$

It has also been shown (C13) that for dispersions of random particle shapes, if no concave surfaces are present, $a = 4a_p$ as in the case of spheres, whence

$$\ln \frac{I_0}{I} = \frac{al}{4} \tag{2}$$

Data which show the validity of Eq. (2) are given in Fig. 3.

This light-obscuration method is valid for dispersions of transparent, colored, or opaque particles in fluids which are sufficiently transparent to light to make measurement possible. It has also been used by Rose (R2) to determine the properties of powders dispersed in liquids. For dispersions of transparent particles, the minimum angle of the forward scattered light decreases as the particle size decreases, and to avoid scattered light reaching the photocell the latter is moved further away from the region of scattering so as to subtend a solid angle less than the minimum angle of forward scattering as shown in Fig. 9. The higher the refractive index ratio of the phases of the dispersion the larger the minimum angle of forward scattering and in practice it is a simple matter to adjust the position of the photocell so that it is sufficiently distant from the dispersion to record an intensity which is unaltered by placing it at a still more distant position. Calderbank (C2) connected the photocell to a light quantity meter and electric timer and measured the time for a given quantity of light to be received by the photocell when the light passed through the continuous phase alone (t_0) and when it passed through the dispersion (t). In this way, time-average interfacial areas could be determined in dynamically maintained dispersions,

$$\ln \frac{t}{t_0} = \frac{al}{4} \tag{3}$$

Interfacial areas up to 20 cm.$^{-1}$ were measured reliably, the optical path length being chosen to give conveniently measurable values of the time, t.

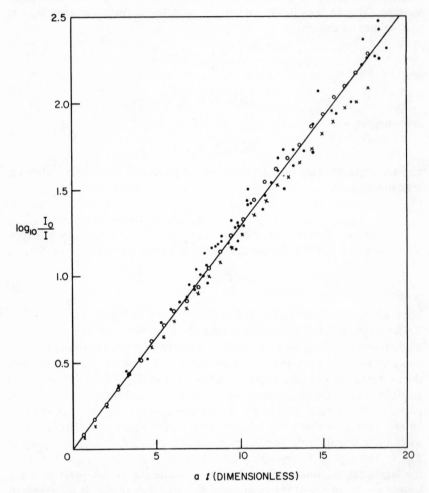

FIG. 3. Test of Eq. (2) with dispersion of polystyrene spheres and gas bubbles in water (C2). KEY:

● Air bubbles in water
× Polystyrene spheres in water, $D_p = 0.1$ mm.
○ Polystyrene spheres in water, $D_p = 0.75$ mm.

Neutral density filters were used to measure the time when the continuous phase only was present, this time being divided by the known filter factor to give the time, t_0. Calibration of the equipment with dispersions of known interfacial area established a high order of accuracy for the method which holds for values of $al < 25$. This latter criterion sets a limit to the values of interfacial area which can be measured since the optical path length cannot be reduced indefinitely as it must always include a statistically representative sample of the dispersion and although the cross-sectional area of the light beam can be made as large as desired, the optical path length must obviously be equal to an indefinite, though probably large, number of particle diameters. A further limitation of the method is experienced when the particle size is reduced below about 50 μ when the particle approaches the same order of size as the wavelength of light, and due to diffraction the scattering cross-sectional area of the particles is no longer the same as the geometric cross-sectional area. In these case, the interfacial area appearing in Eq. (3) has to be multiplied by a scattering factor, k. Values of k are given in Fig. 4.

Figure 4 was deduced from the electromagnetic theory of light by Mie (M8) and shows that as the particle size becomes large compared with the wavelength of light, the particle scattering cross section is twice its projected area. This result may also be deduced on the Babinette principle which states that a disk scatters by diffraction an amount of light equal to that incident upon it. However, for the comparatively large particles of interest in this discussion,

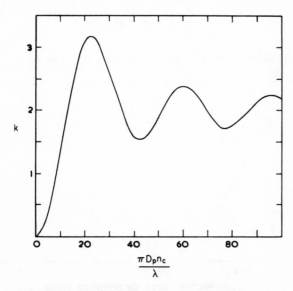

FIG. 4. Light-scattering factors for small particles for a refractive index ratio of 1.09, where k = scattering factor, D_p = particle diameter, n_c = refractive index of continuous phase, λ = wavelength of light in air (M8).

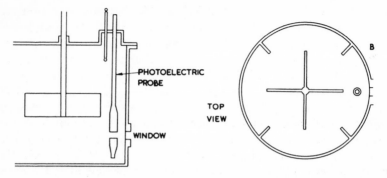

FIG. 5. Light-transmission probe (V4).

diffracted light is scattered at such a small forward angle that it is practically impossible to avoid its reception by a photocell placed even at vast distances from the dispersion. This fact, which has been discussed by Walton (W1), leads to an observed scattering factor of unity for large particles.

Vermeulen (V4) and Roger *et al.* (R1) placed their photocell close to the region of light scattering and consequently measured both directly transmitted and forward scattered light (Fig. 5). They calibrated the apparatus by observations using dispersions of known interfacial area, and Vermeulen concluded that for his apparatus

$$\frac{I_0}{I} - 1 = \beta a \qquad (4)$$

where β is a function of the refractive-index ratio (see Fig. 6).

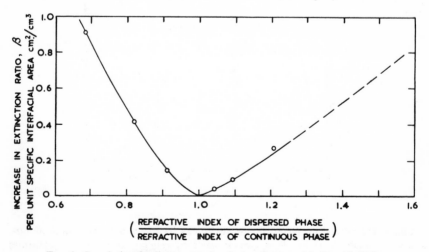

FIG. 6. Correlation for light-scattering factor, β, as used in Eq. (4) (V4).

Calderbank and Evans (C4) measured the optical reflectivity or backward scattered light from dispersions (Fig. 7), since this method is particularly suitable for optically dense dispersions, and concluded that

$$\frac{R_\infty}{R} - 1 = \frac{\beta'}{a} \qquad (5)$$

where β' is a constant, R is the intensity of light reflected back by the dispersion, and R_∞ is the intensity of light reflected back when the interfacial area, a, is infinite (as obtained by extrapolation of a plot of $1/R$ against $1/a$ to $1/a = 0$) (see Fig. 8). It was found that R_∞ was a function of the refractive index ratio and a method of predicting R_∞ for a given dispersion was presented.

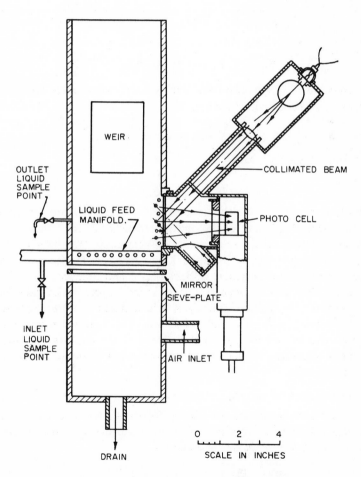

FIG. 7. Light reflectivity probe attached to sieve-plate column (C4).

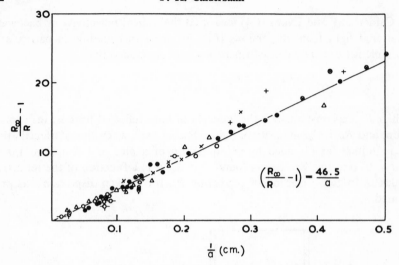

$$\left(\frac{R_\infty}{R} - 1\right) - \frac{46.5}{a}$$

$$\frac{1}{a} \text{ (cm.)}$$

Fig. 8. Correlation for R_∞ in Eq. (5) (C4).

Churchill *et al.* (C17) have given a theoretical treatment of light scattering by dense dispersions which reduces to

$$\frac{I_0}{I} - 1 = \beta'' \frac{al}{4} \tag{6}$$

for forward scattering, and

$$\frac{I_0}{I} - 1 = \frac{4}{\beta'' al} \tag{7}$$

for back scattering, where the constants β'' in Eqs. (6) and (7) are similar though different functions of the optical properties of the fluids.

These equations confirm the results of Vermeulen (V4) who used a fixed optical path length and of Calderbank and Evans (C4) who also employed a fixed geometrical arrangment of light source and detector.

Figure 9 illustrates the principles of the several optical techniques described in this chapter, for the case of a ray of light incident on a transparent sphere and undergoing reflection and refraction. The light-scattering lobes are drawn so that the intensity of light scattered at any angle is proportional to the length of the line at this angle from the center of the sphere to the periphery of the lobe.

2. Chemical Method

In a later part of this chapter [Eq. (155)] it is shown that when a gas combines with a liquid by a very rapid first-order irreversible reaction, the rate of gas absorption, r is given by the equation

$$\frac{r}{V_D} = C_i a \sqrt{k_1 D_L} \tag{8}$$

where C_i is the solubility of the gas in the exhausted liquid reagent, which is taken to be the constant concentration of gas at the gas-liquid interface during the absorption process; k_1 is the first-order chemical rate constant; D_L is the liquid-phase diffusivity of the gaseous reactant; and V_D is the volume of the gas-liquid dispersion. Thus, the rate constant of Eq. (8) is $\sqrt{k_1 D_L}$ and is independent of hydrodynamic factors. This fact makes it possible to determine the rate constant by measuring the rate of gas absorption in any convenient apparatus for which the area of contact between gas and liquid is known precisely, such as a wetted-wall or liquid-jet contactor. The rate constant so determined is then applicable to situations where the interfacial area is un-

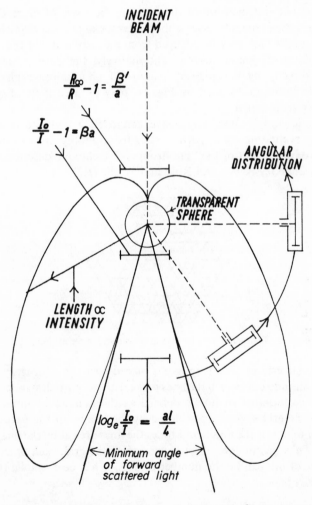

FIG. 9. Schematic representation of the light-scattering pattern from a transparent sphere.

known and can be used to determine the latter in, for example, aerated mixing vessels and bubble columns from measurements of gas absorption rates in these devices. Thus, if a gas-liquid reaction can be found to fulfill the requirements detailed in Section IV,C, and Eq. (8) is applicable, the interfacial area in dispersions of the gas in the liquid reagent may be found from measurements of the gas absorption rate in the dispersion and in a wetted-wall or liquid-jet apparatus. Westerterp (W4), by using oxygen gas and catalyzed aqueous solutions of sodium sulfite, has applied this technique to the measurement of interfacial areas in aerated mixing vessels as described later.

B. The Particle Size

The most relevant measure of the mean particle size of dispersions from the point of view of mass transfer is the Sauter mean particle diameter defined by Eq. (1), and which may be deduced from measurements of the interfacial area and dispersed-phase hold-up. Alternatively, the particle size may be evaluated directly by a statistical analysis of photomicrographs or high-speed flash photographs such as Fig. 11, may be used if the dispersion is dynamically maintained.

Calderbank and Rennie (C11) have employed a convenient and rapid method for evaluating mean particle sizes from photographs of dispersions which is a development of the "pin-dropping" technique described by Crofton (C20), Chalkley (C14), and Rose and Wyllie (R3).

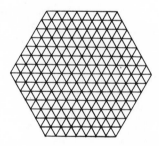

Fig. 10. Triangular grid for use in pin-dropping technique (C11).

A triangular grid of lines (Fig. 10) is placed over the photograph and the number of hits and cuts which the lines make with the particles are counted. In this sense, the number of hits is defined as the number of times the ends of lines on the grid are totally enclosed by the images of the particles while the number of cuts is the number of times the lines are cut by the images of the particles. If n is the number of lines comprising the grid, each of length, L, h the number of hits, and c the number of cuts, it has been shown (C11) that

$$D_{\text{S.M.}} = \frac{3Lh}{2c}; \qquad H = \frac{h}{2n}; \qquad a = \frac{2c}{nL} \qquad (9)$$

where H is the mean volume fraction of the dispersed phase and a is the interfacial area per unit volume of dispersion. Using the triangular grid, hits are thus recorded in groups of 6, 3, and 2, within the grid, on the sides of the triangle, and at the corners of the triangle, respectively.

Sullivan and Lindsey (S10) have measured the angular distribution of scattered light (see Fig. 9) in a fine-grained liquid-liquid dispersion, and applied the theory developed by Sloan (S6) to the evaluation of the average particle size and the distribution of sizes. The apparatus used was similar to that described by Stein and Keane (S7) and a sample of the dispersion, generated in a mixing vessel, was continuously pumped through the cell of the light-scattering assembly. The need to sample in this manner clearly introduces some uncertainty in the method, if further dispersion or coalescence in the pump and sample lines can be anticipated. The principal conditions which must be fulfilled in using this technique are that the particle diameter must be in the region of 0.2 to 200 μ and the transmission of the incident light beam must be between 40 and 80% of the incident intensity. Under these conditions, the intensity of the light scattered at a given angle is a function of the size of the particle causing scattering.

C. THE DISPERSED-PHASE HOLD-UP

With solid-fluid and liquid-liquid dispersions there is generally no difficulty in measuring the volume fraction of the dispersed phase. In batch operations, the dispersed-phase hold-up is determined by the proportions of the two phases charged, whereas in continuous flow operations samples of the dispersion have to be taken and the hold-up measured after phase separation.

With gas-liquid dispersions of bubbles, the gas hold-up has been determined by sampling the dispersion into an evacuated receiver or by measuring the volume of the dispersion and that of the liquid after phase separation. Calderbank and Rennie (C11) have employed a gamma-ray transmission method for measuring the hold-up in sieve-tray distillation columns (Fig. 11) which permits accurate measurements at varying levels in a foam. Further applications of the same technique are described by Cameron (C12).

The principles of the application of gamma-ray absorption to density measurements depend on the use of the relationship:

$$\ln \frac{I_0}{I} = \alpha \rho l \tag{10}$$

where α is the mass absorption coefficient, values of which for most atoms have been published (C12), ρ is the density to be measured, and l is the radiation path length. Typical data are shown in Fig. 12 and Table I.

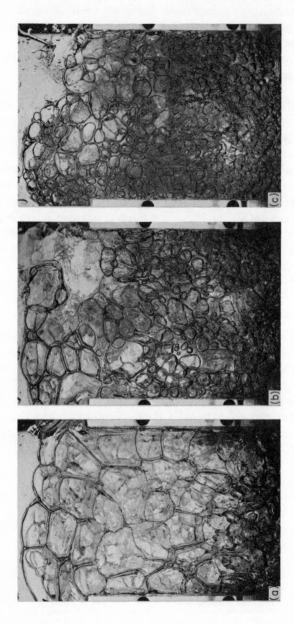

Fig. 11. Flash photograph of gas dispersion on a sieve plate showing transition from foam to froth. System: air-water, 4 by $3\frac{1}{2}$ in. sieve plate. Fifty-three $\frac{1}{8}$-in. diam. holes. (a) $N_{Re} = 2200$ (for $d = \frac{1}{8}$ in.); (b) $N_{Re} = 2800$; and (c) $N_{Re} = 3600$.

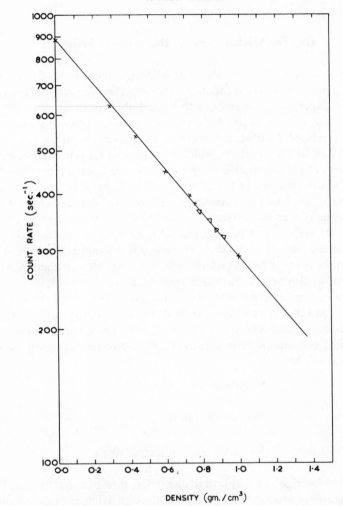

FIG. 12. Test of Eq. (10) with gas-liquid dispersions and pure liquids (C12).
KEY: × Air-water foams   ▽ Alcohol-water solutions

Table I

Mass Absorption Coefficients for Selected Liquids for a
Photon Energy of 0.681 m.e.v. as Given by Cesium-137[a]

Liquids	α
Water	0.085
Methyl alcohol	0.085
Glycerol	0.083
Benzene	0.082

[a] From Cameron (C12).

III. The Mechanics of the Dispersion of Fluids

The interdispersion of immiscible fluids is brought about by fluid dynamical forces which have to overcome the static force of surface tension. Such surface forces resist dispersion by attempting to retain bubble or drop sphericity and prevent gross distortion leading to break-up. This static force of surface tension may be reduced if surface active agents are present due to the adsorption of charged ions at the interface, which by their mutual repulsion cause an outward force of electrostriction in opposition to the inward force of surface tension. The dynamic forces which bring about dispersion may be due to buoyancy or induced fluid flow creating viscous or inertial forces which, if they do not act equally over the surface of a drop or bubble, may cause it to deform and eventually break-up. A consequence of these dynamic forces acting unequally over the surface of the drop or bubble is internal circulation of the fluid within the drop or bubble which induces viscous stresses therein. These internal stresses also oppose distortion and break-up.

If we represent the external shear stress acting on a drop or bubble of diameter D_p as τ, and if μ, ρ, and σ are the fluid viscosity, density, and interfacial surface tension, respectively, the subscripts d and c being used to denote the dispersed and continuous phases, then the three stresses acting on the dispersed phase will be

$$\text{Shear stress} \quad \propto \tau$$

$$\text{Surface tension} \propto \frac{\sigma}{D_p}$$

$$\text{Viscous stress in dispersed phase} \propto \frac{\mu_d}{D_p} \sqrt{\tau/\rho_d}$$

where the first force, which may lead to dispersion, is resisted by the other two.

In many cases of practical interest such as the dispersion of gases in liquids, the viscous stress in the dispersed phase may be ignored so that for these cases, a dispersed-phase particle size equilibrium is reached when the ratio of shear to surface tension stresses has a particular value which may be characteristic of the dispersion equipment and perhaps also of the physical properties of the dispersion. Thus the dimensionless Weber number is defined as

$$N_{We} = \tau D_p / \sigma \tag{11}$$

If one evaluates the Weber number for dispersions in dynamic equilibrium in a particular piece of dispersion equipment where the shear stress is known, the maximum drop or bubble size will thereby be related to the interfacial tension. Thus, D_p will be the maximum diameter of the drop or bubble which can survive at dynamic equilibrium in a flow or turbulent field of shear stress,

τ. Hinze (H9) has suggested that different flow arrangements may exhibit a varying effectiveness for dispersion so that the Weber number may be characteristic of the dispersion equipment. Taylor (T2) has also shown that in dispersions produced by laminar shear, the Weber number is a function of the ratio of the viscosities of the two phases.

A simple application of the above principles is to be found in the break up of liquid drops falling through other immiscible liquids or gases. Here the shear stress due to buoyancy is $\frac{1}{6}(D_p \Delta \rho g)$ and the Weber number is therefore $(D_p^2 \Delta \rho g)/6\sigma$. If a critical and definite value of the Weber number defies the point of break-up of the drop,

$$D_p \propto \left(\frac{\sigma}{\Delta \rho}\right)^{1/2} \tag{12}$$

This relationship has been experimentally tested by Hu and Kintner (H11) who found $N_{We} = 2\cdot 4$. This result has been confirmed by Schotland (S3) in a study of the impact between water drops and a large liquid hemispherical target.

A. Dispersion in Mixing Vessels

In mixing vessels, mechanical agitation is employed to create shear stress by means of turbulence. In order to apply the concept of a critical Weber number which determines the grain of the dispersion, it is necessary to evaluate the shear stress due to this turbulence.

Turbulent flow produces primary eddies which have a wavelength or scale of similar magnitude to the dimensions of the main flow stream (see Chapter 2 by Brodkey, in Vol. I). These large primary eddies are unstable and disintegrate into smaller eddies until all their energy is dissipated by viscous flow. When the Reynolds number of the main flow is high, most of the kinetic energy is contained in the large eddies, but nearly all of the dissipation occurs in the smallest eddies. If the scale of the main flow is large compared with that of the energy dissipating eddies, a wide spectrum of intermediate eddies exist which contain and dissipate little of the total energy. The large eddies transfer energy to smaller eddies in all directions, and the directional nature of the primary eddies is gradually lost. Kolmogoroff (K4) concludes that all eddies which are much smaller than the primary eddies are statistically independent of them and that the properties of these small eddies are determined by the local energy dissipation rate per unit mass of fluid. Thus, if a small volume of fluid is considered whose dimensions are small compared with the scale of the main flow, the fluctuating components of the velocity are equal, and this so-called local isotropic turbulence exists even though the turbulent motion of the larger eddies may be far from isotropic. For local isotropic turbulence the smallest eddies are responsible for most of the energy dissipation and their scale is given by Kolmogoroff as

$$l = \frac{\mu_c^{3/4}}{\rho_c^{1/2}} \left(\frac{P}{V}\right)^{-1/4} \tag{13}$$

The mean-square fluctuating velocity over a distance, d in a turbulent fluid field where $L \gg d \gg l$ (L being the scale of the primary eddies and l that of the smallest eddies) is given by Batchelor (B6) as

$$\overline{u^2} \propto \left(\frac{P}{V}\right)^{2/3} \left(\frac{d}{\rho_c}\right)^{2/3} \tag{14}$$

The condition for local isotropy where the above equation applies is frequently encountered. For example, in a fluid mixing vessel, operating at an energy input of 10 hp./1000 gallons of water, l is 25 μ. If the agitator has blades 5 cm. wide, the scale of the main fluid flow from it is approximately 5 cm. Thus, $L/l = 2000$ which is large enough to assume local isotropy.

Turbulence in the immediate vicinity of a particle of a dispersion affects heat and mass-transfer rates between the particle and the fluid, and if the particle is itself fluid may lead to its break up. The theory of local isotropy may be used here since it gives information on the turbulent intensity in the small volume around the particle. Thus $\sqrt{\overline{u_{D_p}^2}}$ is a statistical parameter describing the flow of fluid around a particle of diameter, D_p, and may be used in place of the velocity of the particle in the Reynolds and Weber numbers of the particle. It may be concluded that the shear stress due to turbulence is given by

$$\tau = C_1 \rho_c \left(\frac{PD_p}{V\rho_c}\right)^{2/3} \tag{15}$$

where P/V = power input per unit volume of fluid and C_1 is a dimensionless number to which Batchelor (B6) assigns a value of 2. Thus from Eqs. (11) and (15), taking N_{We} as a constant, characteristic of the mixing equipment,

$$D_p \propto \frac{\sigma^{0.6}}{\rho_c^{0.2}(P/V)^{0.4}} \tag{16}$$

Equation (16) may be transformed by using the following groups conventionally employed in mixing vessel studies,

$$\text{Power number } N_P = \frac{P}{N^3 D^5 \rho_c}$$

$$\text{Weber number of impeller } (N_{We})_I = \frac{N^2 D^3 \rho_c}{\sigma}$$

to give

$$\frac{D_p}{D} \propto (N_{We})_I^{-3/5} \, N_P^{-2/5} \left(\frac{D^3}{V}\right)^{-2/5}. \tag{17}$$

where N is the speed of the impeller in, for example, r.p.m., D is the impeller diameter, and V is the volume of the mixing vessel containing the agitated fluid. Thus, for a particular geometry of agitator and mixing vessel, in which (D^3/V) is constant with scale-up and in which the power number is also constant as it most frequently is in the agitation of mobile liquids in fully baffled vessels,

$$\frac{D_p}{D} \propto (N_{We})_I^{-3/5} \tag{18}$$

Hinze (H9) has pointed out that the maximum drop diameter observed by Clay (C18) in the dispersion of oil in water between rotating and stationary coaxial cylinders conforms with the two-fifths exponents of the power input as predicted by Eq. (16).

Vermeulen (V4) dispersed a number of two liquid systems in a fully baffled tank with a four-blade paddle stirrer under conditions where N_P and V/L^3 were both constant and found

$$\frac{D_p}{D} \propto f_d \, (N_{We})_I^{-3/5} \tag{19}$$

thus confirming the 3/5- exponent of the Weber number predicted by Eq. (19). The factor f_d in Vermeulen's empirical equation was found to depend on the dispersed-phase hold-up and is given by

$$f_d = 2.5H + 0.75 \tag{20}$$

Writing Eq. (16) as

$$D_p = C_2 \left[\frac{\sigma^{0.6}}{\rho_c^{0.2}(P/V)^{0.4}} \right] \tag{21}$$

the value of C_2 from the data of Clay (C18) is 0.725 whereas Vermeulen's observations lead to a value of C_2 of about 0.03. However, the drop diameters quoted in Clay's work were maximum values while Vermeulen measured the Sauter mean diameter. Other more cogent reasons for this apparent discrepancy are discussed later.

Endoh and Oyama (E2) dispersed very small quantities of one liquid in a large body of another agitated liquid and found

$$D_p = 0.0169 \left[\frac{\sigma^{0.6}}{(P/V)^{0.4} \rho_c^{0.2}} \right] \tag{22}$$

Calderbank (C2) carried out measurements of drop and bubble sizes in mixing vessels using six-blade turbine impellers in geometrically similar vessels of a range of sizes (see Fig. 13). He found for liquid-liquid dispersions that

$$D_p = 0.224 \left[\frac{\sigma^{0.6}}{(P/V)^{0.4} \rho_c^{0.2}} \right] H^{0.5} \left(\frac{\mu_d}{\mu_c} \right)^{0.25} \tag{23}$$

for dispersions of gas bubbles in solutions of electrolyte that

$$D_p = 2.25 \left[\frac{\sigma^{0.6}}{(P/V)^{0.4}\rho_c^{0.2}} \right] H^{0.4} \left(\frac{\mu_d}{\mu_c}\right)^{0.25} \tag{24}$$

and for dispersions of gas bubbles in aqueous solutions of alcohols that

$$D_p = 1.90 \left[\frac{\sigma^{0.6}}{(P/V)^{0.4}\rho_c^{0.2}} \right] H^{0.65} \left(\frac{\mu_d}{\mu_c}\right)^{0.25} \tag{25}$$

(The constants in the above equations are dimensionless.)

Bubble sizes produced by the aeration of pure liquids in mixing vessels were considerably greater than when solutions of electrolytes or other hydrophilic solutes were aerated. This was attributed to the greater ease of bubble coalescence in the former case.

It seems evident that Eq. (16) which does not take account of coalescence phenomena, has to be modified by hold-up and viscosity-ratio parameters for

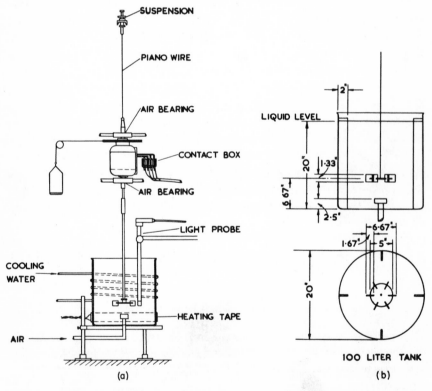

FIG. 13. Apparatus for study of dispersion properties as a function of the power dissipated by an impeller: (a) Torque measuring equipment; (b) impeller and vessel geometry.

practical cases where the drops or bubbles are in close proximity. Also it is likely that the critical Weber number for drop break-up is a function of (μ_d/μ_c) as was found by Taylor (T2) and Hinze (H9) in experiments on single liquid drops.

For the dispersion of gases in pure agitated liquids, Vermeulen (V4) finds

$$\frac{D_p}{D} = f_d \, (N_{We})_I^{-3/4} \, (N_{vi})_I^{-1/2} \left(\frac{\mu_d}{\mu_c}\right)^{-3/4} \tag{26}$$

where

$$(N_{We})_I = \frac{N^2 D^3 \rho_c}{\sigma}$$

and

$$(N_{vi})_I = \frac{\mu_c}{\sqrt{D\sigma\rho_c}}$$

Calderbank (C2) concluded that Vermeulen's measurements, which were made close to the tip of the impeller, were not representative of mean values for the whole tank contents and proposed the equation

$$D_{S.M.} = 4.15 \left[\frac{\sigma^{0.6}}{(P/V)^{0.4} \rho_c^{0.2}}\right] H^{1/2} + 0.09 \text{ cm.} \tag{27}$$

which was obtained by measuring the interfacial area at many points in the body of a mixing vessel, calculating an integral mean area and combining this with a measured over-all gas hold-up in the vessel. A typical interfacial area distribution for aerated water in a mixing vessel is shown in Fig. 14. These extreme variations in the properties of gas-liquid dispersions throughout a mixing vessel largely disappear when the liquid viscosity is increased and when hydrophylic solutes are added to water to reduce coalescence and produce smaller bubbles.

Calderbank (C2) also proposed the equation

$$a = 1.44 \left[\frac{(P/V)^{0.4} \rho_c^{0.2}}{\sigma^{0.6}}\right] \left(\frac{v_s}{v_t}\right)^{1/2} \tag{28}$$

where v_t = terminal velocity of bubbles in free rise, and
v_s = superficial gas velocity in mixing vessel

for the aeration of pure liquids in mixing vessels, so that combining Eqs. (27) and (28), a correlation for the gas hold-up is obtained.

$$H = \left(\frac{v_s H}{v_t}\right)^{1/2} + 0.0216 \left[\frac{(P/V)^{0.4} \rho_c^{0.2}}{\sigma^{0.6}}\right] \left(\frac{v_s}{v_t}\right)^{1/2} \tag{29}$$

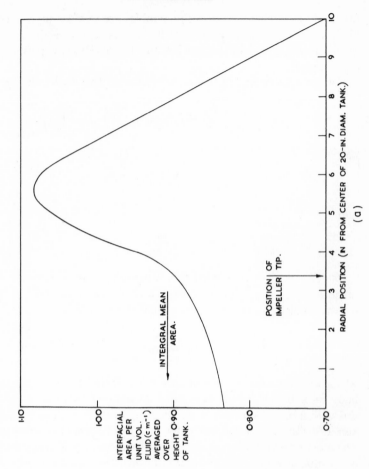

FIG. 14. Distribution of air in water in an aerated mixing vessel (C2).

(a) Radial distribution for 400 r.p.m. Air at 1 c.f.m. in water in 20 in. diam. tank.

(b) Vertical distribution in 20 in. diam. tank at mid-radial position. Air flow at 1 c.f.m.

Run	r.p.m.
1	200
2	250
3	300
4	350
5	400

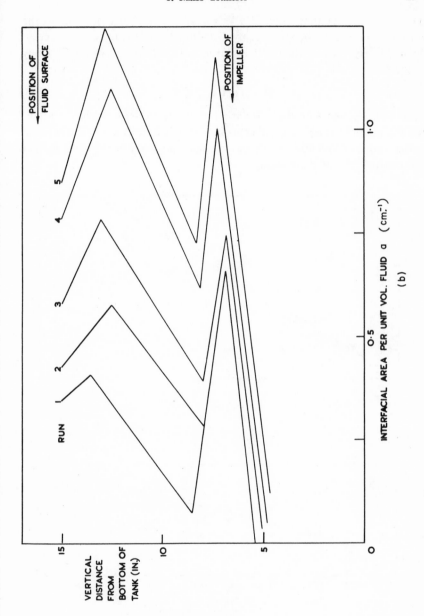

(b)

Equation (29) shows that when $(P/V) \to 0$, $H = v_s/v_t$, corresponding to the condition when the bubbles are evenly distributed over the tank cross section and rising freely. Again, when (P/V) is large and the first term of Eq. (29) may be neglected,

$$H \propto (P/V)^{0.4} \, v_s^{0.5} \tag{30}$$

Equation (30) agrees with the correlation of Rushton (R6) which was established for mixing vessels of large capacity, while Eq. (28) may be compared with the results of Valentin and Preen (V1) who find $a \propto P^{0.35} \, v_s^{0.6}$ for constant physical properties of the system.

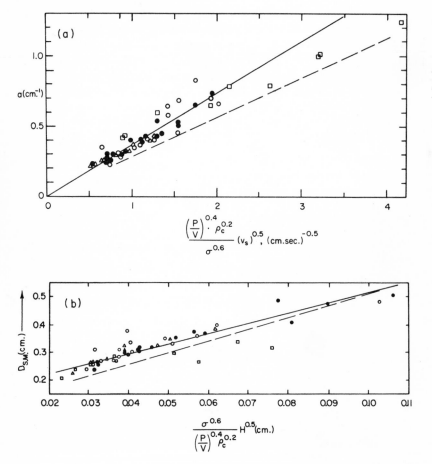

FIG. 15. Mean interfacial areas (a) and bubble sizes (b) in an aerated mixing vessel (C2, M9).

KEY: - - - Calderbank (C2) [(a) Eq. (28); (b) Eq. (27)]
        ○ Air-water   □ Air-ethanol   ● O_2-water   △ CO_2-water

Moo-Young (M9) has carried out measurements of the interfacial area and Sauter mean bubble size in a mixing vessel in which pure liquids were aerated. Although the mixing vessel, in this case, was of unusual geometry and both phases were in continuous flow and undergoing mass transfer, the results obtained were only slightly different from those predicted by Eqs. (27) and (28) as shown in Fig. 15.

Calderbank (C3) showed that Eq. (28) seriously underestimated the interfacial area in aerated mixing vessels operating at high impeller Reynolds numbers. He used mass-transfer experiments to demonstrate that this departure was due to aeration from the turbulent free surface of the tank contents under these conditions. The results of the above investigation are shown in Fig. 16, in which b is the ratio of total gas flow (above and below the tank) to that fed to the sparger and a/a_0 is the ratio of the observed mean interfacial area to that predicted by Eq. (28). These quantities are plotted against the parameter $(N_{Re})_I^{0.7} (ND/V_s)^{0.3}$ where $(N_{Re})_I$ is the impeller Reynolds number $(ND^2 \rho_c/\mu_c)$. It may be seen that when the parameter above is less than 15,000, no surface aeration takes place, but when it is greater than 60,000 the whole of the tank contents are largely affected by aeration from the free surface.

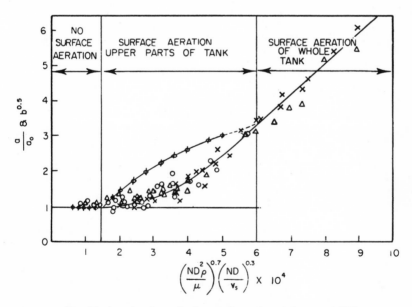

FIG. 16. Aeration from the free surface of a mixing vessel (C3)

Tank Diam. in.	a/a_0	$b^{0.5}$
12	×	
15	△	ϕ
20	○	

These results were obtained in mixing vessels geometrically similar to that depicted in Fig. 13 and in which the clear liquid height was equal to the vessel diameter. In commercial vessels, the height to diameter ratio is often large and surface aeration may not play such an important role as is suggested by these results obtained in comparatively shallow liquid pools.

Westerterp (W4) has carried out measurements of the interfacial area and mean bubble size in an aerated mixing vessel using the "chemical" method previously described. These data are shown in Fig. 17 where they are compared with Eq. (28) to which the surface aeration correction has been applied.

Westerterp employed a range of gas hold-ups which was exceptionally high compared with other work and it may be seen from Fig. 17 that the bubble size is substantially independent of impeller speed under these conditions, when the agitator is functioning primarily as a distributor. Only at very high impeller speeds does an increase in impeller speed begin to decrease the bubble size, when the data have converged to the results predicted by Eq. (28), which was derived under these conditions. Westerterp presents his results as

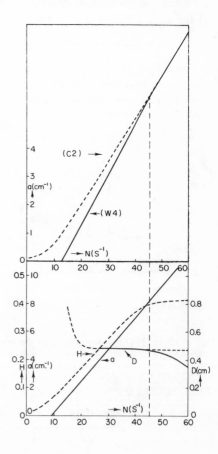

FIG. 17. Properties of gas-liquid dispersions in mixing vessels with high gas hold-up: - - - Calderbank (C2); —Westerterp (W4).

follows: For 16 geometrically similar turbine impellers of different sizes in five vessels

$$\frac{ah}{1-H} = (N - N_0) D (\rho_c T/\sigma)^{1/2} \tag{31}$$

For nine geometrically similar 4-blade paddles of different sizes in three · different vessels,

$$\frac{ah}{1-H} = (N - N_0) D (\rho_c T/\sigma)^{1/2} (D/T)^{1/3} \tag{32}$$

where h = the height of liquid in the tank and
T = the tank diameter.

The minimum agitator speed N_0, at which the agitator begins to have an influence on the dispersion is given by

$$\frac{N_0 D}{(\sigma g/\rho_c)^{1/4}} = A + B \left(\frac{T}{D}\right) \tag{33}$$

where $A = 1.22$; $B = 1.25$ for the turbines and
$A = 2.25$; $B = 0.68$ for the paddles.

Calderbank (C2) studied the effect of impeller geometry on liquid-liquid dispersion in a mixing vessel by comparing a 4-blade paddle, geometrically similar to that employed by Vermeulen (V4) with a 6-blade turbine. The results are shown in Fig. 18 which is a plot of the following equation:

$$D_p = 0.06 f_H (N_P)^{0.4} \left(\frac{D^3}{V}\right)^{0.4} \left[\frac{\sigma^{0.6}}{(P/V)^{0.4} \rho_c^{0.2}}\right] \tag{34}$$

or,

$$\frac{D_p}{D} = 0.06 f_H \left(\frac{\sigma}{\rho_c N^2 D^3}\right)^{0.6} = 0.06 f_H (N_{We})^{-0.6} \tag{35}$$

where $f_H = (1 + 3.75H)$ for the 4-blade paddle ($D/T = \frac{2}{3}$), and
$f_H = (1 + 9H)$ for the 6-blade turbine ($D/T = \frac{1}{3}$).

Equation (35) for paddles agrees satisfactorily with the results of Vermeulen (V4) who reported

$$D_p/D = 0.063 (1 + 3.3H) (N_{We})^{-0.6} \tag{36}$$

Vanderveen and Vermeulen (V2), in subsequent work, report some coalescence in the dispersion and by making a correction for this, find that the bubble diameter in the immediate vicinity of the impeller tip is given by

$$D_p/D = 0.069 (1 + 2.5H) (N_{We})^{-0.6} \tag{37}$$

30   **P. H. Calderbank**

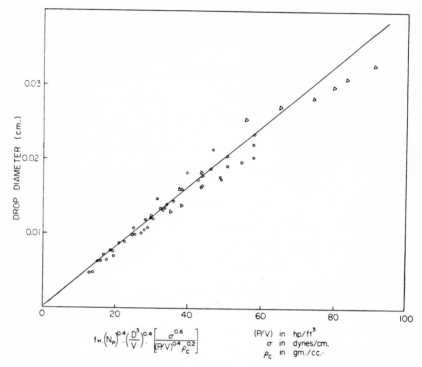

$$f_H \cdot (N_P)^{0.4} \cdot \left(\frac{D^3}{V}\right)^{0.4} \cdot \left[\frac{\sigma^{0.6}}{(P/V)^{0.4} \rho_c^{0.2}}\right]$$

(P/V) in hp/ft³
σ in dynes/cm.
ρ_c in gm./cc.

Fig. 18. Liquid-liquid dispersion in mixing vessels using different agitators (C2, V4). Key:
△ 10-in. diam. four-blade paddle   ○ 5-in. diam. flat-blade turbine

Calderbank (C2) did not observe coalescence in liquid-liquid dispersions, presumably because he employed hard water as the continuous phase instead of the distilled water used by Vermeulen. The presence of dissolved electrolytes in the water evidently creates an electrical double layer at the interface which reduces the rate of coalescence as has been reported by Benton and Elton (B8) for aerosols. Equation (35) for turbines should be compared with Eq. (23) which relates to the same situation. Taking $\mu_d/\mu_c = 1$ in Eq. (23), it is found that agreement with Eq. (35) is obtained at a dispersed-phase hold-up of 10%, which is the mean value employed in the investigation. In order to decide which correlation is the more accurate, reference may be made to the work of Endoh and Oyama [Eq. (22)] in which the dispersed-phase hold-up was 0.2%. When this is done, it is found that Eq. (35) predicts a constant of 0.029 and Eq. (23), a constant of 0.01 compared with Endoh and Oyama's value of 0.0169, so that no conclusion can be arrived at.

Reference to Eq. (35) shows that the mean drop size in noncoalescing liquid-liquid dispersions, as the dispersed-phase hold-up approaches zero, is primarily determined by the impeller tip speed. Thus, for a given system of fluids, the influence of agitator properties on the mean drop size is given by

$$D_p \propto \frac{D^{0.4}}{(ND)^{1.2}} \tag{38}$$

Bates (B7) studied the dispersion of liquid sodium in oil by means of a wide variety of agitators and concluded that

$$D_p \propto \frac{1}{(ND)^{1.8}} \tag{39}$$

as shown in Fig. 19.

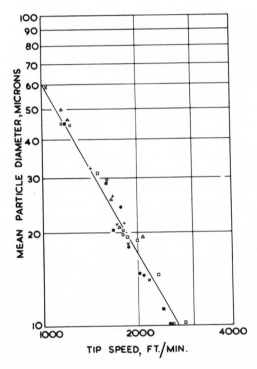

FIG. 19. Liquid-liquid dispersion in mixing vessels using different agitators (B7).

In both cases, the important conclusion is reached that the impeller or paddle width has no influence on the ultimate drop size, so that a flat disk produces the same drop size as a broad-blade paddle operating at the same peripheral speed, despite the fact that the power drawn by these agitators will be very different. Vanderveen (V2) also found that doubling the width of a paddle had no effect on the drop size at constant speed. Bates (B7) however discovered that the rate at which dispersion equilibrium was reached depended on the impeller width or power dissipated, so that practical considerations demand a compromise between high tip speed and a reasonable flow from the impeller, in order that the whole of the tank contents may be circulated through the high shearing zone around the impeller at a rate which is large compared with the rate of settling or coalsecence of the dispersed phase. Since the rate of coalescence will increase with the dispersed-phase hold-up, it becomes increasingly important to achieve a rapid turnover of the tank contents as the hold-up is increased. In this connection, Eq. (23) leads to the suggestion that at high hold-ups, the drop size is approximately determined by the power dissipated per unit volume of dispersed phase. This conclusion is in accord with the previous observation that the power dissipation may be reduced to vanishingly small values as the hold-up and coalescence rate become very small, without affecting the ultimate drop size, provided the impeller tip speed is maintained constant by reducing the blade width.

A somewhat more fundamental analysis of these questions may be attempted if Eq. (15) is considered in the form,

$$D_p = \left(\frac{N_{We}}{2}\right)^{3/5} \left[\frac{\sigma^{0.6}}{(P/V)^{0.4} \rho_c^{0.2}}\right] \tag{40}$$

This equation relates the particle size in a fluid dispersion to the power dissipation per unit volume of liquid if the latter may be considered to be in a state of homogeneous isotropic turbulence which, as previously discussed, may be assumed to be the case, for the small scale of reference associated with particles which are small compared with the leading dimensions of the agitator. In a mixing vessel, it is reasonable to expect that the power dissipated by the agitator will not be equally distributed over the whole volume of the tank contents, being greatest near to the agitator and declining with distance from it. Thus, in coalescing systems, a variation in the dispersed-phase particle diameter with position in the tank will be found, as has been reported by Vanderveen and Vermeulen (V2) and by Calderbank [(C2) and Fig. 14]. In noncoalescing systems, no such variation is observed because the condition of dynamic equilibrium set up near the impeller is preserved in the body of the tank, provided an adequate circulation of the tank contents is maintained. These are the conditions observed by Bates [(B7) and Fig. 19] and Calderbank [(C2) and Fig. 18]. Thus, using Eq. (40), it is instructive to consider the small volume of liquid in the immediate vicinity of the impeller where dispersion is most effective, this state of dispersion being preserved over a much wider area, in the case of noncoalescing systems. If this volume is taken to be a fraction, C_3, of the volume of liquid swept by the agitator, V_I, Eq. (40) becomes

$$D_p = (C_3)^{2/5} \left(\frac{N_{We}}{2}\right)^{3/5} \left[\frac{\sigma^{0.6}}{(P/V_I)^{0.4} \rho_c^{0.2}}\right] \tag{41}$$

and combining this with Eq. (35) when $H \to 0$ or $f_H \to 1$, to eliminate the influence of coalescence,

$$N_{We} = \frac{2(0.06)^{5/3}}{(C_3)^{2/3}} \left[N_P \left(\frac{D^3}{V_I} \right) \right]^{2/3} \tag{42}$$

For 4-blade paddles operating in the turbulent region of mixing, Vanderveen and Vermeulen (V2) report

$$N_P = \left(\frac{81.6}{\pi} \right) \left(\frac{V_I}{D^3} \right) \tag{43}$$

Therefore, from Eqs. (42) and (43),

$$N_{We} = \frac{0.16}{(C_3)^{2/3}} \tag{44}$$

Again, for 6-blade turbines operating in the turbulent region of mixing, Calderbank and Moo-Young (C9) report

$$N_P = \left(\frac{160}{\pi} \right) \left(\frac{V_I}{D^3} \right) \tag{45}$$

and from Eqs. (42) and (45),

$$N_{We} = \frac{0.25}{(C_3)^{2/3}} \tag{46}$$

Hinze (H9) has analyzed the data of Clay (C18) in which liquid-liquid systems were sheared between rotating coaxial cylinders, and has found that Eq. (40) applies with

$$\left(\frac{N_{We}}{2} \right)^{3/5} = 0.725$$

or

$$N_{We} = 1.18 \tag{47}$$

In these experiments, the volume of liquid undergoing turbulent shear was small enough to allow the assumption that the power was uniformly distributed throughout the whole of the liquid. Further, the conclusion that $N_{We} = 1.18$ is supported by Taylor's observations (T2) on dispersion by laminar shear fields, using liquids of viscosity ratio close to one. Combining Eqs. (44) and (47) for paddles and Eqs. (46) and (47) for turbines, it is seen that C_3 for paddles is 6.5% and C_3 for turbines is 12.5% indicating that the fraction of the swept volume which is effective in dispersion differs by a factor of nearly two when a 4-blade paddle is compared with a 6-blade turbine. It may be significant to note that this ratio is the same as the ratio of the number of blade edges in the two cases considered.

Thus one can visualize dispersion in mixing vessels as occurring in the region of intense turbulent shear around the impeller edges or tips with the tip speed primarily determining the local turbulent intensity and particle size. The blade area determines the flow of liquid from the impeller and most of the power drawn by the agitator. Flow from the impeller enables the whole of the tank contents to be circulated through the small dispersion zone. A proper rate of circulation will depend on the rates of coalescence and settling which the dispersion may undergo in the body of the fluid.

The above conclusions, arrived at by a consideration of liquid-liquid dispersion, are almost certainly applicable to gas-liquid dispersions also since the correlations of Westerterp (W4) and Eqs. (31) and (32) may be put in the following form if N_0, the minimum agitator speed required to influence the properties of the dispersion, is neglected and the physical properties of the fluids are constant.

For turbines,

$$D_p \propto \left(\frac{1}{ND}\right) \frac{h}{T^{1/2}} \tag{48}$$

For paddles,

$$D_p \propto \left(\frac{1}{ND}\right) \frac{1}{D^{1/3}} \frac{h}{T^{1/6}} \tag{49}$$

The paramount influence of the impeller tip speed and the absence of an effect of blade width is again evident, while the influence of liquid height and tank dimensions indicates the importance of coalescence in gas-liquid systems.

B. Dispersion in Turbulent Pipe Flow

Again applying the principles of local isotropic turbulence and noting that in pipe flow $P/V \propto [(fv^3 \rho_c)/(d_p)]$, Eq. (15) becomes

$$\tau = \text{const. } \rho_c \left(\frac{fv^3 D_p}{d_p}\right)^{2/3} \tag{50}$$

where f = Fanning friction factor,
  D_p = particle diameter, and
  d_p = pipe diameter.

If the grain of the dispersion is determined by a balance between surface tension and dynamic forces, a critical value of the Weber number characterizes the dispersion. Whence

$$\frac{D_p}{d_p} \propto \left(\frac{\sigma}{\rho_c v^2 d_p}\right)^{3/5} \left(\frac{1}{f}\right)^{2/5} \tag{51}$$

or since in turbulent pipe flow $f \propto (N_{Re})^{-1/4}$

$$\frac{D_p}{d_p} \propto (N_{We})^{-3/5} (N_{Re})^{-1/10} \tag{52}$$

where

$$(N_{We}) = [(\sigma/\rho_c v^2 d_p)]^{-1} \text{ and } (N_{Re}) = [(d_p v \rho_c / \mu_c)]$$

Equation (52) agrees with the experimental findings of Rushton and Roy (R8) and Baranayev (B4) who injected oils into water in turbulent pipe flow.

McDonough *et al.* (M3) incorporated an orifice plate of diameter, d_0, in a pipe line of diameter, d_p, and measured the properties of liquid-liquid dispersions produced by flow at a distance 1 ft. downstream from the orifice plate. These results were correlated as

$$\frac{D_{S.M.}}{d_p} = 21.6 \left(\frac{d_o}{d_p}\right)^{3.73} H^{0.121} (N_{Re})^{-0.065} (N_{We})^{-0.722} \tag{53}$$

where H is the volume fraction dispersed-phase hold-up, and $D_{S.M.}$, as before, is the Sauter mean drop diameter.

C. Gas Dispersion with Sieve Plates

Measurements of the physical properties of foams and froths formed in sieve-tray columns under conditions simulating those obtaining in gas absorption and distillation have been carried out by Calderbank and Rennie (C11) using the techniques of optical reflectivity, gamma-ray absorption, and high-speed flash photography previously described. These results, shown in Figs. 20 to 25, were obtained with the air-water system but are believed to be of more general applicability for the reasons discussed below.

Figure 20 shows that, at low gas velocities, the Sauter mean bubble size

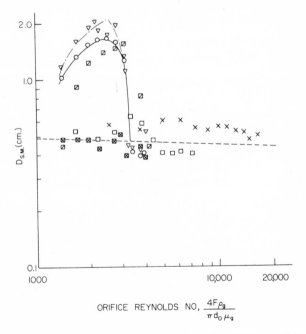

Fig. 20. Sauter mean bubble diameters in a sieve-tray colum (C11); - - - - Leibson (L3).

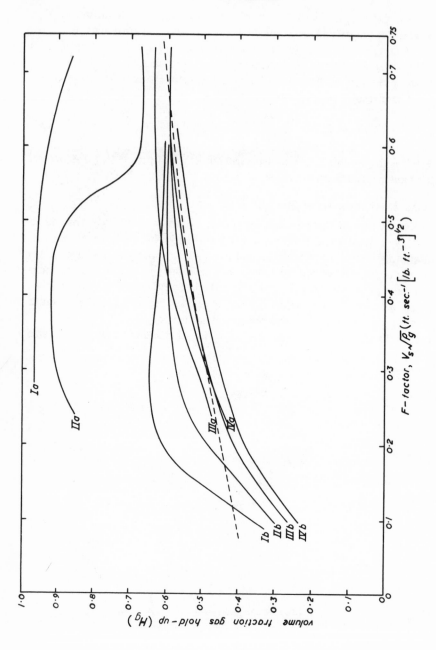

Hold-up F-factor

FIG. 21. Gas hold-up in sieve-tray columns (C11, C21). - - - Crozier (C21). Curves a: $53 \times \frac{1}{8}$-in. diam. holes; curves b: $12 \times \frac{1}{8}$-in. diam. holes; both on a 4 by $3\frac{1}{2}$-in. sieve plate.

Curve	Liquid flow rate, (liters/min.)	Curve	Liquid flow rate (liters/min.)	Type of dispersion[a]
Ia	0.25	Ib	1.0	Cellular foam
IIa	2.8	IIb	2.8	Cellular foam
IIIa	7.3	IIIb	7.3	Froth
IVa	9.5	IVb	9.5	Froth

[a] See Fig. 22.

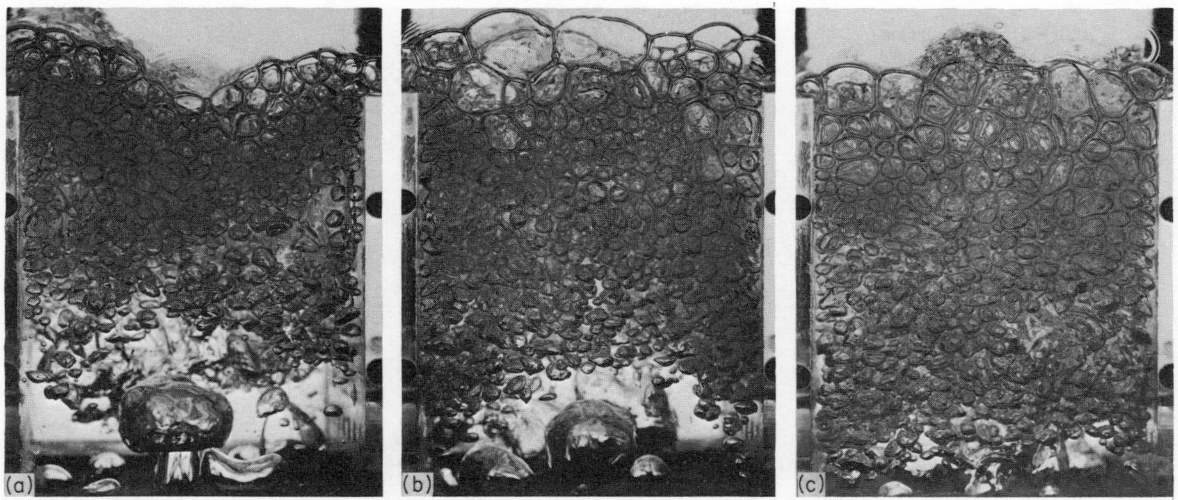

FIG. 22. Stages in the formation of a froth. System: air-water on a 4- by $3\frac{1}{2}$-in. sieve plate with nine $\frac{1}{4}$-in. diam. holes (C11). (a) $N_{Re} = 1500$(for $d_0 = \frac{1}{4}$-in.; (b) $N_{Re} = 2250$; and (c) $N_{Re} = 3000$.

increases with gas flow rate up to an orifice Reynolds number of about 2500 when much smaller bubbles are produced, their mean diameter being about $4\frac{1}{2}$ mm. and substantially unaffected by further increase in gas velocity. This observation was also made by Leibson et al. (L3) using a single orifice to pass air into various liquids and the following correlation for high orifice Reynolds numbers was proposed:

$$D_{\text{S.M.}} = 0.713 \, (N_{\text{Re}})_0^{-0.05} \tag{54}$$

Figure 21 shows values of the gas hold-up observed with various sieve plates. It may be seen that the data for higher liquid flow rates and larger orifices is in reasonable agreement with the correlation proposed by Crozier (C21) who worked with several different liquids and reported his results as

$$\ln \left(\frac{1}{1 - H} \right) = 0.715 \, v_s \sqrt{\rho_g} + 0.45 \tag{55}$$

It may also be seen that very low density cellular foams may be formed with sieve plates containing many small holes operating with low liquid flow rates, a condition which was not observed by Crozier and for which his correlation is inapplicable.

Figure 22 shows stages in the formation of a froth.

For conditions under which cellular foams are not produced, Fig. 23 shows gas-liquid interfacial areas plotted against the gas flow rate. The interfacial areas tend to a maximum value of about 8 cm. $^{-1}$ at high gas rates and agree with results previously reported for a variety of liquids by Calderbank (C3) using the light-transmission method.

The foregoing observations on gas dispersion into liquids using perforated plates are reasonably self-consistent and, as will be seen in a later section, when applied to mass-transfer data, they predict mass-transfer coefficients which are in accord with theory and other experimental data.

D. COALESCENCE IN DISPERSIONS

Vanderveen and Vermeulen (V2) have studied coalescence in agitated liquid-liquid dispersions by observing the variation in mean drop diameter with position in a mixing vessel containing distilled water as the continuous phase. They conclude that the results of Rodger et al. (R1), obtained by measurements at the base of a mixing vessel, also with distilled water as the continuous phase, are strongly influenced by coalescence as the drops move from the impeller tip to this location. Rodger's correlation can be expressed as

$$\frac{D_p}{D} \propto (N_{\text{We}})_I^{-0.36} \left(\frac{\mu_c}{\mu_d} \right)^{0.2} f\left(H, \frac{\Delta\rho}{\rho_c}, t \right) \tag{56}$$

where t is an experimentally determined dispersion settling time. It may be

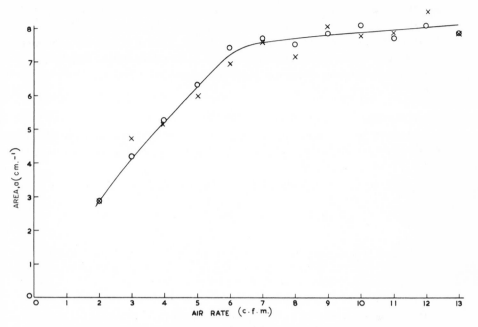

FIG. 23. Interfacial areas in sieve-tray columns (C11). Twelve $\frac{1}{8}$-in. diam. holes in a 4- by 3$\frac{1}{2}$-in. sieve plate. Air-water system. Liquid flow rate is 2.8 liters/min. KEY:

    ○   Light transmission (C3)

  × $\begin{cases} D_p \text{ from photography (C11)} \\ H_g \text{ from gamma-ray transmission (C11)} \end{cases}$

concluded from Eq. (35) or (37), that the mean drop diameter at the impeller tip is given by

$$\frac{D_p}{D} \propto f_H \, (N_{We})_I^{-0.6} \tag{57}$$

Vanderveen has compared Eqs. (56) and (57) to arrive at the change in drop diameter between the impeller tip and the base of Rodger's mixing vessel and also included his own experimental observations to arrive at an expression for the change in mean drop diameter between the impeller tip and any other position in a mixing vessel.

This expression takes the following form for a given system of constant dispersed-phase hold-up,

$$\frac{(D_p)_i - (D_p)_0}{D} \propto (N_{We})_I^{-0.45} \, (N_{Re})_I^{0.25} \tag{58}$$

where $(D_p)_i$ is the mean drop diameter at a particular location in the tank and $(D_p)_0$ is the mean drop diameter in the immediate vicinity of the impeller tip, showing that the extent to which the mean drop diameter changes with position is inversely dependent on the impeller speed to the power 0.65.

Shinnar and Church (S5) propose a semitheoretical argument to predict the effect of coalescence in mixing vessels. Thus the energy of adhesion, E, between two drops of diameter, D_p, is postulated as

$$E = D_p A \qquad (59)$$

where A is the force needed to separate two drops to infinity.

For dispersion equilibrium, the energy of adhesion is equal to the turbulent energy causing dispersion, so that from Eq. (15), which gives the tubulent force per unit area of a drop

$$\frac{\rho_c^{1/3} (P/V)^{2/3} D_p^{11/3}}{D_p A} = \text{const.} \qquad (60)$$

whence

$$D_p \propto (A)^{3/8} (P/V)^{-1/4} \rho_c^{-1/8} \qquad (61)$$

or at constant power number and system properties

$$D_p \propto (1/N)^{3/4} \qquad (62)$$

a result which is in reasonable functional agreement with Eq. (56).

Calderbank (C2) found no measurable rate of coalescence in agitated liquid-liquid dispersions if hard water was used as the continuous phase, while Vanderveen (V2) observed that drop size variation does not occur when (μ_c/μ_d) is close to unity and the interfacial tension is between 35 and 50 dynes/cm. Calderbank (C2) also found that coalescence between gas bubbles in aerated mixing vessels was extensive but considerably reduced when electrolytes or alcohols were added in small amounts to water as the continuous phase.

Calderbank and Moo-Young (C10) measured the rate of coalescence in clouds of carbon dioxide bubbles rising through a 10-ft. high column of water. The transparent rectangular column was traversed vertically by gamma-ray and optical transmission equipment to enable the gas hold-up and interfacial area to be determined at all levels in the column, from which variations in mean bubble diameter with height could be recorded. Experiments were performed when the bubbles were dissolving in the water and when the water was saturated with the gas. In the former case the bubble cloud was subjected to a continuous crossflow of deaerated water by multipoint injection and removal of water through perforated metal screens between which the bubble cloud was contained at two of its opposite extremities, the remaining two confining walls being of transparent plastic sheet and through which the properties of the dispersion were measured. This arrangement of the liquid flow ensured that the bubbles were surrounded by liquid of invariant composition as they progressed through the column. The experimental arrangement is depicted in Fig. 24.

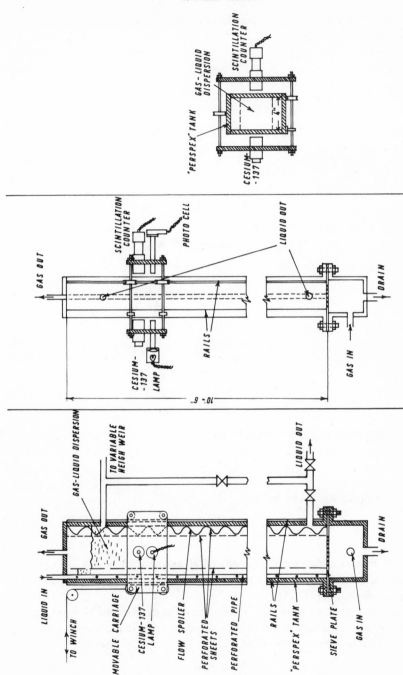

FIG. 24. Apparatus for measuring mass transfer and coalescence rates in bubble clouds rising through deep pools of liquid (C10).

If N_b is the number of bubbles in unit volume of dispersion and f is the coalescence frequency factor

$$\frac{-dN_b}{dt} = f N_b \tag{63}$$

where

$$N_b = \frac{6H}{\pi D_p^3} \tag{64}$$

and

$$t = \frac{h}{v_t} \tag{65}$$

where t = time,
   h = column height, and
   v_t = average velocity of rise of the bubbles;

then, from Eq. (64),

$$\frac{d(\ln N_b)}{dh} = \frac{d(\ln H)}{dh} - \frac{3d(\ln D_p)}{dh} \tag{66}$$

and since

$$a = \frac{6H}{D_p} \tag{67}$$

$$\frac{d(\ln N_b)}{dh} = \frac{3d(\ln a)}{dh} - \frac{2d(\ln H)}{dh} \tag{68}$$

Further, from Eqs. (63) and (65),

$$\frac{d(\ln N_b)}{dh} = \frac{-f}{v_t} \tag{69}$$

and combining Eqs. (68) and (69),

$$f = v_t \left[\frac{2d(\ln H)}{dh} - \frac{3d(\ln a)}{dh} \right] \tag{70}$$

Thus from the measured changes of H and a with height, recorded on semilogarithmic coordinates, the coalescence frequency factor may be evaluated, as shown in Fig. 25.

The average rising velocity of the bubbles was determined by extrapolation of the hold-up plots to both ends of the column where the inlet and outlet gas flow rates were known. Thus,

$$v_t = v_s/H \tag{71}$$

an average of the two determinations being used in Eq. (70).

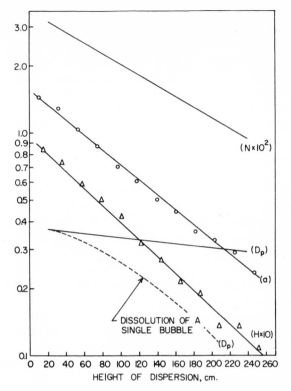

FIG. 25. Typical experimental data for evaluating the coalescence frequency factor in bubble clouds (C10).

For conditions under which mass transfer and coalescence take place simultaneously, Eq. (70) still applies, but in addition, one may write

$$-dH/dt = k_L a \Delta C \qquad (72)$$

where k_L is the liquid-phase mass-transfer coefficient and ΔC is the constant concentration driving force given by $\Delta C = C_i - C$, C_i being the concentration of carbon dioxide at the gas-liquid interface (the solubility of CO_2 in water in cm.³/cm.³), and C the concentration of CO_2 in the bulk liquid which was measured at the liquid outlet from the column.

From Eqs. (65), (67), (72),

$$\frac{-d(\ln H)}{dh} = \frac{6k_L \Delta C}{v_t D_p} \qquad (73)$$

Equation (73) enables the mass-transfer coefficient to be determined from the experimental data to reveal any effect on its value due to coalescence. Surface active material, notably hexanol, was found to completely inhibit bubble coalescence and the mass-transfer coefficient appeared to be little affected by the presence or absence of coalescence.

The work described above was conducted with bubbles of an average diameter of approximately 3 mm., the dispersion at the base of the column possessing a hold-up of about 10% of gas. In the experiments where mass transfer and coalescence occurred simultaneously, the bubble size did not change substantially, although the number of bubbles per unit volume of dispersion decreased by a factor of about 10 in the 10 ft. of column height provided. This circumstance made it possible to use a constant value of D_p in Eq. (73) for these conditions under which dissolution and coalescence rates combined to produce a balance, in which D_p was constant. If D_p is constant, it may be seen from the foregoing equations that $f = [(6k_L \Delta C)/(D_p)]$, showing that the collision frequency depends on the rate of dissolution, and its value was found in all cases to be such that the mean constant bubble diameter corresponded to the transition region between spherical and elipsoidal bubbles. This observation may be explained as follows: The velocities of rise of elipsoidal bubbles are almost independent of their size, so that even if their sizes differ they cannot overtake each other and collide; however, as they dissolve they move into the spherical bubble region where the velocity is strongly dependent on size and rapid overtaking and coalescent is possible; thus, further reduction in size by dissolution is resisted by coalescence and the mean size remains constant.

For fluid particles which rise or fall in the Stokes regime of flow, the larger particles overtake the smaller ones and collide with them to produce even larger and more rapidly moving particles so that coalescence is a self-accelerating process. These conditions have been investigated by Benton et al. (B8) with particular reference to the coalescence of water droplets descending in air, and are also applicable to bubble swarms rising in viscous liquids, where there is no plateau in the D_p/V_t curve (H1). Thus when the experiments described above were repeated, using 80% aqueous glycerol in place of water, coalescence was so pronounced that spherical-cap bubbles left the top of the 10-ft. column of such a size that they bridged the column cross section.

The foregoing treatment of the mechanics of dispersion processes provides a useful basis for appreciation of the factors which govern the physical properties of dispersions, and, as can be seen, receives substantial support from experimental observations. The importance of impeller tip speed in determining the grain of dispersions in mixing vessels is established while the need to maintain a sufficient liquid flow rate from the agitator to reduce coalescence and settling is revealed. The influence of these factors has been quantitatively evaluated in a number of cases but more work is needed to achieve a wider generality in their application to industrial processes. In particular, it is necessary to obtain a better understanding of coalescence, and at the present time experimental data are too sparse to permit this. However, it is clear that there is a large effect of even mildly surface active solutes on coalescence rates and these subtle properties may defy prediction in many cases of practical importance. In this situation, it is expedient to carry out measurements of the properties of particular dispersions and dispersion equipment on the laboratory scale, using the experimental tools described at the beginning of this chapter. The problem is thus reduced to that of scaling-up and this matter is returned to in the latter part of the chapter.

For gas-liquid dispersions produced in sieve-tray columns, one can say, as a first approximation, that the Sauter mean bubble diameter at high gas flow rates is $4\frac{1}{2}$ mm., the corresponding gas hold-up is about 60% and the interfacial area 8 cm.$^{-1}$, and that these results are independent of the fluid properties. These are the conditions under which sieve plates operate in distillation and gas absorption practice, and in a following section, continuous phase mass-transfer coefficients deduced from these results and measured point plate efficiencies are discussed in relation to mass-transfer coefficients obtaining in other kinds of dispersions. Here again, the effect of coalescence becomes important when deep pools of liquid are aerated and the properties of the dispersion may be very different near to and far from a multiple orifice disperser.

IV. Theory of Mass Transfer in Continuous Phases

A. Equilibrium and Operating Relationships

The rate of mass transfer in continuous phases has been successfully represented by an equation which for batch contacting between two phases x and y takes the form

$$- (dn/dt)_x = (dn/dt)_y = r = K_{ox}A(C_x - C_x^*) = K_{oy}A(C_y^* - C_y) \qquad (74)$$

where K_{ox} and K_{oy} are the over-all mass-transfer coefficients in cm./sec., A is the interphase contact area in cm.^{2}, C_x is the concentration of transferring solute in one phase in moles/cm.^{3}, C_y^* or C_x^* are the concentration of solute in the same phase which would be in equilibrium with the second phase, and r is the mass-transfer rate in moles/sec. More simply stated, this equation proposes that the rate of mass transfer is directly proportional to the interphase contact area and the uncompleted concentration change The constant of proportionality, known as the mass-transfer coefficient, depends on the fluid dynamics obtaining in the phase contacting equipment and on the physical properties of the system of fluids and transferring solute. The significance of Eq. (74) is clearer when the rates of mass transfer in the individual phases designated x and y are considered and represented as

$$r = k_x A (C_x - C_{xi}) = k_y A (C_{yi} - C_y) \qquad (75)$$

the suffix i representing concentrations at the interface at which point the phases are considered to be in a state of equilibrium. Thus $C_{yi} = mC_{xi}$ (if m is, for example, an equilibrium or partition constant) and also by definition of C^*, $C_y = mC_x^*$.

Combining Eqs. (74) and (75) it is seen that

$$\frac{1}{K_{ox}} = \frac{1}{k_x} + \frac{1}{mk_y} \tag{76}$$

and K_{ox} is recognized as being defined in order to secure the additivity of resistances implied by Eq. (76).

If both sides of Eq. (75) are divided by the volume of fluid V_x, or V_y, one obtains

$$-dC_x/dt = k_x a_x (C_x - C_{xi}) \tag{77}$$

and

$$dC_y/dt = k_y a_y (C_{yi} - C_y) \tag{78}$$

where a_x is the interfacial area per unit volume of phase, x, and a_y is the interfacial area per unit volume of phase, y, and integrating these equations to obtain the amount of mass transfer in time, t,

$$-\int \frac{dC_d}{C_x - C_{xi}} = k_x a_x t = N_x \tag{79}$$

$$\int \frac{dC_y}{C_{yi} - C_y} = k_y a_y t = N_y \tag{80}$$

where N_x and N_y are the dimensionless number of transfer units for phases x and y, respectively, k has the units of velocity, a of reciprocal length, and t of time. It is the frequent purpose of chemical engineering design calculations to evaluate the residence time necessary to secure a desired number of transfer units in an item of equipment, from experimental values of the interfacial area, and the mass-transfer coefficients in each phase, the total mass transfer resistance being deduced as

$$-\int \frac{dC_x}{C_x - C_x^*} = K_{ox} a_x t = N_{ox} \tag{81}$$

where

$$\frac{1}{N_{ox}} = \frac{1}{N_x} + \frac{V_x/V_y}{m N_y} \tag{82}$$

V_x and V_y being the volumes of phases x and y, respectively.

Turning to Eq. (75), the rates of mass transfer in the individual phases are proposed as being proportional to the solute concentration difference between the bulk of the fluid and the interface. This result is founded on a large body of experimental evidence and has been explained in terms of the two film theory originally proposed by Whitman (W5) and by the penetration theory developed by Danckwerts (D1) and others.

The individual mass-transfer coefficients k_x and k_y for many systems have been correlated in terms of hydrodynamic and physical property parameters such as

$$\frac{k_x d}{D_L} \propto \left(\frac{dv\rho}{\mu}\right)^{n_1} \left(\frac{\mu}{\rho D_L}\right)^{n_2} \tag{83}$$

or

$$N_{Sh} \propto (N_{Re})^{n_1} (N_{Sc})^{n_2} \tag{84}$$

where d and v are a diameter and velocity characterizing the fluid motion, D_L is the diffusion coefficient of the solute in the fluid, and ρ and μ are the density and viscosity of the fluid. For example,

$$N_{Sh} = 1.13 \, (N_{Re})^{1/2} (N_{Sc})^{1/2} \quad [\text{Higbie (H8)}] \tag{85}$$

$$N_{Sh} = 2.0 + 0.552 \, (N_{Re})^{1/2} (N_{Sc})^{1/3} \quad [\text{Froessling (F5)}] \tag{86}$$

are correlations for mass transfer in fluids flowing at velocity, v, around bubbles and solid spheres of diameter $d = D_p$. Evidence for decisions about the mechanism of mass transfer is being sought and judged on the ability of the various models to predict experimental correlations such as those above.

The majority of equipment used for mass transfer operates with continuous flow of the phases and in these cases it is sometimes misleading to think in terms of residence times. Here the rate of mass transfer $r = G \, dy = -L dx$ where G and L are the constant molar flow rates of phases y and x and the symbols y and x are most conveniently expressed as mole fractions.

Thus

$$-L dx = K_{ox} a \, (x - x^*) \, dV \tag{87}$$

and

$$-\int \frac{dx}{x - x^*} = K_{ox} a \frac{V}{L} = N_{ox} \tag{88}$$

in which K_{ox} has the units of moles/cm². sec.

The units of velocity for the mass-transfer coefficient may be preserved by multiplying V/L by ρ_x the density of phase x in moles per unit volume. Thus,

$$K_{ox} a \left(\frac{V \rho_x}{L}\right) = N_{ox} \tag{89}$$

where a = interfacial area per unit volume of mixed phases, and
V = total volume of equipment occupied by phases x and y.

The use of ρ_x, as above, changes the mole fraction driving force into concentration units as used in Eq. (74) and results in preserving the units of velocity for the mass-transfer coefficient. However, this may be expected to lead to difficulties in cases where ρ_x changes appreciably in the passage of phase x through the equipment as it will in some cases of liquid extraction. It is expedient to use mole fractions as mass-transfer driving forces because equilibrium and operating relationships most often employ these units and also have to be combined with mass-transfer rate equations in design calculations. However, in these calculations it is generally necessary to deduce a mass-transfer coefficient from a correlation or from a theoretical relationship which in either case will give a mass-transfer coefficient in velocity units and imply the use of the volumetric concentration units C_x or C_y, leading to the employment of a mean value of ρ_x or ρ_y as above.

If the mass rate of flow varies appreciably due to mass transfer, it may be shown as in Ref. (S4) that Eq. (88) has to be modified to give

$$N_{ox} = \tfrac{1}{2}\ln\left[\frac{1-x_2}{1-x_1}\right] + \int \frac{dx}{x^* - x} \qquad (90)$$

and similarly

$$N_{oy} = \tfrac{1}{2}\ln\left[\frac{1-y_2}{1-y_1}\right] + \int \frac{dy}{y^* - y} \qquad (91)$$

where the suffixes 1 and 2 represent terminal conditions. A frequent problem in chemical engineering design calculations is to evaluate the number of transfer units required for a given separation by evaluating the integral

$$\int \frac{dy}{y^* - y} \qquad \text{or} \qquad \int \frac{dx}{x - x^*}$$

by the use of the equilibrium relationship, a material balance (the operating relationship), and the appropriate boundary conditions. If the operation is not isothermal it will also be necessary to employ an enthalpy balance so that variations in temperature may be known and used in order to employ the proper equilibrium constant and fluid physical properties. By way of illustration one may consider the case of the countercurrent contacting of two fluids where the equilibrium and operating relationships are linear.
Rate equation

$$\int \frac{dy}{y - y^*} = N_{oy} \qquad (92)$$

Material balance (operating relationship)

$$y = (L/G)x + b \qquad (93)$$

Equilibrium relationship

$$y^* = mx + c \tag{94}$$

Boundary conditions

When $y = y_1$, $x = x_1$ and when $y = y_2$, $x = x_2$

where L = molar flow rate of phase x,
$\quad\quad G$ = molar flow rate of phase y, and
$\quad\quad m$ = equilibrium constant in terms of mole fractions
$\quad\quad\quad\quad$ (e.g., Henry's or Raoult's law).

It is convenient to define two kinds of mass-transfer efficiencies for flow processes.

(1) The Murphree efficiency or fractional approach to equilibrium of one stream with the other stream *leaving* the equipment. The meaning of the symbols is shown in the diagram below.

$$E_{My} = \frac{y_1 - y_2}{y_1 - y_1^*} \quad \text{and} \quad E_{Mx} = \frac{x_1 - x_2}{x_2^* - x_2} \tag{95a}$$

$$
\begin{array}{cc}
x_2 & y_2 \\
\uparrow & \uparrow \\
L \;\big|\; x & y \;\big|\; G \\
\downarrow & \\
x_1 & x_1
\end{array}
$$

$$\xleftarrow{\hspace{3cm}}$$

Direction of

mass transfer

This efficiency is most significant and numerically less than one where one of the streams is perfectly mixed in its flow through the apparatus.

(2) The fractional approach to equilibrium of one stream with the other stream *entering* the equipment.

$$E_y = \frac{y_1 - y_2}{y_1 - y_2^*} \quad \text{and} \quad E_x = \frac{x_1 - x_2}{x_1^* - x_2} \tag{95b}$$

Substituting for y^* in Eq. (92) by means of Eqs. (93) and (94) and integrating over the boundary conditions given,

$$N_{oy} = \frac{1}{1 - (1/A)} \ln \left[\frac{1 - (1/A)E_y}{1 - E_y} \right] \tag{96}$$

where $A = L/mG$ is called the separation factor. It may be readily shown that Eq. (96) can also be derived by the employment of a logarithmic mean driving force for mass transfer which may therefore be conveniently used when the equilibrium and operating lines are straight.

Also for phase x, the equation analogous to Eq. (96) becomes

$$N_{ox} = \frac{1}{1-A} \ln \left[\frac{1-AE_x}{1-E_x} \right] \tag{97}$$

where $N_{oy} = K_{oy}\,a\,(V\rho_y/G)$ and
$\quad\quad N_{ox} = K_{ox}\,a\,(V\rho_x/L)$,

V being the volume of the contactor and a the interfacial area per unit volume of contactor.

From Eq. (96) when $1/A$ approaches zero (high solubility of solute in x-phase, and x-phase flow rate $\gg$ y-phase flow rate),

$$N_{oy} \approx N_y = -\ln(1-E_y) = -\ln(1-E_{My}) \tag{98}$$

since under these conditions the composition of the x-phase changes but slightly and y^* is almost constant. In other words, the driving force in the x-phase is large compared with that in the y-phase which therefore imposes the major resistance to mass transfer. Also when $1/A$ approaches unity

$$N_{oy} = \frac{E_y}{1-E_y} = E_{My} \tag{99}$$

corresponding to the condition when the equilibrium and operating lines are parallel and the driving forces in both phases remain constant. Again when $1/A$ approaches infinity (low solubility of solute in x-phase and x-phase flow rate $\ll$ y-phase flow rate).

$$N_{ox} \approx N_x = -\ln(1-E_x) = -\ln(1-E_{Mx}) \tag{100}$$

where a large y-phase driving force exists and the major resistance to mass transfer is in the x-phase.

Other situations of practical interest arise when either or both of the phases are well mixed. Thus, when perfect mixing is represented by the symbol, $\xrightarrow{\;\varrho\;}$,

$$-\ln(1-E_{Mx}) = N_{ox} \tag{101}$$

where all of the mass transfer occurs at a constant value of y equal to y_2.

$$\frac{E_{Mx}}{1-E_{Mx}} = N_{ox} \tag{102}$$

$$\frac{E_{My}}{1-E_{My}} = N_{oy} \tag{103}$$

where all of the mass transfer occurs at constant values of both x and y equal to x_1 and y_2.

Also, when the solute is consumed by a rapid chemical reaction in the x-phase for semibatch operations,

$$x = 0 \qquad \begin{array}{c} y_2 \\ \uparrow \\ \vert \\ y_1 \end{array} \qquad -\ln(1 - E_y) = N'_{oy} = -\ln(1 - E_{My}) \qquad (104)$$

$$x = 0 \qquad \begin{array}{c} y_2 \\ \uparrow \\ \circ \\ \vert \\ y_1 \end{array} \qquad \frac{E_{My}}{1 - E_{My}} = N'_{oy} \qquad (105)$$

where in Eqs. (104) and (105)

$$N'_{oy} = K'_{oy} a \frac{V \rho_y}{G} \qquad (106)$$

$$\frac{1}{K'_{oy}} = \frac{1}{k_y} + \frac{m}{k'_x}$$

and where k'_x is the equivalent mass-transfer coefficient for mass transfer accompanied by chemical reaction, the concentration of solute in the bulk x-phase being taken as zero and the interface concentration as that corresponding to the solubility of the solute in the exhausted reagent. If k_x is the mass-transfer coefficient in the x-phase, without chemical reaction, k'_x/k_x has been evaluated (S4) for gas-liquid systems where the hydrodynamic situation is simple (laminar liquid jets, wetted-wall columns) and has also been experimentally observed (S8) in more complex items of equipment (packed beds).

1. Cascade of Well-Mixed Stages

Analytical solutions for the over-all mass-transfer efficiency of a series of mixing stages are available if the equilibrium and operating relationships are linear. Thus, for countercurrent contacting of phases and where each stage is a theoretical stage giving rise to mass-transfer equilibrium, it has been shown (S4)

$$n = \frac{\ln\left[\dfrac{1 - (E_y/A)}{1 - E_y}\right]}{\ln A} \qquad (107)$$

where E_y has been previously defined and n is the number of theoretical stages. Comparing this result with Eq. (96) it is seen that the relationship between the number of theoretical stages and the number of transfer units for this case is

$$n = N \frac{1 - (1/A)}{\ln A} \tag{108}$$

whence $n \to N$ as $A \to 1$. Also for crossflow of the y-phase (96) where the mass-transfer efficiency of each stage is given by E_p (the point efficiency), and n is now the number of mixing stages,

$$E_{oy} = A\left[\left(1 + \frac{E_{py}}{nA}\right)^n - 1\right] \tag{109}$$

where

$$E_{oy} = \frac{y_{Av} - y}{y_n^* - y}$$

and

$$E_{py} = \frac{y_1 - y}{y_i^* - y} \quad \text{(referring to the } i\text{th stage)}$$

$y_{Av.}$ being the mixed mean composition of the outlet y-phase and y_1 the composition of the inlet y-phase. When $n \to \infty$, Eq. (109) reduces to

$$E_{oy} = A[e^{(E_{py}/A)} - 1] \tag{110}$$

as deduced by Lewis (L4) for plug flow of the x-phase. Also, it is seen that $E_{oy} = E_{py}$ when $n = 1$ (complete mixing). Again it may be shown (F2) that when all of the resistance to mass transfer is in the x-phase, Eq. (109) reduces to,

$$1 - E_{Mx} = \frac{1}{\left(1 + \dfrac{N_x}{n}\right)^n} \tag{111}$$

Equation (111) becomes, for $n = \infty$ (plug flow),

$$-\ln(1 - E_{Mx}) = N_x \tag{112}$$

and for $n = 1$ (complete mixing)

$$\frac{E_{Mx}}{1 - E_{Mx}} = N_x \tag{113}$$

as previously employed [see Eqs. (98) to (103)].

Equation (109) is a particularly useful form for application to the problem of liquid mixing on large diameter gas absorption and distillation trays where the liquid and gas streams are in cross flow and the liquid mixing can be conveniently represented as due to a number of equivalent well-mixed pools in series. Foss *et al.* (F2) have shown experimentally that the variance of the liquid residence time distribution function divided by the mean liquid residence time is approximately equal to the reciprocal of the number of equivalent well-mixed pools on gas-liquid plate contactors and have correlated the variance with the froth height, froth density, and length of tray.

For cases where the equilibrium and or operating lines are nonlinear the mass-transfer equations are best solved by graphical means to give the number of transfer units or theoretical stages for a desired separation as described in textbooks devoted to mass transfer, e.g., (S4). The use of a variety of operating relationships resulting from the use of reflux and other equilibrium laws peculiar to different systems is also discussed fully in these places.

B. Mass Transfer in the Continuous Phase around an Axisymmetric Body of Revolution

The rates of mass transfer from small drops or bubbles into a continuous phase can be analyzed rigorously if the drops or bubbles are spherical and rise or fall in straight paths (G8, H1, R4). The shapes and paths of larger drops or bubbles are generally irregular and vary rapidly with time, making an exact theoretical treatment impossible. A reasonably satisfactory approach to the problem is to assume that the perturbations in shape and path affect the transfer rates to a minor extent only. Approximate mass-transfer rates can then be derived for large bubbles and drops by supposing them to have well-defined shapes and to rise in straight lines. Since deformed bubbles are generally classified (H1, R4, T1) as ellipsoids of revolution or spherical caps, it is convenient to assume that all bubbles are symmetrical about the axis in the direction of rise. Mass-transfer relations are derived below for any such axisymmetric body of revolution when both the Peclet and Schmidt numbers are large, as in the case of rising bubbles. These general equations are then used to predict the rates of dissolution of solid and fluid spheres, fluid oblate spheroids, and fluid spherical caps.

When the thickness of the velocity boundary layer is far less than the equivalent spherical radius of the dissolving body, a dimensionless continuity equation for the continuous phase fluid can be written (B9), by employing dimensionless coordinates $U_1 = u/U$, $v_1 = v/U$, $r_1 = r/R_e$, $x_1 = x/R_e$ and $y_1 = y/R_e$, where R_e is the radius of a sphere having the same volume as the axisymmetric body of revolution. (Symbols are defined in Fig. 26.) Thus,

$$\frac{\partial}{\partial x_1}(u_1 r_1) + \frac{\partial}{\partial y_1}(v_1 r_1) = 0 \tag{114}$$

Similarly, for a thin concentration boundary layer (F5),

$$u_1 \frac{\partial C_1}{\partial x_1} + v_1 \frac{\partial C_1}{\partial y_1} = \frac{2}{N_{Pe}} \frac{\partial^2 C_1}{\partial y_2} \tag{115}$$

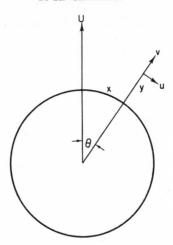

FIG. 26. The coordinates of an axisymmetric body of revolution (taken as a sphere for simplicity) rising through a continuous phase with velocity, U.

where $C_1 = C/C^*$, C^* being the concentration of solute in the solvent at saturation.

It has been suggested (F4) that the thin concentration boundary layer approximation is valid for N_{Pe} greater than roughly 10^2. Since $N_{Pe} = N_{Re}N_{Sc}$ and N_{Sc} is roughly 10^3 for gases dissolving in liquids, the approximation applies to gas bubbles moving in the regime $N_{Re} > \frac{1}{10}$.

Expanding u_1 in a Taylor series gives

$$u_1 = (u_1)_{y_1=0} + \left(\frac{\partial u_1}{\partial y_1}\right)_{y_1=0} y_1 + \left(\frac{\partial^2 u_1}{\partial y_1^2}\right)_{y_1=0}\left(\frac{y_1^2}{2!}\right) + \qquad (116)$$

If the concentration boundary layer is far thinner than the velocity boundary layer, all the penetration of the solute into the solvent will occur in a layer of fluid in which the velocity can be represented by the first two terms of Equation (116) abreviated to read

$$u_1 = u_{10} + u'_{10} y_1 \qquad (117)$$

Since the Schmidt number is a measure of the ratio of the thicknesses of the velocity and concentration boundary layers, this simplification is possible only when $N_{Sc} \gg 1$.

The tangential velocity given by Eq. (117) can be used in Eq. (114) to find the radial velocity, and substitution of the two velocities into Eq. (115) leads to

$$(u_{10} + u'_{10} y_1)\frac{\partial C_1}{\partial x_1} - \frac{1}{r_1}\frac{\partial}{\partial x_1}\left\{\left(u_{10}y_1 + \frac{u'_{10} y_1^2}{2}\right)r_1\right\}\frac{\partial C_1}{\partial y_1} = \frac{2}{N_{Pe}}\frac{\partial^2 C_1^2}{\partial y_1^2} \qquad (118)$$

This is the general equation describing the dissolution of an axisymmetric body of revolution at high N_{Pe} and N_{Sc}. The boundary conditions which must be applied to any concentration distribution resulting from its solution are

$$\left.\begin{array}{lll} C_1 = 1, & y_1 = 0 & \text{all } x_1 \\[2mm] C_1 = 0, & y_1 = \infty & \text{all } x_1 \end{array}\right\} \tag{119}$$

Equation (118) can be solved analytically in two limiting situations.

Case 1: Immobile Interface $u_{10} = 0$.

Equation (118) becomes

$$u_{10}' y_1 \frac{\partial C_1}{dx_1} - \frac{1}{r_1} \frac{\partial}{\partial x_1} \left(\frac{u_{10}' y_1^2 r_1}{2} \right) \frac{\partial C_1}{\partial y_1} = \frac{2}{N_{Pe}} \frac{\partial^2 C_1}{\partial y_1^2} \tag{120}$$

and can be solved (L7) to give the following general mass-transfer equation when the boundary conditions (119) are applied.

$$N_{Sh} = 0.641 \left(\frac{4\pi R_e^2}{A} \right) \left[\int_0^{x_1} (u_{10}' r_1)^{1/2} r_1 \, dx_1 \right] N_{Pe}^{1/3} \tag{121}$$

where A is the total area of the body. Knowledge of the shapes and flow patterns outside particular bodies allows Eq. (121) to be reduced to more convenient forms.

Case 2: Mobile Interface $u_{10} \gg u_{10}' y_1$

In this case Eq. (118) reduces to

$$u_{10} \frac{\partial C_1}{\partial x_1} - \frac{1}{r_1} \frac{\partial}{\partial x_1} (u_{10} y_1 r_1) \frac{\partial C_1}{\partial y_1} = \frac{2}{N_{Pe}} \frac{\partial^2 C_1}{\partial y_1^2} \tag{122}$$

Solution of this equation (L7) with the same boundary conditions as before gives

$$N_{Sh} = \left(\frac{2}{\pi} \right)^{1/2} \left(\frac{4\pi R_e^2}{A} \right) \left[\int_0^{x_1} u_{10} r_1^2 \, dx_1 \right]^{1/2} N_{Pe}^{1/2} \tag{123}$$

for $N_{Sh} \gg \mu_{10}'/\mu_{10}$. Some particular solutions of the general mass transfer equations will now be given.

1. Solid Spheres

For a sphere, $r_1 = \sin \theta$, $dx_1 = d\theta$, and $A = 4\pi R_e^2$, so that Eq. (121) can be reduced to

$$N_{Sh} = 0.641 \left[\int_0^\pi (u_{10}' \sin \theta)^{1/2} \sin \theta \, d\theta \right]^{2/3} N_{Pe}^{1/3} \tag{124}$$

This equation has been derived by Baird and Hamielec (B3) from a simplified differential equation in which the transfer contribution due to radial convection was neglected. The problem of predicting N_{Sh} is therefore reduced to one of finding the radial gradient at the interface of the tangential velocity.

(a) $N_{Re} < 1$. By neglecting the inertial terms appearing in the Navier-Stokes equation, Stokes (S9) derived an analytical solution for the stream function around solid spheres, from which

$$u'_{10} = \tfrac{3}{2} \sin \theta \tag{125}$$

Substitution of this expression into Eq. (124) and evaluation of the integral term leads to

$$N_{Sh} = 0.99 \, N_{Pe}^{1/3} \qquad N_{Re} < 1 \tag{126}$$

Friedlander (F4) has also derived this result from a solution of Eq. (115) (in spherical coordinates) in which the velocity components were taken from Stokes' stream function. The form of Eq. (126) has also been verified for the dissolution of small bubbles (C8), liquid drops (W2), and solid spheres (A1).

(b) $N_{Re} > 1$. In this regime, the Navier-Stokes equation cannot be solved explicitly for u'_{10}, as the inertial terms cannot be ignored. However, the Navier-Stokes equation can be modified (B9) to apply in the boundary layer which forms over the moving sphere.

$$u\frac{\partial u}{\partial x} + v\frac{\partial u}{\partial y} = U_\infty \frac{dU_\infty}{dx} + v_c \frac{\partial^2 u}{\partial y^2} \tag{127}$$

where v_c is the kinematic viscosity of the continuous phase and U_∞ the velocity at the edge of the velocity boundary layer.

By assuming that the velocity distribution in the boundary layer can be expressed by a "similar" equation

$$\frac{u}{U_\infty} = f(\eta) \tag{128}$$

where $\eta = y/\delta$, δ being the velocity boundary layer thickness which can be found (L7) from an integrated form of Eq. (127). Substitution of this thickness into the assumed velocity profile allows u'_{10} to be found and substituted into Eq. (124) for N_{Sh}.

In particular, for the quartic profile,

$$\frac{u}{U_\infty} = 4\eta - 6\eta^2 + 4\eta^3 - \eta^4 \tag{129}$$

$$N_{Sh} = 0.62 \, N_{Re}^{1/2} N_{Sc}^{1/3} \tag{130}$$

Equation (130) was deduced on the assumption of no flow separation at the sphere's surface: in fact, separation occurs (T3) at $\theta \approx 108°$. Estimation of the transfer occurring over the front surface of the sphere up to this separation angle gives the following result, again for the quartic distribution.

$$N_{Sh} = 0.56 \, N_{Re}^{1/2} N_{Sc}^{1/3} \tag{131}$$

This relationship was first used by Froessling (F5) to correlate the rates of evaporation of naphthalene spheres, and drops of aniline, benzene, and water, into an air stream and has since been used by many workers to describe heat and mass transfer from solid spheres. Small bubbles also dissolve (C8, H4) according to Eq. (131).

Froessling (F6) studied mass-transfer rates theoretically by expanding the stream function and the concentration distribution outside the sphere into complex series functions. This approximate approach resulted in a constant. of 0.59 in Eq. (131). The method used above to derive Eq. (131) therefore gives a satisfactory result, and has the advantage of relative simplicity.

2. Fluid Spheres

For a mobile sphere, Eq. (123) becomes

$$N_{Sh} = (2/\pi)^{1/2} \left[\int_0^\pi u_{10} \sin^2 \theta d\theta \right]^{1/2} N_{Pe}^{1/2} \tag{132}$$

Precise equations for N_{Sh} are possible if the interfacial velocity is known.

(a) $N_{Re} < 1$. From Hadamard's (H2) stream function for flow around a fluid sphere,

$$u_{10} = \left(\frac{\mu_c}{\mu_c + \mu_d} \right) \frac{\sin \theta}{2} \tag{133}$$

Use of this equation in Eq. (123) gives N_{Sh} as

$$N_{Sh} = 0.65 \left(\frac{\mu_c}{\mu_c + \mu_d} \right)^{1/2} N_{Pe}^{1/2} \tag{134}$$

For Eq. (134) to apply,

$$N_{Pe} \gg 9.4 \left(\frac{\mu_d}{\mu_c} + 1 \right) \left(3\frac{\mu_d}{\mu_c} + 1 \right)^2$$

Hammerton's results for the dissolution of bubbles in glycerol (H4) follow Eq. (134) reasonably well, while Ward (W2) has also derived this equation, with a constant of 0.61, by means of an approximate integral method due to Friedlander (F3) and has used it to correlate data for the dissolution of drops of water in cyclohexanol at $N_{Re} < 10$.

(b) $N_{Re} > 1$. Chao (C15) has analyzed the boundary layer around fully circulating spheres moving at high N_{Re}, and has expressed the velocity distributions both inside and outside the sphere as dimensionless perturbations from the potential-flow fields. These equations lead to

$$u_{10} = \tfrac{3}{2} \sin \theta - \frac{2\sqrt{3}}{N_{Re}^{1/2}} \frac{2 + 3\mu_d/\mu_c}{1 + (\rho_d \mu_d/\rho_c \mu_c)^{1/2}} \frac{[\int_0^\theta \sin^3 \theta d\theta]^{1/2}}{\sin \theta} \frac{1}{\sqrt{\pi}} \tag{135}$$

The dissolution of mobile spheres can therefore be described (L7) by

$$N_{Sh} = 1.13 \left[1 - \frac{2 + 3\mu_d/\mu_c}{1 + (\rho_d\mu_d/\rho_c\mu_c)^{1/2}} \frac{1.45}{N_{Re}^{1/2}} \right]^{1/2} N_{Pe}^{1/2} \qquad (136)$$

when the following approximate condition applies:

$$N_{Pe} \gg \pi \left[\frac{4\mu_d/\mu_c - (\rho_d\mu_d/\rho_c\mu_c)^{1/2}}{1 + (\rho_d\mu_d/\rho_c\mu_c)^{1/2}} \right]^2 \left[1 + \frac{4.35}{N_{Re}^{1/2}} \frac{2 + 3\mu_d/\mu_c}{1 + (\rho_d\mu_d/\rho_c\mu_c)^{1/2}} \right] \qquad (137)$$

Equation (136) correlates data for drops of water dissolving in iso butanol (G8, H6) and drops of ethyl acetate dissolving in water (G8) reasonably well. For gas bubbles, both μ_d/μ_c and $(\rho_d\mu_d/\rho_c\mu_c)^{1/2}$ are far less than one so that Eq. (136) becomes

$$N_{Sh} = 1.13 \left[1 - \frac{2.96}{N_{Re}^{1/2}} \right]^{1/2} N_{Pe}^{1/2} \qquad (138)$$

This equation should apply for gas bubbles moving in pure liquids when N_{Re} lies in the approximate range $100 \leqslant N_{Re} \leqslant 400$.

When N_{Re} is very high (potential flow)

$$N_{Sh} = 1.13 \, N_{Pe}^{1/2} \qquad (139)$$

which has been used (H4, H8, W3) in attempts to correlate the rates of dissolution of large bubbles.

3. Fluid Oblate Spheroids

Gas bubbles in the range $0.2 < d_e < 1.8$ cm. are deformed, but can be assumed to be oblate spheroids (H1, R4, T1). The Reynolds numbers of such bubbles are very high, so that the flow is potential and can be described by equations published by Zahm (Z1). From these equations

$$u_{10} = (1 + k) \sin \theta \qquad (140)$$

where k is a complex function of e_c the eccentricity of the spheroid defined as the ratio of its width to height.

Analysis of the shape of the spheroid, and use of Eq. (140) leads to a solution (L7) of the general Eq. (123) which applies to dissolving oblate spheroids of different eccentricities.

$$N_{Sh} = \{\tfrac{2}{3}(1 + k)\}^{1/2} \left\{ \frac{2.26 \, e_c^{1/3} (e_c^2 - 1)^{1/2}}{e_c(e_c^2 - 1)^{1/2} + \ln [e_c + (e_c^2 - 1)^{1/2}]} \right\} N_{Pe}^{1/2} \qquad (141)$$

From Eq. (141) it may be shown that flattening a spherical bubble does not influence the transfer coefficient to an important degree, which explains the successful use (H4, H8, W3) of Eq. (139) in describing transfer around oblate bubbles.

4. Fluid Spherical Caps

Bubbles greater than approximately 1.8 cm. in diameter assume mushroomlike or spherical-cap shapes, and undergo potential flow (D5). The interfacial velocity is therefore given by the potential value for a sphere.

$$u_{10} = \tfrac{3}{2} \sin \theta \tag{142}$$

Neglecting transfer from the rear of the bubble, Eq. (142) can be used (L7) to give

$$N_{Sh} = 1.79 \frac{(3e_c^2 + 4)^{2/3}}{e_c^2 + 4} N_{Pe}^{1/2} \tag{143}$$

Here, e_c is the ratio of the bubble width to its height, and is therefore equivalent to the eccentricity of an oblate spheroid.

Equation (143) can be simplified if necessary by eliminating the rising velocity since (H1)

$$U = 1.02 (g R_e)^{1/2} \tag{144}$$

Rosenberg (R4) and Takadi and Maeda (T1) have suggested that $e = 3 \cdot 5$ for all spherical-capped bubbles.

Equation (143) therefore takes the final form

$$N_{Sh} = 1.28 N_{Pe}^{1/2} \tag{145}$$

which is nearly the same as Eq. (139) for spheres in potential flow.

C. Mass Transfer with Chemical Reaction

The important problem of mass transfer of a substance from one phase into another phase where chemical reaction takes place, cannot be fully treated in this section, where attention will be confined to the case where the chemical reaction is first order and irreversible. For the treatment of other types of reaction, reference may be made to (S4).

Suppose that a substance A diffuses from one phase into a second continuous phase where it is consumed by a first-order irreversible reaction having a rate constant, k. Under steady state conditions, a material balance at distance, x, from the interface in the diffusion zone gives

$$D_A \frac{d^2 c_A}{dx^2} = k c_A \tag{146}$$

where D_A is the diffusion coefficient of A in the phase where reaction takes place. Applying the boundary conditions, when $x = 0$, $C_A = C_{Ai}$ (equilibrium at the interface) and when $x = l$, $C_A = C_{Al}$ (l being the thickness of the diffusion zone) integration of Eq. (146) gives

$$C_A = \frac{C_{Al} \sinh bx + C_{Ai} \sinh b(l - x)}{\sinh bl} \tag{147}$$

where $b = \sqrt{k/D_A}$.

From the film theory of mass transfer,

$$l = D/k_1 \quad \text{or} \quad bl = \sqrt{kD_A}/k_1 = \phi \tag{148}$$

where k_1 is the mass-transfer coefficient of A in the reacting phase, and ϕ is a dimensionless number defined by Eq. (148). Thus, the rate of material transfer between the two phases at steady state in moles per unit time is found by differentiating Eq. (147) to give

$$r = -AD_A \left(\frac{dC_A}{dx}\right)_{x=0} = \frac{A\sqrt{KD_A}}{\sinh \phi} \, (C_{Ai} \cosh \phi - C_{Al}) \tag{149}$$

Here, A is the total interphase contact area and is equal to aV_D where a is the area per unit volume of dispersion or mixed phases of total volume, V_D. Also,

$$r = H_1 V_D k C_{Al} = -AD_A \left(\frac{dC_A}{dx}\right)_{x=l} = \frac{A\sqrt{kD_A}}{\sinh \phi} \, (C_{Ai} - C_{Al} \cosh \phi) \tag{150}$$

where H_1 is the volume fraction of the phase in which reaction occurs. Eliminating C_A from the two previous equations, one finds

$$\frac{r}{V_D} = \frac{aC_{Ai}\sqrt{kD_A}}{\tanh \phi} \left[1 - \left(\frac{H_1 k_1}{aD_A} \phi \sinh \phi \cosh \phi + \cosh^2 \phi \right)^{-1} \right] \tag{151}$$

This equation simplifies for restricted ranges of the value of ϕ. Thus when $\phi < 0.3$,

$$\frac{r}{V_D} = \left[\frac{1}{k_1} + \frac{a}{H_1 k} \right]^{-1} a \, C_{Ai} \tag{152}$$

and as a further corollary, when $k \ll k_1(a/H_1)$ (slow chemical reaction rate),

$$\frac{r}{V_D} \approx H_1 k C_{A1} \tag{153}$$

and the chemical reaction rate is rate-determining. Also, when $k \gg k_1(a/H_1)$ (fast chemical reaction rate),

$$\frac{r}{V_D} = k_1 a C_{Ai} \tag{154}$$

and diffusion controls the over-all rate. However when $\phi > 2$,

$$\frac{r}{V_D} = \sqrt{kD_A} \, a C_{Ai} \tag{155}$$

and comparing Eqs. (154) and (155) it is seen that $\sqrt{kD_A}/k_1 = \phi$ is the "enhancement factor" which represents the extent to which material transfer is increased due to chemical reaction in the diffusion zone, above that which would occur if the process were physical rate-controlled and $C_{Ai} = 0$. Since the kinetic rate constant $\sqrt{kD_A}$ in Eq. (155) is independent of the ·hydrodynamic behavior of the fluids, it may be determined in any convenient apparatus in which the area of contact between the fluids is known, such as a liquid jet or wetted surface device. The kinetic rate constant determined in this way may then be used to predict the rate of material transfer when the same process occurs in more complex contacting equipment if the area of contact can be assessed or the latter may be evaluated from experimental observations of the rate of transfer r/V_D in commercial equipment. This is the principle of the "chemical method" for determining interfacial areas, referred to in the first part of this chapter. Since Westerterp (W4) has shown that $\phi \approx 6$ for the absorption of oxygen in catalyzed sodium sulfite solutions, this system is an attractive one for evaluating gas-liquid interfacial areas in a variety of equipment such as aerated mixing vessels, bubble reactors, and packed columns. Some caution must however be exercised when gas bubble dispersion equipment is evaluated since the presence of electrolytes in water is known to increase the interfacial area above that obtaining when pure liquids are aerated. In gas-liquid contacting, it is worthwhile noting that the use of the chemical method for determining interfacial areas does not require a knowledge of the Henry constant, since this is the same in both the dispersion equipment and in the liquid or wetted surface apparatus used to evaluate $\sqrt{kD_A}$ if the temperature is the same in both cases. Thus a value of $B = \sqrt{kD_A}/m$ may be found from the jet or similar experiments and employed in Eq. (155) as

$$\frac{r}{V_D} = \frac{BaP}{RT} \tag{156}$$

where P is the partial pressure of the absorbing gas, and m (the Henry constant) is the reciprocal of the solubility of the gas in the liquid, in, for example, cm.^{3}gas/cm.^{3} liquid.

Westerterp (W4) finds for the absorption of oxygen in aqueous sodium sulfite solution catalyzed by cupric ions of concentration $(0.5$ to $4.0) \times 10^{-3}$ moles-liter, $B = (7.3 \pm 0.9) \times 10^{-3}$ cm./sec. at 30°C.

V. Continuous Phase Heat and Mass-Transfer Properties of Dispersions

The mass-transfer properties of dynamically maintained dispersions have only recently received systematic attention with the advent of ways of measuring the interphase contact area as previously described. Hitherto, one had to accept transfer unit correlations for particular items of equipment or make the assumption that the particles of the dispersion obey semitheoretical mass-transfer equations of the type proposed by Froessling (F5) and Higbie (H8).

Calderbank (C3) and Calderbank and Moo-Young (C8) have measured

Table II

Fluid Systems Investigated and Physical Properties[a]

Solute	Solvent	$D_L \times 10^5$ at 20°C. (cm.²/sec.)			Henry constant at 20°C. (atm.)	Vapor pressure of pure solute at 20°C. (atm.)	Activity coeff. of solute in solvent at 20°C.
		Measured C(3)	Literature values	Ref.			
Benzene	Water	1.13	1.06	(W7)	—	0.100	2410
Ethyl bromide	Water	1.29	1.24	(W7)	—	0.513	662
Carbon disulfide	Water	1.36	1.34	(W7)	—	0.397	580
Methyl iodide	Water	1.18	1.40	(W7)	—	0.44	1950
Ethylene	Water	—	1.51	(W7)	10.18×10^3	—	—
Propylene	Water	—	1.20	(W7)	6.32×10^3	—	—
Butadiene	Water	—	1.05	(W7)	3.0×10^3	—	—
Oxygen	Water	—	2.0	(D3)	4.0×10^4	—	—
Carbon dioxide	Water	1.92^b	1.92^b	(D4)	1.63×10^{3b}	—	—
Sulfur dioxide	Water	—	1.46	(D3)	Varies with concn.	—	—
Benzene	Glycol	0.135	—		—	0.100	26.7
Oxygen	Glycol	0.238	—		1.5×10^{4b}	—	—
Butadiene	Glycol	0.125	—		—	—	—
Carbon dioxide	Glycol	0.229^b	—		4.5×10^{2b}	—	—

[a] From Calderbank (C8).

[b] At 25°C.

mass-transfer rates in the continuous phase of dispersions. Experiments were performed in which sparingly soluble volatile or gaseous solutes were desorbed from or dissolved in liquid solvents by a gaseous phase dispersed in the liquid as a bubble cloud. The rate of material transfer and the interfacial area were simultaneously observed, the experimental equipment being such as to ensure that the liquid phase was perfectly mixed. In this way values of the liquid-phase mass-transfer coefficients were determined in aerated mixing vessels and in sintered and sieve-plate columns. This work covered a wide range of bubble sizes and liquid-phase diffusion coefficients, extreme variations in the latter being achieved by the use of aqueous glycol and glycerol solutions, and small variations by the use of various solutes, as shown in Tables II and III.

The gas hold-up was varied over a considerable range so that the results could be applied to the processes of fermentation, gas absorption, and distillation. The data for bubble swarms were extended, first by the use of an aqueous thickening agent (polyacrylamide), which enabled the viscosity to be independently and largely increased without affecting the diffusion coefficient. Solutions having viscosities up to 87 cp. were used and found to be Newtonian in rheological behavior, and to have the same diffusion coefficient for carbon dioxide as pure water. The data were also extended to include the high diffusion coefficients obtained using hydrogen as solute. This was achieved by

Table III

Fluid Systems Investigated and Properties—Carbon Dioxide[a]

Solvent (wt.—%)	ρ_c (gm./cc.)	μ_c (c.p.)	$D_L \times 10^6$ at 25°C. (measured) (cm²/sec.)
94 Glycol-water	1.11	15.4	3.5
85 Glycol-water	1.10	12.0	4.35
79 Glycol-water	1.10	9.75	5.4
72 Glycol-water	1.09	7.1	7.08
58 Glycol-water	1.07	4.15	8.05
41 Glycol-water	1.046	2.35	10.6
70.4 Glycerol-water	1.081	19.2	1.52
67.6 Glycerol-water	1.174	15.4	2.32
64.4 Glycerol-water	1.166	12.0	2.88
60.4 Glycerol-water	1.155	9.2	3.88
54.0 Glycerol-water	1.138	6.2	4.99
45.0 Glycerol-water	1.114	4.0	6.05
28.0 Glycerol-water	1.107	2.1	9.34
19.3 Glycerol-water	1.1046	1.5	13.40
9.5 Glycerol-water	1.022	1.1	17.10

[a]From Calderbank and Moo-Young (C8).

measuring the rates of catalytic hydrogenation reactions in various solvents containing an excess of catalyst, and through which hydrogen bubble swarms were passed. The excess of catalyst ensured that the chemical reaction rate was fast and not rate-determining, and the progress of the reaction was determined by the rate of solution of hydrogen, and followed by the infrared absorption characteristics of the product of reaction.

Finally, other published data on continuous-phase heat and mass transfer to dispersions of solids and to single bubbles and drops, as shown in Table IV, were combined with the above to achieve a comprehensive and convincing correlation.

The above work showed that the liquid-phase diffusion coefficient was the major factor which influenced the value of the mass-transfer coefficient. It was clear that agitation intensity, bubble size, and free rising velocity had no effect on the mass transfer coefficient, but that large bubbles (>2.5 mm. diam.) gave greater mass transfer coefficients than small bubbles (<2.5 mm. diam.), whose mass-transfer coefficients correlated well with those observed for solid-liquid dispersions. It was concluded that small "rigid sphere" bubbles experience friction drag, causing hindered flow in the boundary layer sense, and that under these circumstances the mass-transfer coefficient is proportional to the two-thirds power of the diffusion coefficient as found by Froessling (F5) and others. For large bubbles (>2.5 mm. diam.) form drag predominates, and the condition of unhindered flow envisaged by Higbie (H8) is realized and as postulated by Higbie, the mass-transfer coefficient is proportional to the half-power of the diffusion coefficient. It is important to appreciate that the Froessling and Higbie equations had never hitherto been tested with systems of gas bubbles for extreme variations of diffusion coefficients and liquid viscosities. Although the experimental results supported these equations as regards the effect of the diffusion coefficient on the mass-transfer coefficient, they also showed that the effects of bubble size and slip velocity were such as to be mutually compensating, so that these variables were replaceable by physical property parameters.

Figure 27 shows two correlations for liquid-phase mass-transfer coefficients for bubble swarms passing through liquids in sieve and sintered-plate columns and in aerated mixing vessels. The upper correlation line refers to bubble swams of average bubble diameter greater than about $2\frac{1}{2}$ mm., such as are produced when pure liquids are aerated in mixing vessels and sieve-plate columns, and leads to the result

$$N_{Sh} = 0.42\, N_{Sc}^{1/2}\, N_{Gr}^{1/3} \tag{157}$$

or

$$k_L (N_{Sc})^{1/2} = 0.42 \left(\frac{\Delta\rho\mu_c g}{\rho_c^2} \right)^{1/3} \tag{158}$$

Table IV

Range of Variables for Fig. 27

Ref.	k_L(cm./sec.)	h_c(cal./cm.²sec.°C.)	$D_L \times 10^5$(cm.²/sec.)	λ(cal./cm.sec.°C.)	$\Delta\rho$(gm./cc.)	ρ_c(gm./cc.)	μ_c(cp.)	D(cm.)
Large bubbles (C8)	0.0024-0.010	—	0.19-4.8	—	1-1.178	1-1.178	0.6-87	0.20-0.80
Small bubbles and rigid spheres	0.0024-0.010	—	0.34-1.78	—	1.11-1.160	1.11-1.160	1.13-10.5	0.02-0.06
(C8)	0.07-0.122	—	5-26	—	0.698-0.772	0.698-0.772	1.99-8.9	—
(C8)	0.007	—	1.13	—	1.00	1.00	1.00	0.075
(C8)	0.00122-0.00187	—	6.5-9.3	—	0.174-0.177	1.0	0.89-1.52	0.09
(C7)	—	0.039-0.459	—	$1.43-66.7 \times 10^{-3}$	12.27-12.55	1.0-1.249	1.08-3.95	0.225
(M2)	0.0253-0.048	—	4.75-14.25	—	0.867	0.867	0.28-0.90	0.14
(P2)	0.011	—	1.35	—	0.89	1.178	1.55	0.01-0.065
(K2)	0.0059-0.138	—	1.06-1.84	—	1.30-1.98	1.0	0.4-1.0	0.09-1.2
(R5)	0.0127-0.0215	—	2.8-4.3	—	1.165	1.26	0.371-0.535	0.11
(O1)	0.00435	—	0.45	—	0.588	1.0	1.0	0.33-0.415
(H10)	0.0042-0.00627	—	1.15-1.75	—	0.266-0.36	0.79-1.11	0.55-1.0	0.45-0.65
(H13)	0.00291	—	0.62	—	0.69	1.0	1.0	0.14-0.47

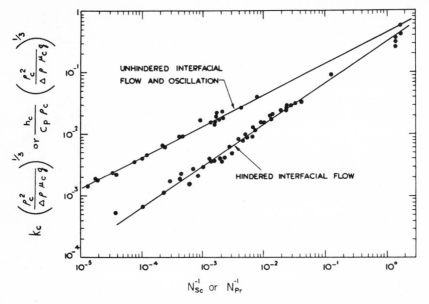

FIG. 27. Continuous phase heat and mass-transfer coefficients for freely suspended dispersions and for dispersed-phase rising or falling freely (C8). (For a key to the symbols, see the accompanying tabulation.)

The lower correlation line refers to bubble swarms of average bubble diameter less than about $2\frac{1}{2}$ mm. Such bubbles are produced when aqueous solutions of hydrophylic solutes are aerated in mixing vessels, or when liquids generally are aerated with sintered plates, or plates containing very small perforations. This leads to the result

$$k_L(N_{Sc})^{2/3} = \frac{h_c}{C_p\rho_c}(N_{Pr})^{2/3} = 0.31\left(\frac{\Delta\rho\mu_c g}{\rho_c^2}\right)^{1/3} \tag{159}$$

or

$$N_{Sh} = N_{Nu} = 0.31\, N_{Ra}^{1/3} \tag{160}$$

where N_{Sh} = Sherwood number, N_{Sc} = Schmidt number, N_{Gr} = Grashoff number, N_{Nu} = Nusselt number, N_{Ra} = Rayleigh number, and N_{Pr} = Prandtl number. It will be seen that both correlations show that the liquid-phase mass-transfer coefficients are independent of bubble size and slip velocity, and depend only on the physical properties of the system. In mixing vessels the mass-transfer coefficients are independent of the power dissipated by the agitator, and in sieve-plate columns they are independent of the fluid flow rates.

F.R.—Free rise; F.F.—Free fall; S.T.—Stirred tank; F.B.—Fluidized bed

System	Process	Ref.
Aqueous glycerol solutions–CO_2	Absorption (F.R., S.T.)	(C8)
Aqueous glycol solutions–CO_2	Absorption (F.R., S.T.)	(C8)
Glycol–CO_2	Absorption (F.R., S.T.)	(C8)
Water–CO_2	Absorption (F.R., S.T.)	(C8)
Water–O_2; 15°C.	Absorption (F.R., S.T.)	(C8)
Water–O_2; 45°C.	Absorption (F.R., S.T.)	(C8)
Brine–O_2	Absorption (F.R., S.T.)	(C8)
Polyacrylamide solutions–CO_2	Absorption (F.R., S.T.)	(C8)
Wax–H_2	Absorption (F.R.)	(C8)
Aqueous ethanol solution–C_6H_6/air	Desorption (F.R.)	(C8)
Water–resin beads	Ion exchange (S.T.)	(C6)
Aqueous glycerol solutions–Hg	Heat transfer (F.F.)	(C7)
Water–O_2	Absorption (F.R.)	(C19)
Water–O_2	Absorption (F.R.)	(H3)
Water–O_2; CO_2	Absorption (F.R.)	(S4)
Toluene–H_2	Absorption (F.R.)	(M2)
Brine–NaCl	Dissolution (S.T.)	(P2)
Water–KCl	Dissolution (F.B., S.T.)	(K2)
Water–urea	Dissolution (F.B., S.T.)	(K2)
Brine–NaCl	Crystallization (F.B.)	(R5)
Water–sucrose	Dissolution (S.T.)	(O1)
Solvents–benzoic acid	Dissolution	(H10)
Water–sodium thiosulfate	Dissolution (S.T.)	(H13)
Water–air	Large rain drops evaporating (F.F.)	(K1)
Water–air	Small rain drops evaporating (F.F.)	(K1)
Air-sand ⎫ Air-caborundum ⎬ Air-aloxite ⎭	Small particles falling through hot air undergoing heat transfer (F.F.)	(J3)
Water–air	Dissolution in centrifugal apparatus	(W8)
Water–H_2	Absorption (F.R.)	(H3)
Water–CO_2	Absorption (F.R.)	(B2)
Water-benzoic acid ⎫ Water-benzoic acid ⎬ Water-benzoic acid ⎭	Dissolution by forced convection at fluid flow equivalent to particle terminal velocity	(G2) (L5) (L6)

A. Small Bubble Correlation

The data for small bubbles are in agreement with those reported by other workers for both heat and mass transfer with solid in liquid and liquid in liquid dispersions as shown in Fig. 27.

Figure 28 also shows that Eq. (159) is in agreement with data on heat and mass transfer in free turbulent convection from stationary spheres, if the term $\Delta\rho$ is taken as the difference in density between the fluid at the interface and the bulk in free convection as compared with the difference in density

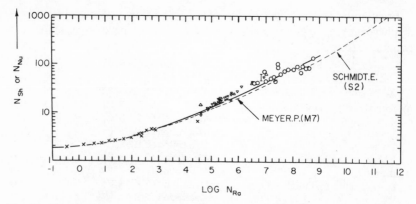

Fig. 28. Heat and mass transfer in natural convection at spherical surfaces and in free rise or fall of dispersed phase. KEY: ○ (G2); × (M1); I (M2); △ (C7); ▽ (C8).

between the dispersed and continuous phases which is employed when the former is free to move under gravity. This result is obviously due to the fact that in both cases the relative rate of flow between the phases is induced, and its magnitude determined, by the appropriate value of the density difference.

It may be noted that Morgan and Warner (M11) have deduced the equation $N_{Nu} = 0.55\ N_{Ra}^{1/4}$ for natural convection heat transfer from laminar boundary layer theory. This result is not significantly different from Eq. (160) at low values of the Rayleigh number but shows a more marked departure at high values where the theory is less certain. Figure 28 demonstrates the fact that as the density difference between the phases diminishes, the Sherwood and Nusselt numbers approach the value of 2 appropriate to molecular diffusion. Allowing for this, Eq. (160) in more precise form becomes

$$N_{Sh} = N_{Nu} = 2.0 + 0.31\ N_{Ra}^{1/3} \tag{161}$$

It may be shown that this relationship is formally identical with the Froessling equation when the drag coefficient for spheres is inversely proportional to the square root of the Reynolds number, as proposed by Allen (A2) for small gas bubbles and solid spheres. Thus,

$$v_t = \frac{1}{4}\left(\frac{\Delta\rho^2 g^2}{\rho_c \mu_c}\right)^{1/3} D_p \quad \text{(Allen)} \tag{162}$$

and

$$\frac{k_L D_p}{D_L} = 0.552\ N_{Re}^{1/2}\ N_{Sc}^{1/3} \quad \text{(Froessling)} \tag{163}$$

Combining these results it is seen that

$$k_L (N_{Sc})^{2/3} = 0.28\left(\frac{\Delta\rho g \mu_c}{\rho_c^2}\right)^{1/3} \tag{164}$$

Friedlander (F3) treated the problem of diffusion in the fluid surrounding a sphere moving in the Stokes region by solving the linearized Navier-Stokes equations to give

$$N_{Sh} = 0.89 \, N_{Pe}^{1/3} = 0.89 \left(\frac{D_p v_t}{D_L}\right)^{1/3} \tag{165}$$

Where from Stokes' law

$$v_t = \frac{g D_p^2 \Delta\rho}{18\mu_c} \tag{166}$$

Thus,

$$k_L(N_{Sc})^{2/3} = 0.34 \left(\frac{g\Delta\rho\mu_c}{\rho_c^2}\right)^{1/3} \tag{167}$$

which also compares reasonably with the experimental result, Eq. (159).

Bowman et al. (B10) have extended the calculations to include the effect of internal circulation on the continuous-phase transfer coefficients, by employing the Hadamard (H2) solution of the linearized Navier-Stokes equation and find that for large fluid spheres undergoing internal circulation, the continuous-phase mass-transfer coefficients are greater than those for rigid spheres by a factor of about 3. It would be inappropriate to compare this result directly with the behavior of large bubbles, which do not obey Stokes' law, but directionally the conclusion is supported by experimental evidence (see below).

It is significant to note that the experimental correlation [Eq. (159)] shows reasonable agreement with the theoretical predictions in the Stokes region [Eq. (167)] and also with the experimental observations of Allen and Froessling at higher Reynolds numbers [Eq. (164)]. The experimental constant proposed is 0.31 which is exactly the mean of 0.34 and 0.28 given by Eqs. (167) and (164), respectively, and is probably a more significant compromise figure than either.

In summary, Figs. 27 and 28 show that a comprehensive correlation for heat and mass transfer in turbulent free convection and for suspended or free falling or rising particles, bubbles, and drops is obtained. This correlation holds equally for dispersions and for single particles, and shows that small bubbles move in liquids as rigid spheres under conditions of hindered flow in the boundary layer sense. The correlation, originally applied to small bubbles, has been tested for heat and mass transfer in turbulent free convection, for mass transfer in mixing vessels where suspended solids and gases are undergoing dissolution, for mass transfer in liquid fluidized beds of crystallizing solids, for heat transfer when liquid drops fall through other liquids, and for several other similar processes as shown in Table IV.

It may also be noted that Zdanovskii (Z2) has proposed a correlation for the dissolution of solid suspensions in mixing vessels which is of the same form as Eq. (160), while Schmidt (S2) has proposed a similar equation for turbulent free convective heat transfer.

B. LARGE BUBBLE CORRELATION

If the experimental correlation for large gas bubbles [Eq. (158)] is compared with that proposed by Higbie

$$k_L = 2\sqrt{D_L/\pi t_e}$$

where $t_e = D_p/v_t$ (the exposure time of liquid elements at the gas-liquid interface), it may be shown that these two equations become identical in form when the drag coefficient is inversely proportional to the square root of the Reynolds number and the equations are strictly identical if

$$v_t = 0.14 \left(\frac{\rho_c g^2}{\mu_c}\right)^{1/3} D_p \tag{168}$$

Here, v_t is the boundary layer velocity, taken by Higbie as equal to the bubble free rising velocity. The latter equation is formally the same as that proposed by Allen (A2) but has a smaller value of the constant and indicates lower rising velocities for gas bubbles of diameters between $2\frac{1}{2}$ and 4 mm. than are observed. Evidently, the Higbie model, while giving order of magnitude agreement with the experimental observations for large bubbles, is by no means completely satisfactory because it applies for fully developed potential flow of liquid round a spherical-cap bubble (diameter > 5 mm.) and is not applicable to ellipsoidal bubbles (diameter $2\frac{1}{2}$ to 5 mm.) which lie in the transition region between hindered and unhindered flow.

No adequate theory is available for the transition region but it is perhaps significant that Eq. (158) can be derived by considering the hypothetical case of diffusion into liquid flowing over a hemispherical cap, the flow rate of liquid being taken as the volumetric rate of displacement of liquid by the hemispherical cap if it were a bubble rising in potential flow. Thus Haberman and Morton (H1) and Davies and Taylor (D5) have shown that the rising velocity of spherical-cap bubbles v_t is given by

$$v_t = 2\sqrt{Rg/3.9} \tag{169}$$

where R is the radius of the sphere of volume equal to the spherical cap. Thus the volumetric rate, L, of liquid displaced by the rising bubble is

$$L = \frac{4\pi}{3} \left(\frac{g}{3.9}\right)^{1/2} R^{2.5} \tag{170}$$

Davidson and Cullen (D4) have considered the problem of diffusion of gas into liquid flowing over solid spheres and vertical cylinders and their theoretical equation, for spherical surfaces, may be put in the form

$$k_L = (12 \times 1.68)^{1/2} \left(\frac{2\pi}{3}\right)^{1/6} \frac{1}{4\pi} (N_{Sc})^{-1/2} \left(\frac{\mu_c g}{\rho_c}\right)^{1/3} \frac{D^{1/3}}{g^{1/6} R^{5/6}} \qquad (171)$$

Combining Eqs. (170) and (171), one finds

$$k_L(N_{Sc})^{1/2} = 0.52 \left(\frac{\mu_c g}{\rho_c}\right)^{1/3} \qquad (172)$$

which is formally the same as the experimental correlation [Eq. (158)], but contains the constant 0.52 instead of 0.42. This difference is reduced if it is noted that large gas bubbles take the shape of spherical caps and not spheres and that the factor 1.68 introduced by Davidson and Cullen in Eq. (171), represents a conversion factor such that $1.68R$ is the height of a cylinder equivalent to a sphere of radius R. If it is assumed that the gas bubble is a hemisphere, then its height or radius will be $2^{1/3}R$ where R is the radius of the equivalent sphere, and the height of the equivalent cylinder will be $\frac{1}{2} \times 1.68 \times 2^{1/3} \times R = 1.06R$. Thus using the factor 1.06 instead of 1.68 in Eq. (171), one obtains

$$k_L(N_{Sc})^{1/2} = 0.41 \left(\frac{\mu_c g}{\rho_c}\right)^{1/3} \qquad (173)$$

which is in good agreement with the experimental result [Eq. (158)]. This rather arbitrary derivation assigns to the fluid moving round the bubble a flow rate characteristic of unhindered potential flow, while the residence time of elements of fluid moving round the bubble is taken as being dependent on the viscosity, in the same way as if the moving fluid were constrained by a solid boundary such as exists in liquid film flow over a solid surface. The fact that this apparent contradiction is used to predict the experimental results indicates that, in the transition region between hindered and unhindered flow, neither extreme exists but the state of the system may be represented as a combination of the two extremes. Alternatively, it could be argued that the experimental correlation, in attempting to encompass the transition region, has produced a compromise between the two extreme states, without revealing the finer detail between them.

Baird (B2) has studied the extreme case of large spherical-cap bubbles of carbon dioxide dissolving in water and using potential flow theory combined Eq. (169) with the equation for diffusion into a semi-infinite medium (the Higbie equation). Thus, from

$$v_t = 2\sqrt{Rg/3.9} \quad \text{(D5)} \qquad (169)$$

and

$$k_L = 2\sqrt{D_L v_t/2\pi R} \quad \text{(H8)} \qquad (174)$$

$$k_L = 0.975 \, D_L^{1/2} (g/D_p)^{1/4} \qquad (175)$$

where D_p is the diameter of the equivalent spherical bubble, and this equation and Eq. (158) are compared with Baird's experimental data in Fig. 29.

It may be seen that the experimental data are not able to discriminate between the two correlations although it is fairly clear that Baird's equation overemphasizes the effect of bubble diameter if such an effect exists.

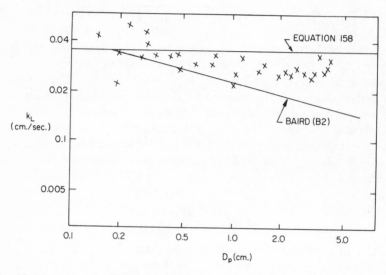

FIG. 29. Mass-transfer coefficients for large spherical-cap bubbles (CO_2 in water) (B2).

C. TRANSITION REGION BETWEEN SMALL AND LARGE BUBBLES

Figures 30 and 31 show reported mass-transfer data for bubbles of a size intermediate between those previously classified as large and small, exhibiting unhindered and hindered flow, respectively. Similar results have been observed by Garner (G1) for gas bubbles, and by Griffiths (G8) for liquid drops. It has also been observed by the above workers that surface active agents and trace impurities inhibit the transition from the hindered to the unhindered flow condition.

In aerated mixing vessels, as the agitator power is increased, the gas-liquid interfacial area increases and the bubble size diminishes, while the mass-transfer coefficient remains constant until a critical point is reached at which the transfer coefficient begins to fall rapidly as the bubbles pass through the transition size region, when ultimately the transfer coefficient again becomes constant, while the interfacial area continues to increase with increasing power dissipation. A similar behavior is observed with sieve-plate columns as the gas flow rate is increased.

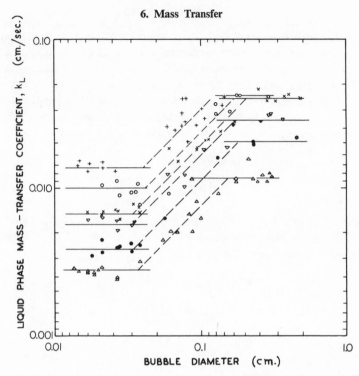

FIG. 30. Transition region between small and large bubbles in mixing vessels (C8).

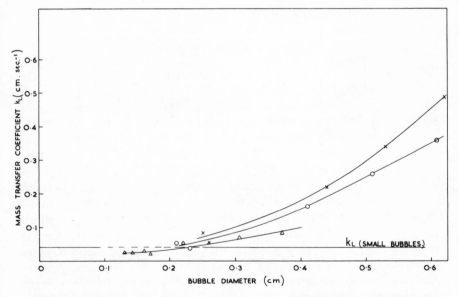

FIG. 31. Transition region between small and large bubbles in "deep pool" sieve-plate columns (C5). KEY: × 2 ft. column;   ○ 4 ft. column;   △ 15 ft. column.

Figure 32 shows the results of observations carried out on sieve trays operating under simulated distillation conditions, for all of which the data have been found to lie in the transition region. It was found that under conditions producing a large vapor hold-up in the foam, the transition from unhindered to hindered flow occurred with bubbles of greater size than in aerated mixing vessels, where the gas hold-up was comparatively low.

Gerster (G7) reports the results of extensive measurements of the liquid phase mass transfer resistance in plate columns for oxygen desorption from water with air at high gas flow rates as

$$\frac{k_L a}{1 - H} = 0.26F + 0.15 \tag{176}$$

where $k_L a$ is in sec.$^{-1}$, F (the F-factor) is $v_s \sqrt{\rho_g}$, v_s is the superficial gas velocity in ft./sec., and ρ_g the vapor density in lb./ft.^{3}. Crozier (C21) correlates the gas hold-up on sieve plates as

$$\ln\left(\frac{1}{1 - H}\right) = 0.715F + 0.45 \tag{177}$$

The interfacial area per unit volume of dispersion a may now be computed as

$$a = \frac{6H}{D_p} \tag{178}$$

taking $D_p = 0.45$ cm. (see Fig. 20) and calculating H from Eq. (177).

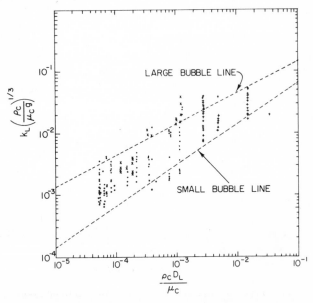

FIG. 32. Mass-transfer coefficients in "shallow pool" sieve-plate columns (C8).

Alternatively, a may be read from Fig. 23 extrapolated as shown in Fig. 33. Results of such calculations are given in the accompanying tabulation.

F factor	H_g [Eq. (177)]	k_La(sec.$^{-1}$) [Eq. (176)]	a(cm.$^{-1}$) [Eq. (178)]	k_L (cm./sec.)	a(cm.$^{-1}$) (Fig. 33)	k_L (cm./sec.)
0.5	0.554	0.125	7.28	0.0169	7.9	0.0158
1.0	0.629	0.1275	8.4	0.0152	8.75	0.0146
1.5	0.782	0.118	10.4	0.0114	9.6	0.0123
2.0	0.847	0.1025	11.3	0.0091	10.5	0.0098

These results may now be compared with the value of k_L, deduced from Eq. (159) which applies to rigid sphere bubbles, taking D_L for oxygen in water as 2×10^{-5} cm.^{2}/sec. and giving $k_L = 0.011$ cm./sec. It is evident that the values of k_L decrease with gas flow rate, approaching the value for rigid sphere bubbles at $F > 1.5$ or at gas velocities greater than 5 ft./sec.

Much work remains to be done in order to throw light on this interesting region of bubble behavior, but some relevant observations are worth making at this stage.

(a) The high mass transfer coefficients observed with large bubbles are only obtained in shallow liquid pools. With deep pools, a progressive decrease in the point values of the mass-transfer coefficient with height or residence time is observed and these values finally decline to those obtained with small rigid sphere bubbles. These observations have been made by Griffiths (G8), Baird (B2), Downing (D7), and Deindoerfer (D6) among others.

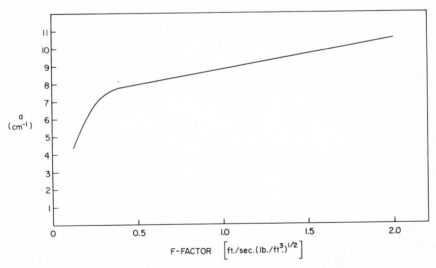

FIG. 33. Gas-liquid interfacial areas in sieve-plate columns (C11).

(b) Surface active agents cause the high mass-transfer coefficients for large bubbles to be reduced to those obtained with rigid spheres.

(c) High rates of gas flow in sieve-plate columns cause the mass-transfer coefficients of large gas bubbles to approach the rigid sphere values. The unhindered flow situation appears to be most readily realized with large bubbles at low rates of liquid flow around them [free rise in viscous liquids or flow round trapped bubbles as studied by Higbie (H9)].

As shown in Fig. 27, the mass-transfer coefficients for large and small bubbles approach each other at low values of the Schmidt group. This fact agrees with the conclusion of Toor and Marchello (T4) who predict that the boundary layer and penetration mechanisms of mass transfer lead to identical results at small values of the Schmidt group.

An important practical conclusion from these observations is that the high mass-transfer coefficients for large bubbles in unhindered flow can only be realized in special circumstances. In industrial situations of high gas flow rates, surface active impurities, and deep liquid pools, the mass-transfer coefficients will approach the lower values obtaining for rigid spheres.

D. Dispersed Phase under the Influence of Turbulence

The subject matter dealt with in the previous section refers to dispersions in which the density difference between phases is sufficiently large to cause the force of gravity to determine exclusively the rate of mass transfer. Such dispersions include those in which solid particles are just completely suspended in mixing vessels. It is known that an increase in agitation intensity above the level needed for complete suspension of particles results in a small increase in the mass-transfer rate as shown in Fig. 34 (K2).

When the particles in a mixing vessel are just completely suspended, turbulence forces balance those due to gravity, and the mass-transfer rates are the

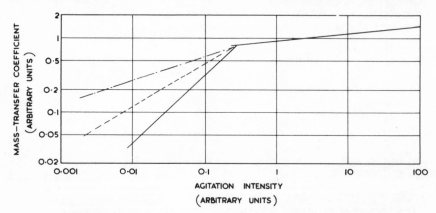

Fig. 34. Effect of agitation intensity on mass-transfer coefficients for solid particles in agitated liquids (K2).

same as for particles moving freely under gravity. At higher agitation levels, turbulence forces are greater than those due to gravity and can no longer be equated with them so that a separate treatment, described below, has to be applied. This approach, which treats turbulent forces as those which determine mass-transfer rates, has also been applied to fixed bodies submerged in mixing vessels. In the case of gas-liquid dispersions it is impracticable to exceed gravitational forces by mechanically induced turbulence since agitators operate poorly in gas-liquid dispersions.

Turbulence in the immediate vicinity of a liquid or solid particle in a dispersion affects heat and mass-transfer rates between the particle and the fluid and the theory of local isotropy may be used to give information on the turbulent intensity in the small volume around the particle. Thus, $\sqrt{u^2_{(D_p)}}$ is a statistical parameter describing the variation in fluctuating velocity over a distance comparable to the particle diameter D_p, and may be used in place of the velocity of the particle in correlations of heat and mass-transfer rates. Further, following Batchelor (B6) the Reynolds number for local isotropy is given by

$$\left(\frac{\sqrt{u^2_{(D_p)}} D_p \rho_c}{\mu_c} \right) = \frac{D_p^{4/3} \rho_c^{2/3} (P/V)^{1/3}}{\mu_c}$$

or, alternatively,

$$N_{Re} = \frac{\rho_c^{1/3} (P/V)^{1/6} D_p^{2/3}}{\mu_c^{1/2}} \tag{179}$$

The turbulence Reynolds number above was apparently first employed for correlation of mass-transfer rates in turbulent fluids by Oyama (O1) and subsequently by Kolar (K3) and the present author (C8). If the usual functional relationship between the Sherwood, Schmidt, and Reynolds numbers for mass transfer is assumed, one obtains the relationship

$$\frac{k_L D_p}{D_L} \propto \left(\frac{\mu_c}{\rho_c D_L} \right)^x \left(\frac{\rho_c^{1/3} (P/V)^{1/6} D_p^{2/3}}{\mu_c^{1/2}} \right)^y \tag{180}$$

Experiments on mass transfer in mixing vessels (H10, J1) have led to the conclusion that $k_L \propto D_L^{2/3}$ and k_L does not depend on D_p. In order to satisfy these conditions it is necessary that $x = \frac{1}{3}$ and $y = \frac{3}{2}$, whence

$$\frac{k_L D_p}{D_L} \propto \left(\frac{\rho_c^{1/3} (P/V)^{1/6} D_p^{2/3}}{\mu_c^{1/2}} \right)^{3/2} \left(\frac{\mu_c}{\rho_c D_L} \right)^{1/3}$$

or

$$k_L \propto \left(\frac{(P/V) \mu_c}{\rho_c} \right)^{1/4} \bigg/ \left(\frac{\mu_c}{\rho_c D_L} \right)^{2/3} \tag{181}$$

the analogous expression for heat transfer being,

$$\frac{h}{C_p\rho_c} \propto \left(\frac{(P/V)\mu_c}{\rho_c^2}\right)^{1/4} \bigg/ \left(\frac{C_p\mu_c}{\lambda}\right)^{2/3} \tag{182}$$

Following Eq. (181) and (182), heat and mass-transfer data for mixing vessels observed by many workers are plotted in Fig. 35.

Figure 35 reveals a correlation, which despite a considerable scatter in the data, extends for a 10^7-fold change in the variables and therefore merits serious consideration. It may be argued that some of the scatter of the experimental data is due to uncertainties in the values of the interfacial areas for suspended particles as obtained by screen analysis, and to uncertainty concerning the state of suspension of the particles in the mixing vessel. The equation of the regression line drawn through the data gives

$$k_L(N_{Sc})^{2/3} = \frac{h_c}{C_p\rho_c}(N_{Pr})^{2/3} = 0.13\left(\frac{(P/V)\mu_c}{\rho_c^2}\right)^{1/4} \tag{183}$$

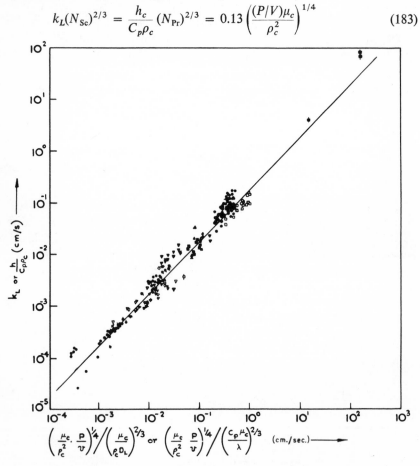

FIG. 35. Heat and mass-transfer coefficients in mixing vessels for suspended and submerged bodies. (For data references, see the accompanying tabulation.)

DATA FOR FIG. 35

Heat transfer

Water   (R6)
$\left.\begin{array}{l}\text{Water}\\\text{Glycerol}\\\text{L.M. oil}\\\text{A-12 oil}\\\text{Hot water}\end{array}\right\}$ (C15)

Nitrating liquor   (B11)

Mass transfer

Solid dissolution
Sucrose-water   (O1)
Sodium thiosulfate–water   (H13)
$\left.\begin{array}{l}\text{Benzoic acid–sperm oil}\\\text{Benzoic acid–cotton seed oil}\\\text{Benzoic acid–rape seed oil}\\\text{Benzoic acid–glycol}\end{array}\right\}$ (H10)

$\left.\begin{array}{l}\text{Succinic acid–acetone}\\\text{Succinic acid–}n\text{-butanol}\\\text{Salicylic acid–benzene}\\\text{Salicylic acid–water}\\\text{Benzoic acid–water}\end{array}\right\}$ (J1)

Liquid-liquid extraction
Benzoic acid–kerosene–water   (F1)

with a standard deviation of 66%. This result may be compared with the most recent and extensive investigation by Harriott (H5) of the dissolution rates of particles suspended in mixing vessels. For particles large compared with the scale of the smallest eddies, Harriott found

$$k_L \propto (P/V)^{0.17}; \quad k_L \propto D_L^{0.6-0.8}; \quad k_L \propto \mu_c^{-0.37}$$

while k_L was also independent of the particle size and difference in density between the two phases.

Figure 35 includes data on heat and mass transfer to fixed bodies submerged in agitated liquids (heat transfer jackets and coils and solids cast into the base of mixing vessels). For heat transfer to coils in mixing vessels it is hard to avoid the conclusion that the correlation has concealed some real trends in the scatter of the data taken as a whole, since Rushton (R7) has shown that the size of the impeller has a small effect on heat transfer to coils in mixing vessels even when the power per unit volume of fluid is kept constant.

It appears probable that the type of agitator used may also have a minor influence on the mass-transfer coefficients which is not revealed in Eq. (183). Thus, when the agitator power is increased, the particles become completely suspended when the mass transfer coefficient is given by Eq. (159). As the power is further increased, Eq. (183) is followed, but with the practical limitations of allowable power dissipation the increase in mass-transfer coefficient may not be easily observed. However, it is almost certain that agitators producing high ratios of flow to shear for a given power are most effective for suspending particles, so that particles suspended with these agitators come under the influence of power at a lower value of power dissipation than with agitators producing high shear and low flow rates. This question is complicated by the fact that small particles are most easily suspended, and if their size is less than the scale of the smallest eddies, the mass-transfer coefficients will remain independent of power dissipation until the latter is raised considerably above that needed for complete suspension. This behavior has been observed by Piret (P2). As a result of these considerations some scatter of the results shown in Fig. 35 is to be expected due to variations in agitator characteristics in the several experiments reported. It should also be made clear that the range of variables in Fig. 35 is largely made up of variations in the Schmidt group and that variations in the power dissipation factor for a given system are comparatively small, so that the figure confirms the two-thirds exponent of the Schmidt group but does not provide precise evidence for the form of the power variable used.

It has been shown (C8) that Eq. (183) also holds reasonably well for turbulent mass transfer in packed beds if the power dissipation per unit volume of fluid is computed from the flow work in this case. It should also be noted that Eq. (183) was derived by combining the theory of local isotropic turbulence with a conventional mass-transfer correlation and by using some experimental findings with mixing vessels to evaluate the constants in this correlation. The result has been found to apply reasonably well to turbulent flow in packed beds and pipes where the theory of local isotropic turbulence could not be supposed to apply so that it would appear that a more general law operates to interrelate energy dissipation and the specific rates of turbulent transport phenomena.

In this connection it is of particular interest to compare Eq. (183) with the work of Norman (N1) who derived a relationship between the heat or mass-transfer coefficients and the shear stress exerted by fluids passing over surfaces in turbulent flow. Norman derived his equations from classical boundary layer theory and successfully applied them to mass transfer in tubes roughened with turbulence promoters and also to mass transfer in packed towers. Thus, if we note that the power dissipation per unit volume of fluid equals the shear rate times the shear stress and also that the viscosity is defined as the shear stress divided by the shear rate, Eq. (183) may be restated as

$$k_L(N_{Sc})^{2/3} = \frac{h_c}{C_p \rho_c}(N_{Pr})^{2/3} = 0.13 \left(\frac{\tau_0}{\rho_c}\right)^{1/2} \tag{184}$$

where τ_0 is the viscous shear stress. The above equation is identical in form to that deduced by Norman who, however, suggests a constant of 0.072 instead of 0.13. It is clear, however, that local isotropic turbulence and boundary layer theories lead to the same formal relationship [Eq. (184)].

E. IMPELLER REYNOLDS NUMBER

Most of the published work on heat and mass transfer in mixing vessels makes use of the impeller Reynolds number as a correlating parameter. Equation (183) may also be put in this form to yield

$$\frac{k_L}{ND}(N_{Sc})^{2/3} = 0.13 \left[\frac{(N_P)}{(N_{Re})_I}\left(\frac{D^3}{V}\right)\right]^{1/4} \tag{185}$$

or

$$\frac{k_L D}{D_L} = 0.13(N_P)^{1/4}(N_{Re})_I^{3/4} N_{Sc}^{1/3} \left(\frac{D^3}{V}\right)^{1/4} \tag{186}$$

or

$$\frac{k_L T}{D_L} = 0.13(N_P)^{1/4}(N_{Re})_I^{3/4} N_{Sc}^{1/3} \left(\frac{D^3}{V}\right)^{1/4}\left(\frac{T}{D}\right) \tag{187}$$

where (N_P) = power number = $\dfrac{P}{\rho_c N^3 D^5}$,

$(N_{Re})_I$ = impeller Reynolds number = $\dfrac{ND^2 \rho_c}{\mu_c}$,

D = impeller diameter,

T = mixing vessel diameter,

N = impeller speed,

and V = volume of tank contents.

Thus it is seen that in scaling-up in the fully turbulent region of constant power number while preserving geometric similarity

$$\frac{k_L T}{D_L} \propto (N_{Re})_I^{3/4} \tag{188}$$

Typically, experimentally determined exponents of the impeller Reynolds number have varied considerably. Thus Hixson and Baum (H10) report a value of 0.6 and Barker and Treybal (B5) a value of 0.83 compared with the value of 0.75 given by Eq. (188).

Again various workers report values of the exponent of the Schmidt group of one-third and two-thirds compared with one-third given by Eq. (187). No systematic investigation has been carried out to discover the effect on the mass-transfer coefficient of variations in impeller and tank geometry. Practical experience suggests that scaling-up at constant power dissipation per unit volume of fluid, as suggested by Eq. (183), leads to satisfactory results, but further work on this problem is needed.

F. POWER REQUIRED TO SUSPEND SOLIDS IN MIXING VESSELS

As previously discussed, the rate of solution of particles just completely suspended in a mixing vessel is dependent on factors which do not include the agitator power, although as the power dissipation is further increased a comparatively small increase in solution rate may be realized. This fact is demonstrated by Kneule (K2) as shown in Fig. 34. At the point of complete suspension of solids, the mass-transfer coefficient is given by either Eqs. (159) or (183), so that on equating these relationships in order to eliminate the mass-transfer coefficient, we arrive at an expression for the power required to just suspend the particles, which takes the form

$$\frac{P}{V} = \text{const.} \ \frac{(g\Delta\rho)^{4/3} \mu_c^{1/3}}{\rho_c^{2/3}} \tag{189}$$

This result may be compared with that proposed by Kneule (K2) who finds

$$\frac{P}{V} = \text{const.} \ \frac{(g\Delta\rho)^{1.5} D_p^{1/2} H^{1/2}}{\rho_d^{1/2}} \tag{190}$$

and with that proposed by Zwietering (23) who finds

$$\frac{P}{V} = \text{const.} \ \frac{(g\Delta\rho)^{1.35} \mu_c^{0.3} D_p^{1/2} H^{0.39}}{\rho_c^{0.35}} \tag{191}$$

while Oyama and Endoh (O2) report

$$\frac{P}{V} = \text{const.} \ \frac{(g\Delta\rho)^{3/2} D_p^{1/2}}{\rho_c^{1/2}} \tag{192}$$

The most obvious discrepancy between the deduced and the experimental results is that Eq. (189) does not contain the particle size as a variable, and it would seem reasonable to suppose that it should. However, if the square root of the particle diameter is included in Eq. (189) as suggested by the experimental correlation above, this might imply that Eq. (183) should include a factor of $D_p^{1/8}$. It is unlikely, with the present precision of experimental mass-transfer data obtainable with solid dispersion, that this factor could be identified with certainty. The same argument suggests that it is only slightly more likely that an effect of solids hold-up on the mass-transfer coefficient could be easily identified, although in this connection both Piret (P2) and Wilhelm (W6) report a small influence. A further test of the validity of Eqs. (159) and (183) may be made by considering the problem of the suspension of dispersions of solids whose free falling velocities are in the Stokes regime. Thus Eq. (159) has been shown to predict the mass-transfer coefficients for rigid particles which obey the free falling or free rising relationship given by Allen (A2).

$$v_t = \frac{1}{4}\left(\frac{\Delta\rho^2 g^2}{\rho_c\mu_c}\right)^{1/3} D_p \qquad (193)$$

If this equation is combined with Eq. (159), a more general relationship is obtained which includes the particle diameter and velocity and is not restricted to a particular kind of fluid flow around the particle.

Thus,

$$k_L(N_{Sc})^{2/3} = 0.62\left(\frac{v_t\mu_c}{\rho_c D_p}\right)^{1/2} \qquad (194)$$

which is one form of the Froessling equation (F5). Also as has already been demonstrated, Eq. (183) may be expressed in the alternative form

$$k_L(N_{Sc})^{2/3} = 0.13(\tau_0/\rho_c)^{1/2} \qquad (195)$$

where τ_0 is the viscous shear stress. On equating the two relationships above one finds the minimum viscous shear stress which has to be generated in a mixing vessel in order to cause the free suspension of particles whose free falling velocities are determined by viscous and gravitational forces. Thus,

$$\tau_0 = 22.7\mu_c(v_t/D_p) \qquad (196)$$

which may be compared with Stokes' law

$$\tau_0 = 12\mu_c(v_t/D_p) \qquad (197)$$

The formal identity of this result with Stokes' law appears to be significant. The fact that the total shear stress which has to be generated in a mixing vessel is some 1.9 times that acting vertically on the particle is not unexpected since much of the flow from a mixing impeller is radial and ineffective in suspending particles. The equation given by Norman (N1) for mass transfer in flow through packed beds, corresponding to Eq. (195), when treated in the above way gives a constant of 12.8 in Eq. (196) and indicates the directional nature of the flow in this case.

Much of the difficulty attending the correlation of the dissolution rates of suspended solids, such as has been attempted in Eq. (183), is due to uncertainty as to whether the solid particles are completely suspended in the solvent. Evidently from Fig. 34 failure to secure complete suspension can lead to results which are greatly in error.

Calderbank and Jones (C6) used a light-transmission method to determine the power needed to obtain a uniform suspension of ion exchange resin beads in water in their experiments on mass transfer to suspended solid particles.

The effect of impeller speed on the light obscuration induced by the suspended particles is shown in Fig. 36.

Even though the density difference between solid and liquid was as low as 0.1 gm./cc., surprisingly high impeller speeds were needed to produce a uniform suspension such that further increases in 'speed did not affect the amount of light obscuration. Also the range of power dissipation between complete suspension of particles and entrainment of air from the free liquid surface is seen to be small, so that it is not surprising that published work on the dissolution rates of heavy suspended solutes suffers from considerable scatter of the data because the effect of power dissipation on the mass-transfer coefficients is difficult to identify with certainty due to the small range of power available between complete suspension and air entrainment.

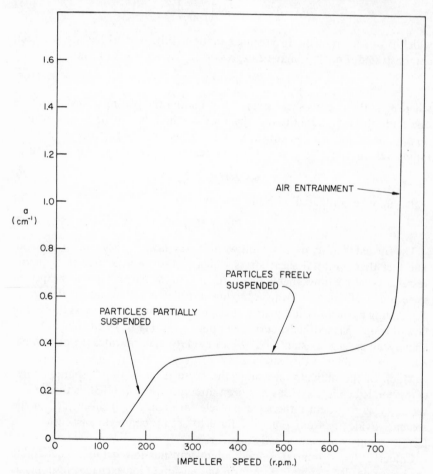

Fig. 36. Effect of impeller speed on suspension of particles in a mixing vessel as measured by an optical obscuration method (C6).

G. CONCLUSIONS OF PRACTICAL INTEREST

One clear conclusion is that when a dispersed phase is suspended or in free rise or fall and under the influence of gravity, the heat and mass-transfer coefficients are almost completely unaffected by mechanical power dissipated in the system. Thus, for example, in dissolving solid particles in liquids, little advantage can be gained by increasing the agitation intensity above the point at which the particles are freely suspended. Clearly, mixing impellers should be designed to secure suspension of particles with a minimum power consumption which should not greatly exceed this level.

For gas-liquid dispersion, power dissipation increases the gas-liquid interfacial area and increases the mass transfer rate for this reason although, as before, little effect on the mass-transfer coefficient can be realized.

Equation (159) shows the important influence of the difference in density between the suspended dispersed and the continuous phases on the transfer coefficients. This means that where this difference in density is small, transfer rates are extremely low. This situation is most likely to be realized in some liquid-liquid extraction processes, and can be countered by increasing the gravitational field strength as in centrifugal extractors. The problem is also met with in dissolving polymers in solvents, where the density difference between polymer solution and solid polymer can be quite small, and the rate of solution is further impeded by the high viscosity of the solution as would be expected from Eq. (159).

VI. Mass Transfer in Dispersed Phases

If the dispersed phase is treated as an assembly of equivalent spheres of diameter equal to the Sauter mean value, and if the fluid comprising these spheres is stagnant, for one such sphere a material balance across a shell of thickness, ∂r, at radial position, r, gives

$$D_D \left[\frac{\partial^2 x}{\partial r^2} + \frac{2}{r} \frac{\partial x}{\partial r} \right] = \frac{\partial x}{\partial t} \tag{198}$$

where x = concentration of diffusing solute at radial position, r,

D_D = molecular diffusivity of solute in drop fluid, and

t = time.

A solution of this equation in terms of the integral mean solute concentrations in the sphere as a function of the time for mass transfer is available provided the concentration of diffusing solute at the interface of the sphere (x_i) is constant.

Thus for x_i = constant and $R > r > 0$

$$E = \frac{x_0 - x_t}{x_0 - x_i} = 1 - \frac{6}{\pi^2} \sum_1^\infty (1/n^2) \exp \{ -n^2 D_D \pi^2 t / R^2 \} \tag{199}$$

where E is the fractional approach to equilibrium of the sphere with its surroundings whose composition is unvarying at x_i expressed as a function of the sphere radius (R), the diffusivity of the solute and the time for mass transfer. The initial uniform solute concentration in the sphere at time zero is x_0, and x_t is the solute concentration at time, t.

Equation (199) is a converging infinite Fourier series, the roots of which have been determined to give

$$E = 1 - \frac{6}{\pi^2} \left[\exp\left\{ -\pi^2 D_D t/R^2 \right\} + \frac{1}{4} \exp\left\{ -4\pi^2 D_D t/R^2 \right\} \right. $$
$$\left. + \frac{1}{9} \exp\left\{ -9\pi^2 D_D t/R^2 \right\} + \ldots \right] \quad (200)$$

For values of $E > 0.4$ (long contact times) the first term only of this series is significant when

$$E = 1 - \frac{6}{\pi^2} \exp\left\{ -\pi^2 D_D t/R^2 \right\} \quad (201)$$

By the method of Laplace transforms, an alternative solution to Eq. (198) may be derived which takes the form of a diverging infinite series. For values of $E < 0.4$ (short contact times) the first term only of this series is significant when

$$E = 6/R \sqrt{D_D t/\pi} = 2a' \sqrt{D_D t/\pi} \quad (202)$$

where $a' =$ the surface area of the dispersion per unit volume of dispersed phase.

Equation (202) applies for short contact times when the concentration gradient of the solute has not penetrated to the center of the sphere which consequently behaves as a semi-infinite body. Equations (201) and (202) may be expressed in terms of the conventional mass-transfer coefficient (k_D) which is defined by the relationship

$$-\ln(1 - E) = k_D a' t = k_D (3/R) t \quad (203)$$

Thus for long contact times $(E > 0.4)$, from Eq. (201)

$$-\ln(1 - E) = \frac{\pi^2 D_D t}{R^2} - \ln\frac{6}{\pi^2} = k_D \frac{3}{R} t \quad (204)$$

for noncoalescing systems, or

$$k_D = \frac{1}{3} \frac{\pi^2 D_D}{R} - \frac{R}{3t} \ln\frac{6}{\pi^2} \quad (205)$$

where for most practical cases the second term of Eq. (205) is insignificant and

$$k_D \approx \frac{1}{3} \frac{\pi^2 D_D}{R} \quad (206)$$

For short contact times ($E < 0.4$), from Eq. (202),

$$-\ln(1 - E) \approx E = 6/R\sqrt{D_D t/\pi} = k_D(3/R)t$$

or

$$k_D = 2\sqrt{D_D/\pi t} \qquad (207)$$

which gives a parabolic rate equation of the form $dx/dt \propto k/x$.

It may be noted that Higbie (H8) deduced a similar equation to the above for the continuous-phase mass-transfer coefficients obtaining for gas bubbles in free rise, where for this case t was the exposure time of elements of the continuous phase with the gas bubble and was taken as $2R/v_t$. This deduction assumed unhindered flow of liquid round the bubble and destruction of concentration gradients in the wake of the bubble. In the derivation above, t is the total residence time of the dispersed phase in the phase contacting equipment if no dispersed-phase coalescence takes place, or the lifetime of a particle of the dispersed phase if coalescence does occur. In practice there is some uncertainty about the degree of circulation of the fluid in the drop or bubble and in addition the results apply only if the continuous phase composition is constant as would be the case, for example, in continuous flow perfectly mixed stage equipment.

Equations (201) and (202) are limiting forms of the general expression given by Eq. (199) which takes a rather intractable form but which, as shown by Vermeulen (V3), can be accurately represented by the empirical result

$$E = (1 - \exp\{-\pi^2 D_D t/R^2\})^{1/2} \qquad (208)$$

For fluid spheres, Kronig and Brink (K5) evaluated the effect of internal circulation on diffusion which for a sphere Reynolds number < 1 and zero interfacial tension gave

$$E = 1 - \frac{3}{8}\sum_{n=1}^{\infty} A_n^2 \exp\{-\mu_n 16 D_D t/R^2\} \qquad (209)$$

where the eigenvalues A_n and μ_n have been tabulated (H6). Danckwerts (D2) has pointed out that this result may not be as limited as the above restrictions suggest and Calderbank and Korchinski (C7) have shown that the equation may be accurately represented by the empirical form

$$E = [1 - \exp\{-2.25 D_D \pi^2 t/R^2\}]^{1/2} \qquad (210)$$

as shown in Fig. 37.

Comparing Eqs. (208) and (210) it is immediately obvious that the Kronig and Brink model of internal circulation results in an effective diffusion coefficient of $2\frac{1}{4}$ times the molecular value.

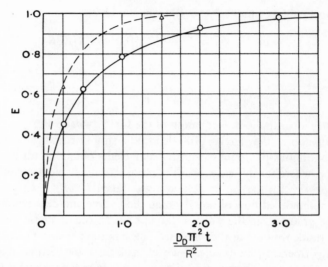

FIG. 37. Equations for diffusion into spherical bodies. KEY:
— Eq. (199)   ○ Eq. (208)   - - - Eq. (209)   △ Eq. (210)

Calderbank and Korchinski (C7) carried out heat transfer experiments on liquid drops falling through various field liquids and concluded that the effective diffusivity for circulating but nonoscillating drops confirmed the above prediction but that drop oscillation gave rise to considerably higher values of the effective diffusivity. A summary of these and other published results is given in Table V. Johnson and Hamelec (J2) have concluded that the effective diffusivity within liquid drops is further and largely increased when interfacial turbulence is induced by mass transfer.

Calderbank (C3) carried out careful measurements of mass-transfer rates within rising bubbles and concluded that the effective diffusivity is many times the molecular value to the extent that mass-transfer resistance in the gas phase compared with that in the liquid may be ignored. However, on distillation and gas absorption plates, at high gas velocities, bubbles may rise at rates far in excess of their free rising velocity, as shown by the fact that the gas hold-up increases but slowly with gas velocity (see Fig. 21) and increasing gas flow has to be accommodated by increasing bubble velocity. Under these circumstances, due to the small bubble residence time in the froth, mass-transfer resistance in the gas phase may become significant. Experimental data presented by Gerster (G7) for ammonia absorption and air humidification in plate columns gives $k_D a = 17.5$ sec.$^{-1}$ over the gas flow range 3.34 to 6.20 ft./sec. where the gas residence time varied from 0.1 to 0.2 sec. Applying Eq. (205) to this result and taking the mean bubble diameter as 4.5 mm. (see Fig. 20), a gas phase diffusivity of about 0.06 cm.^{2}/sec. is deduced, compared with the accepted value of 0.246 cm.^{2}/sec. This result is surprising in view of

Table V

Ratio of Effective to Molecular Diffusivity, D_e/D_M

Source of data	System	D_e/D_M (range)	N_{Re} (range)	Type of drop behavior
(G5)	Cold benzene drops rising in hot water (heat transfer)	1.72–2.1	1500–1780	No drop oscillation
(G5)	Hot benzene drops rising in cold water (heat transfer)	1.6–2.4	730–810	No drop oscillation
(M4)	Cold kerosene drops rising in hot water (heat transfer)	2.25	2000	No drop oscillation
(M4)	Cold xylene drops rising in hot water (heat transfer)	2.25	2400	No drop oscillation
(C7)	Bromobenzene drops falling in hot or cold glycerol-water solutions (heat transfer)	1.8–3.3	10–120	No drop oscillation
(H6)	Isobutanol drops rising in water (mass transfer)	1.9–2.1	20–200	No drop oscillation
(C7)	Bromobenzene drops falling in hot or cold glycerol-water solutions (heat transfer)	7–12	315–620	Drop oscillations present
(H12)	Water drops falling through carbon dioxide atmosphere (mass transfer of carbon dioxide)	20	300–3000	Drop oscillations present
(G4)	Nitrobenzene drops containing acetic acid falling in water (mass transfer of acetic acid)	6.5 10 21	270 370 480	Type of drop behavior not stated by authors. Oscillations indicated by R-values
(P1)	Perchlorethylene drops containing acetic acid falling in water (mass transfer of acetic acid)	25–53 40–61	375 595	Drop oscillations present, although oscillation decay indicated by range of R for constant conditions

the fact that the gas within the bubble will be in a state of circulation and considerably higher values of the effective diffusivity would be anticipated. The explanation probably lies in an effect of gas "bypassing," perhaps through gas "jetting" or "leakage" through bubble chains. This phenomenon is evidently responsible for the finite gas phase mass-transfer resistance encountered in commercial gas absorption and distillation at high vapor loading. Thus an alternative explanation of the slow increase in gas hold-up above 60% with vapor loading is that the bubble rising velocity does not increase, as suggested above, but gas "leaks" through the bubble chains; or, as photographs suggest, leaks through by way of periodic discharges of large volumes or jets of gas. Such leakage may be avoided on sieve plates by using many perforations, when increasing gas flows are accommodated by the formation of cellular foams of very low density (see Figs. 21 and 22).

Garner (G3) has replotted Gerster's results by drawing lines, following Eq. (205), through the data to give an intercept of $-\ln(6/\pi^2)$. This interpretation leads to higher diffusion coefficients as shown in the accompanying tabulation

Gas rate (ft./sec.)	0.5	0.8	1.5	3.5	6.5
Effective diffusivity (cm.²/sec.)	0.15	0.18	0.21	0.34	0.45

The existence of effective diffusivities lower than the molecular value is still evident and gas leakage has to be postulated.

VII. Summation of Transport Resistances in Dispersed and Continuous Phases

Grober (G9) has presented a solution in chart form in the case where a finite resistance to heat transfer is present in a sphere and in the surrounding fluid for the particular case where the bulk fluid temperature is constant. Grober's chart has been replotted in convenient form and is shown in Fig. 38 in terms of both heat and mass-transfer parameters.

Treybal (T6) has pointed out that the Grober chart may be used for continuous flow agitated extractors since the condition of continuous phase composition is fulfilled in this case, and he has shown that design calculations using this chart and an effective diffusivity within liquid drops of $2\frac{1}{4}$ times the molecular value (see above) lead to good predictions for liquid-liquid extraction operations in mixing vessels. Figure 35 shows that the value of the continuous-phase mass-transfer coefficient predicted by Flynn (F1), in this way, from over-all performance data of liquid extractors, is in agreement with solid dissolution data.

Amundsen (A3) has evolved a solution to the problem of the summation of dispersed and continuous phase transport resistances for countercurrent operations where the separation factor is constant but this solution is too

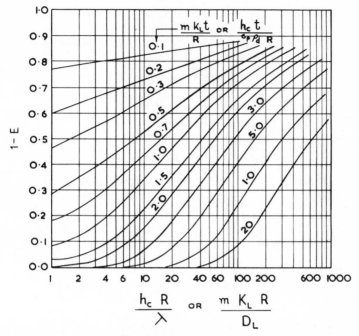

FIG. 38. Grober chart (G9) for summation of heat or mass-transfer resistances in dispersed and continuous phases (constant interfacial temperfacial temperature or concentration).

m = slope of operating line, dC_c/dC_d
λ = thermal conductivity of dispersed phase

laborious to commend itself to common use. Lebeis (L2) has however obtained numerical solutions to Amundsen's equation for a variety of common practical situations and has presented these in graphical form from which reasonably accurate engineering calculations may be made.

Diffusion of substances from a continuous phase into a dispersed phase where chemical reaction occurs is common in the growth of living cells, reactions involving aerosols, and liquid-liquid extraction processes. The differential equations which apply to a spherical particle of the dispersed phase under steady state conditions are

$$D_D\left[\frac{\partial^2 x}{\partial r^2} + \frac{2}{r}\frac{\partial x}{\partial r}\right] - q' = 0 \tag{211}$$

$$D_C\left[\frac{\partial^2 C}{\partial r^2} + \frac{2}{r}\frac{\partial C}{\partial r}\right] = 0 \tag{212}$$

with the boundary conditions,

$$\text{when } r = R, \; D_D\left(\frac{\partial x}{\partial r}\right) = D_c\left(\frac{\partial C}{\partial r}\right) \tag{213}$$

where D_D and x = the diffusivity and concentration of the reacting solute in the dispersed phase,

D_C and C = the diffusivity and concentration of the reacting solute in the continuous phase,

r = the radial distance from the center of the spherical particle

R = the radius of the particle, and

q' = the rate of consumption of the reacting solute by chemical reaction in units, for example, of (moles)/(cm.^{3} of dispersed phase)(sec.).

These equations have been integrated for steady state conditions, various kinds of reactions, and certain limiting conditions. Thus when diffusion in the continuous phase is very rapid compared with diffusion and reaction in the particle and when the reaction is first order and irreversible, the steady state rate of consumption of reactant in moles per unit volume of dispersion per unit time q is given by

$$q = \frac{aC_iD_D}{R}\left[\sqrt{\frac{kR^2}{D_D}}\ \coth\ \sqrt{\frac{kR^2}{D_D}} - 1\right] \qquad (214)$$

where k = the first-order reaction velocity constant,

C_i = the concentration of solute in the dispersed phase at its interface where thermodynamic equilibrium with the continuous phase may be assumed, and

a = the interfacial area per unit volume of dispersion.

When $\sqrt{kR^2/D_D}$ is > 2, Eq. (214) reduces to,

$$q = \sqrt{kD_D}\ aC_i \qquad (215)$$

a result which is strictly comparable with Eq. (155) derived for the continuous phase. Further, when the reaction is of zero order and penetration of the reactant into the particle is small,

$$q = a\sqrt{2C_ikD_D} \qquad (216)$$

and when diffusion in the continuous phase is the slow rate-controlling step,

$$q = \frac{D_c}{R}\ aC_i \qquad (217)$$

corresponding to $N_{Sh} = 2$ as derived by Langmuir (L1) for diffusion into a stagnant medium, a condition which is closely approached when the particle Reynolds number is small (small particles). Cadle and Robbins (C1) discuss situations in which steady state conditions cannot be assumed and present empirical equations for some of these cases, while Essenhigh and Fells (E3) derive equations for the diffusion and reaction of gases into porous solid particles with particular reference to the problem of the combustion of powdered and atomized fuels.

VIII. Practical Applications

Some further examples of the practical application of the foregoing principles to industrial rate processes will now be discussed.

A. AEROBIC FERMENTATION

In principle, the rate of production of aerobic organisms in a fermentation process may be deduced from the rate of consumption of oxygen if one knows the yield of organisms obtained per unit quantity of oxygen consumed. In order to proceed with a prediction of this kind, it is necessary to consider three rate process which operate sequentially. First, oxygen present as air bubbles dissolves in the aqueous nutrient fluid, second, the dissolved oxygen is transported to the surface of the organism, and third, the oxygen reacts with the respiratory enzymes of the organism. One can avoid consideration of mass transfer in the gas (bubble) phase in this sequence since this process has been demonstrated to be relatively very rapid compared with mass transfer in the liquid phase. One need only consider mass transfer from the gas–liquid interface to the bulk liquid, mass transfer from the bulk liquid to the organism interface, and the respiration rate of the organisms. The two mass-transfer rate processes will be considered together, the respiration and reproduction rate of organisms will then be discussed, and finally the equations for the over-all process will be presented.

1. Mass Transfer of Oxygen in Aerobic Fermentations

In treating the above subject, it will be assumed that aeration is carried out by means of bubble clouds in the so-called "deep fermentation" process. Equation (161) has been shown to represent mass-transfer rates to and from both small rigid sphere bubbles and suspended solid particles in liquids. The small rigid sphere bubbles are usually formed when aqueous solutions of hydrophilic solutes such as nutrient broths are aerated with mechanical agitation and if suspended microorganisms are regarded as solid particles, Eq. (161) may be employed to predict oxygen transfer rates both from the bubble interface and also to the organism interface. Thus from Eq. (161),

$$k_L = \frac{2D_L}{D_p} + 0.31(N_{Sc})^{-2/3} \left(\frac{\Delta\rho\mu_c g}{\rho_c^2}\right)^{1/3} \tag{218}$$

where D_p is the diameter of the bubble or organism and D_L is the diffusivity of oxygen in the liquid. This equation has been confirmed by Calderbank and Jones (C6) for the particular case of mass transfer to and from dispersions of low density solid particles in agitated liquids. The latter experiments were designed to simulate mass transfer to microorganisms in fermenters and were carried out using dispersions of spherical ion-exchange resin beads in water contained in a mixing vessel, the rate of mass transfer being followed by means of high-speed pH recording. In particular, the applicability of Eq. (218) to dispersions where the density difference between the dispersed and continuous phases is small was demonstrated, as shown in Fig. 39.

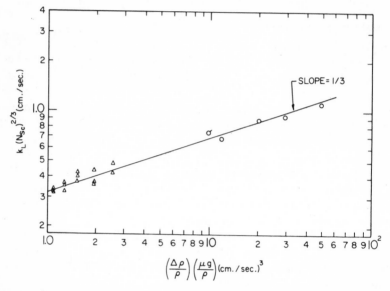

FIG. 39. Application of Eq. (218) to mass transfer to and from rigid sphere bubbles and suspended solid particles in agitated liquids. Results plotted to demonstrate range of values covered (C6). KEY:
$\triangle$ Ion-exchange, resin suspensions
$\bigcirc$ CO_2 bubbles in glycerol solutions
σ O_2 bubbles in brine

For mass transfer from the bubble interface to the bulk liquid, it may be readily shown that the first term of Eq. (218) is insignificant compared with the second term since D and $\Delta\rho$ are both large. For mass transfer from the bulk liquid to the organism interface, however, the reverse is true since $\Delta\rho$ is insignificant and D is very small, and the first term only of Eq. (218) need be considered. To take some typical figures, assuming $D_L = 2 \times 10^{-5}$ cm.^{2}/sec., $\rho_c = 1$ gm./cm.^{3}, and $\mu_c = 1$ cp., for organisms of a few microns in diameter, $(k_L)_{L \to S} \approx 0.1$ cm./sec., whereas $(k_L)_{G \to L} \approx 0.01$ cm./sec. In this case it is evident that mass transfer of oxygen from the liquid to the surface of the organism is very rapid compared with the rate of solution of oxygen from gas bubbles, particularly when it is remembered that the organism-liquid interfacial area is very much larger than the gas-liquid interfacial area for most of the fermentation time. However, if the organisms aggregate together as "clumps" of a few millimeters in diameter, $(k_L)_{L \to S} \approx 0 \cdot 0001$ cm./sec. and a finite resistance to mass transfer from liquid to solid may be anticipated. Thus, although the power dissipated by the agitator in deep fermentation has little or no influence on the values of the liquid phase mass-transfer coefficients from bubbles and to suspended solids, it is usefully employed in dispersion to produce a high gas-liquid and solid-liquid contact area, in the latter case by preventing gross aggregation of the organisms.

To summarize, the mass-transfer coefficient for transfer of solute from the gas-liquid interface to the bulk liquid is given by

$$(k_L)_{G \to L} = 0.31(N_{Sc})^{-2/3} \left(\frac{\mu_c g}{\rho_c} \right)^{1/3} \tag{219}$$

and for transfer from the bulk liquid to the surface of suspended microorganisms by

$$(k_L)_{L \to S} = \frac{2D_L}{D_p} \tag{220}$$

or

$$(k_L a)_{G \to L} = 0.31(N_{Sc})^{-2/3} \left(\frac{\mu_c g}{\rho_c} \right)^{1/3} a \tag{221}$$

where a is the gas-liquid interfacial area per unit volume of dispersion, and

$$(k_L a'')_{L \to S} = 2\pi n D_p D_L \tag{222}$$

where n = number of organisms per unit volume of broth,
D_p = equivalent spherical diameter of organisms, and
a'' = organism-liquid interfacial area per unit volume of broth.

In order to evaluate $(k_L a)_{G \to L}$ from Eq. (221), it is necessary to know a. This may be deduced from the relationship

$$a = \frac{6H}{D_{S.M.}}$$

and

$$D_{S.M.} = 2.25 \left[\frac{\sigma^{0.6}}{(P/V)^{0.4} \rho_c^{0.2}} \right] H^{0.4} \left(\frac{\mu_d}{\mu_c} \right)^{0.25} \qquad [\text{Eq. (24)}]$$

where $D_{S.M.}$ is the Sauter mean bubble diameter and H is the volume fraction gas hold-up. The gas hold-up for aerated solutions of electrolytes in mixing vessels employing flat-blade impellers has been correlated by Moo-Young (M9) following Rushton (R6) as shown in Fig. 40.

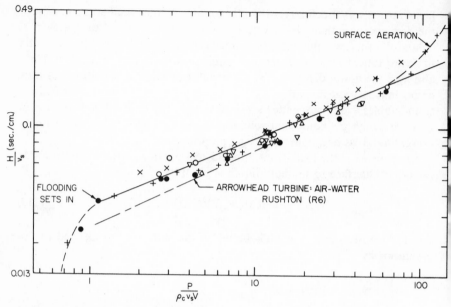

FIG. 40. Correlation of gas hold-up for aerated solutions of electrolytes in mixing vessels employing flat-blade impellers (M9).

T (in.)	D (in.)	Water	Ethanol
20	6.67	+	×
15	5.0	●	○
7.25	2.4	△	▽

In order to predict the power consumption of agitators operating in gas-liquid dispersions and hence to deduce P/V for a given impeller speed, as required by Eq. (24), it is necessary to make use of a further correlation, shown in Fig. 41, which again applies for flat-blade turbine agitators.

This rather circuitous procedure may be avoided by measuring $(k_L a)_{G \to L}$ by direct experiment. This device is commonly practiced in the fermentation industry by observing the rate of oxidation of a copper-catalyzed solution of sodium sulfite, contained in the fermenter under the same conditions of agitation and aeration as is used in the fermentation process itself. The assumption is made that the rate of the chemical oxidation step is very rapid and that the over-all rate is determined by oxygen diffusion from the gas-liquid interface to the bulk solution where the oxygen concentration is maintained at zero by the rapid chemical step.

Thus, $C^*(k_L a)_{G \to L}$ = rate of sulfite oxidation, where C^* is the solubility of oxygen in the solution, representing the concentration of dissolved oxygen at the gas-liquid interface in equilibrium with air of known oxygen partial pressure.

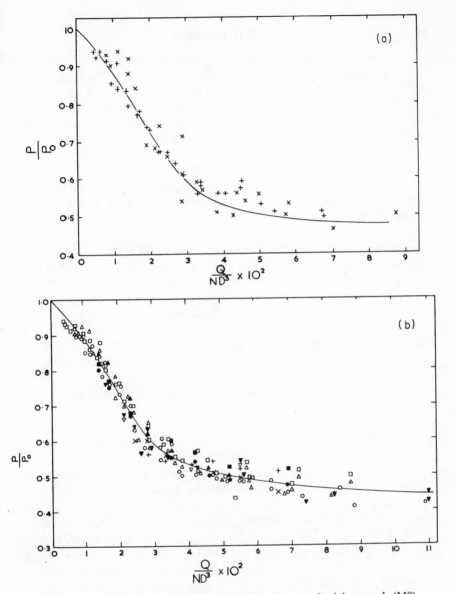

FIG. 41. Power consumption of flat-blade impellers in aerated mixing vessels (M9).

(a)	KEY	T (in.)	D (in.)	Liquid		
	×	20	6.67	5% aqueous ethanol		
	+	15	5.00	Glycerol		
(b)	T (in.)	D (in.)	Water	Glycerol	Ethanol	CCl₄
	20	6.67	○	□	△	
	15	5.0	●	■	▲	
	7.25	2.4	×	+	▽	▼

P_0 = power drawn by agitator with no aeration; P = power drawn by agitator with aeration at same speed; Q = volumetric air flow rate, c.f.m.

Yoshida (Y1) has apparently shown that the value of $(k_L a)_{G \to L}$ obtained by the sulfite oxidation method agrees reasonably well with the corresponding value obtained by measuring the rate of solution of oxygen in the exhausted sodium sulfite reagent, so that the assumptions made in interpreting sulfite oxidation rates appear to be valid. However, Westerterp (W4) reports that the mass-transfer enhancement factor in this system due to chemical reaction, is approximately 6, a contradiction which has so far not been resolved. In either case, the apparent mass-transfer coefficient is constant and the method has the advantage that it allows $(k_L a)_{G \to L}$ or a number proportional to it to be rapidly determined in the actual equipment to be used in fermentation, whereas correlations such as those shown in Figs. 40 and 41 are strictly applicable to a particular mixing vessel and agitator geometry.

In experiments to measure $(k_L a)_{G \to L}$ by the sulfite oxidation method, the air leaving the sulfite solution is depleted in oxygen so that C^* varies as the air passes through the solution. It is usual to assume that the air passes through the solution in plug flow (no back-mixing) when it is proper to employ a value of C^* corresponding to the log mean partial pressure of oxygen in the inlet and outlet air streams. This log mean driving force can be readily deduced from the observed sulfite oxidation rate and the flow rate of air by a material balance.

Thus,

$$-\frac{dC}{dt} = \left(\frac{F}{V}\right) \frac{(P_i - P_0)}{RT} \tag{223}$$

where $-dC/dt$ is the constant rate of oxidation of sodium sulfite expressed in units, for example, of (moles O_2)/(liter) (hr.); F is the air flow rate in liter/hr.; V is the volume of liquid in liters; P_i is the partial pressure of oxygen in air ($\frac{1}{5}$ atm.); and P_0 is the partial pressure of oxygen in the air leaving the solution, also in atmospheres. The gas constant, R, is thus in the units (liter atm.)/(mole.) (°K). Therefore, P_0 is deduced and the log mean driving force calculated as $\{[P_i - P_0]/[\ln(P_i/P_0)]\}$. It is immediately obvious from Eq. (223) that the maximum rate of oxidation of sodium sulfite is obtained when $P_0 = 0$ and that, under these conditions, the rate depends only on the flow rate of air and the volume of solution, or in other words the rate of oxidation is determined solely by the rate of oxygen supply, diffusion and chemical reaction being relatively very rapid.

The assumption of plug flow of gas through aerated mixing vessels may not be completely justified because some gas back-mixing undoubtedly does occur as evidenced by the considerable aeration which can take place at the agitated free liquid surface, as discussed previously (see Fig. 16). Moreover, Westerterp (W4) invokes unpublished work by Piret to propose complete mixing of the gas in an aerated mixing vessel operated at high impeller speeds. The assumption of complete gas mixing would lead to a gas-liquid interfacial oxygen concentration in equilibrium with the oxygen present in the air leaving the fermenter, so that in the sulfite oxidation assessment, the mass-transfer driving force would be $C^* = P_0/m$ where m is the Henry constant instead of $C^* = 1/m \{[P_i - P_0]/[\ln(P_i/P_0)]\}$ as applies for plug flow of gas. In either case P_0 is obtained from the material balance relationship given by Eq. (223).

In commercial practice, the gas flow through liquids in mixing vessels is probably partially mixed and there is a need for measurements of the gas residence time distribution function in order to assess the importance of this factor.

Summarizing,

$$(k_L a)_{G \to L} = \left. \frac{-dC}{dt} \right/ \frac{P_0}{m} \qquad \text{(complete mixing of air)} \qquad (224)$$

or

$$(k_L a)_{G \to L} = \left. \frac{-dC}{dt} \right/ \frac{P_i - P_0}{m \ln (P_i/P_0)} \qquad \text{(plug flow of air)} \qquad (225)$$

where $(k_L a)_{G \to L}$ is obtained by experimental observation of the constant sulfite oxidation rate $(-dC/dt)$ and little uncertainty exists in evaluating the correct driving force if serious depletion of oxygen in the air is avoided.

From Eqs. (223) and (224),

$$\frac{-dC}{dt} = \frac{P_i}{(m/k_L a)_{G \to L} + RTh/v_s} \qquad \text{(complete mixing of air)} \qquad (226)$$

and from Eqs. (223) and (225), taking the arithmetic mean driving force as equal to the log mean (oxygen depletion of air small),

$$\frac{-dC}{dt} = \frac{P_i}{(m/k_L a)_{G \to L} + RTh/2v_s} \qquad \text{(plug flow of air)} \qquad (227)$$

[in Eqs. (226) and (227), h is liquid height]. Thus in either case the rate of sulfite oxidation is time invariant and when the air flow rate is high and the height of solution low, both Eqs. (226) and (227) reduce to

$$(k_L a)_{G \to L} = \left. \frac{-dC}{dt} \right/ \frac{P_i}{m} \qquad (228)$$

where $-dC/dt$ is the constant sulfite oxidation rate. Also,

$$(k_L a'')_{L \to S} = 2\pi n D_p D_L \qquad (229)$$

as derived previously [Eq. (222)]. Equation (229) contains the diameter of the microorganism and the number of these organisms per unit volume of fermentation broth so that attention must now be turned to the microorganisms themselves.

2. The Reproduction and Respiration Rates of Microorganisms

The reproduction rate of microorganisms has been represented by the relationship

$$\frac{dn}{dt} = \frac{knS}{K_m + S} \tag{230}$$

where n is the concentration of organisms, S is the concentration of limiting substrate, and k and K_m are constants at a constant temperature. For the purposes of the following discussion it is assumed that oxygen is the growth rate-limiting substrate and therefore S is the concentration of dissolved oxygen at the cell wall. It may be seen that K_m (the Michaelis constant) is equal to the dissolved oxygen concentration at which the rate of reproduction is one-half the maximum value kn which obtains when $S \gg K_m$. Moreover, when the supply of substrate (oxygen and nutrients) is not limiting, the maximum growth rate is given by

$$\frac{dn}{dt} = kn = \frac{n \ln 2}{t_d} \tag{231}$$

where t_d is the doubling time of the organisms.

Under these conditions the relationship between n and t is given by

$$n = A e^{kt} \tag{232}$$

showing an exponential growth with time.

Longmuir (L8) has measured the Michaelis constant for oxygen uptake by microorganisms and found K_m to be a function of the organism diameter. Thus Longmuir finds $K_m \propto D_p^{2.6}$ for cells ranging in diameter from 0.5 to $4 \cdot 0 \mu$, whereas K_m was approximately constant when cell-free preparations of the various organisms were tested. Moreover, the value of K_m for cell-free preparations was approximately the same as for the smallest organisms investigated so that it appears as if a diffusional barrier to oxygen transport is created with larger organisms.

Baender and Kiesse (B1) have demonstrated that the respiration rate of various organisms including bacteria and yeast is given by

$$\frac{-dS}{dt} = \frac{kS^{1.4}}{K_m + S^{1.4}} \tag{233}$$

a result which has been broadly confirmed by Longmuir (L9) using rat-liver cells. It appears likely that the last equation applies to respiration rates controlled by the enzyme oxidation process itself since comparatively high energies of activation (about 16,000 cal./mole) are associated with it. Since the values of the Michaelis constant in all these experiments are very small (10^{-6} to 10^{-8} M) it is evident that the dissolved oxygen concentration at the cell wall has to be reduced to correspondingly low values before the metabolic process is adversely affected.

Monod (M10) proposed that the growth rate is a constant fraction of the rate of utilization of carbon substrate and presumably of the rate of oxygen utilization also. Thus we may define a quantity Y_s (the yield constant) such that

$$\frac{dn}{dt} = -Y_s \frac{dS}{dt} = \frac{knS}{K_m + S} \tag{234}$$

This equation gives the respiration and reproduction rates of microorganisms and leads to the equations for oxygen supply and demand as shown below.

3. Oxygen Supply and Demand in Fermentation

From the foregoing, and at steady state,

$$\frac{dn}{dt} = Y_s(k_L a)_{G \to L} (C^* - C) = Y_s \pi n D_p D_L (C - S) = \frac{knS}{K_m + S} \tag{235}$$

where C is the oxygen concentration in the bulk solution and the other terms have already been defined. These equations simply state that the rates of oxygen supply from the gas-liquid interface to the bulk solution and from the solution to the cell wall are both equal to the rate of respiration at steady state. In principle, one can eliminate C and S from these equations and by integration obtain n as a function of t in terms of various constants which are determinable. These constants are $(k_L a)_{G \to L}$, which may be determined by a sulfite oxidation test and Y_s, D_p and K_m which are characteristic of the microorganism and can sometimes be obtained from the biochemical literature but generally will have to be measured.

Elsworth (E1) observes that in the growth of *Aerobacter cloacae* the growth rate at the beginning of the fermentation is exponential, implying no substrate or oxygen supply limitations and that at a later time when the cell concentration is about 3 gm./liter, the growth rate becomes constant, suggesting that $C \ll C^*$, when $dn/dt = Y_s(k_L a)_{G \to L} C^*$ and oxygen supply from the gas-liquid interface is rate-limiting. Elsworth confirmed this conclusion by observing that there was a strict proportionality between the constant growth rate and the sulfite oxidation rate at various levels of aeration intensity, this constant of proportionality presumably being proportional to the yield constant, Y_s. The reason for the abrupt change from exponential to constant rate growth is that the rapidly accelerating growth generates new cell-liquid interfacial area so quickly that oxygen depletion is brought about at equal speed. Moreover, since the Michaelis constant is so small, oxygen limitation is only experienced when the dissolved oxygen concentration has fallen to such a low value that $S \ll C^*$. Since the available cell area is rapidly increasing in the exponential phase of growth, $(k_L a'')_{L \to S}$ also rapidly increases since this factor equals $2\pi n D_p D_L$ and so the driving force $(C-S)$ remains small and $C \ll C^*$. Thus, no mass-transfer limitation is to be expected in the diffusion step from the bulk solution to the cell wall unless the cells aggregate into clumps as they may do in mycelial growth. In the absence of cell aggregation therefore, $C \approx S$ and $dn/dt = Y_s(k_L a)_{G \rightarrow L}(C^* - S) = [(kSn)/(K_m + S)]$ so that when S is large, $dn/dt = kn$ and when S is small $dn/dt = Y_s(k_L a)_{G \rightarrow L} C^*$. In the latter case it may be necessary to take into account oxygen depletion in the air stream passing through the fermenter and employing a log mean driving force,

$$\frac{dn}{dt} = [Y_s(k_L a)_{G \rightarrow L}]\left[\frac{P_i - P_0}{m \ln (P_i/P_0)}\right] = \frac{Y_s v_s}{h}\frac{(P_i - P_0)}{RT} \qquad (236)$$

Thus, as a corollary, one finds

$$\ln\left(\frac{P_i}{P_0}\right) = \frac{(k_L a)_{G \rightarrow L}}{m}\frac{RTh}{v_s} \qquad (237)$$

whereby the oxygen utilization of the air may be calculated for constant growth rate conditions. Thus, although good utilization of oxygen is obtained with tall fermenters, a sacrifice in rate of production is entailed and an economic compromise will have to be sought.

B. BUBBLE REACTORS

It is sometimes necessary or advantageous to carry out reactions by contacting a reactant gas or gases with a slurry of solid catalyst in a liquid which may or may not be a participant in the reaction. Liquid-phase catalytic-hydrogenation and oxidation reactions are examples where the liquid participates in the reaction. These processes are becoming of increasing importance in hydrofining, hydroforming, and petrochemical operations. Also, it is sometimes advantageous to conduct gas phase catalytic reactions by bubbling the reactants through a slurry of solid catalyst suspended in an inert liquid, particularly if isothermal operation is required, since the catalyst is, in this way, in intimate contact with a liquid which can be made to serve as a heat transfer medium. In all these examples, the gas is dispersed as a bubble cloud in the slurry by mechanical agitation or by means of an orifice disperser. Evidently, the gaseous reactants have to dissolve in the liquid and diffuse to the catalyst surface before reaction can take place and there are obvious points of similarity between this and the previous example of aerobic fermentation in which three phases also interacted.

An example chosen for more detailed discussion is the Fischer-Tropsch catalyst slurry process in which a mixture of carbon monoxide and hydrogen is bubbled through a slurry of finely divided iron oxide catalyst suspended in molten wax contained in a tall column and in which the liquid circulation induced by the bubble cloud is sufficient to prevent catalyst sedimentation.

The synthesis gas reacts to form a variety of hydrocarbons and oxygenated products, the reaction temperature and pressure being such that all except the heavy waxes are removed as overheads while the wax level in the reactor is maintained by the use of an overflow, incorporated in a sedimentation limb which prevents loss of catalyst from the reaction system. This process has been studied in some detail by the author and co-workers in the course of a pilot scale development program (E4). In this study, use was made of a simpler but analogous reaction scheme in which mixtures of ethylene and hydrogen were bubbled through a slurry of Raney nickel catalyst in inert liquids to form ethane. The kinetics of this reaction could be accurately measured and used to reveal some of the main characteristics of the chemically more complex Fischer-Tropsch catalyst slurry process. Thus, experiments were performed in which the kinetics of ethylene hydrogenation were measured when an excess of catalyst was present, such that further addition of catalyst was without effect on the conversion obtained under otherwise constant reaction conditions. In these circumstances, it was supposed that the rate of the chemical reaction was very rapid and that diffusion of reactants from the gas-liquid interface to the bulk liquid phase was rate-controlling. Further, it was supposed, from a consideration of the solubilities and diffusivities of ethylene and hydrogen in the liquids, that mass transfer of hydrogen in the liquid phase from the gas-liquid interface was rate-controlling and that the catalyst surface was therefore saturated with ethylene but depleted of hydrogen, whose steady state concentration in the bulk liquid could therefore be taken as zero. For this model, the following equations are applicable.

If n represents the moles of a reaction component and x the conversion per unit mass of reactant feed,

	H_2	$+$ C_2H_4	$+$ vapor	$\rightarrow C_2H_6$	$+$ vapor	(total moles)
Reactor inlet	n_1	n_2	n_3	0	n_3	$n_1 + n_2 + n_3$
Reactor outlet	$n_1 - x$	$n_2 - x$	n_3	x	n_3	$n_1 + n_2 + n_3 - x$

where the vapor is present because of the finite vapor pressure of the liquid in which the catalyst is dispersed. From a material balance

$$F\,dx = r\,dV \qquad (238)$$

where F = the mass flow rate of the feed,

$\quad r$ = the reaction rate in moles converted per unit volume of reactor space per unit time, and

$\quad V$ = the volume of reactor space or dispersion.

The rate equation, assuming liquid phase mass transfer of hydrogen to be rate-controlling, is

$$r = k_L a(1 - H)C^* \qquad (239)$$

where H = the volume fraction of gas in the dispersion, and

$\quad C^*$ = the solubility of hydrogen in the liquid in moles per unit volume.

The equilibrium relationship is given by Henry's law.

$$C^* = \frac{N_1 P}{m M_s} \tag{240}$$

where m = the Henry constant,
N_1 = the mole fraction of hydrogen in the gas,
P = the reaction pressure, and
M_s = the molar volume of the liquid.

Thus from the previous three equations,

$$\int_0^x \frac{n_1 + n_2 + n_3 - x}{n_1 - x}\, dx = \frac{k_L a P(1 - H)V}{m M_s F}$$

$$= \frac{k_L a P V_L}{m M_s F} \tag{241}$$

where V_L is the volume of liquid in the reactor.

Solving the above integral, one obtains,

$$x + (n_2 + n_3) \ln\left(\frac{n_1}{n_1 - x}\right) = \frac{k_L a P V_L}{m M_s F} \tag{242}$$

Introducing $N_x = (x)/(n_1 + n_2 - x)$ (the mole fraction of ethane in the vapor-free exit gas)

$$x = \frac{N_x(n_1 + n_2)}{1 + N_x} \tag{243}$$

and

$$\ln\left(\frac{n_1}{n_1 - x}\right) = \ln\left(\frac{1 + N_x}{1 - (n_2/n_1)N_x}\right) \tag{244}$$

Further,

$$F(n_1 + n_2) = F_{VI} \frac{P}{RT} \tag{245}$$

where F_{VI} is the volumetric flow rate of vapor-free gas at the reactor inlet. Also,

$$\frac{n_3}{n_1 + n_2} = \frac{P_s}{1 - P_s} \tag{246}$$

where P_s is the vapor pressure in atmospheres of the liquid at the reaction temperature. Introducing Eqs. (243) to (246) into Eq. (242), one obtains,

$$\frac{N_x}{1 + N_x} + \left[\frac{n_2}{n_1 + n_2} + \frac{P_s}{1 - P_s}\right] \ln\left(\frac{1 + N_x}{1 - (n_2/n_1) N_x}\right) = \frac{k_L a V_L RT}{m M_s F_{VI}} \tag{247}$$

where $(n_2)/(n_1+n_2)$ is the mole fraction of ethylene in the inlet gas and n_2/n_1 is the mole ratio of ethylene to hydrogen in the inlet gas. Equation (247) can be experimentally tested if the gas-liquid interfacial area (a) is evaluated and this was done by using the light-transmission method described earlier, in experiments employing identical equipment and conditions to the kinetic experiments, except that the solid catalyst was omitted from the system. The value of k_L may also be predicted from Eq. (159) since the above experiments showed that small rigid sphere bubbles were produced with the sintered-plate gas disperser employed and Fig. 42 demonstrates that Eq. (247) adequately represents the whole of these data.

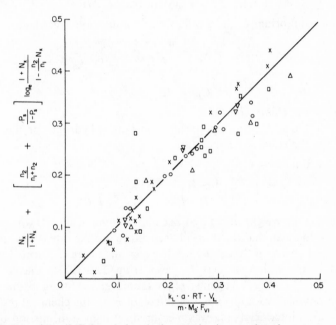

FIG. 42. Test of Eq. (247) for ethylene hydrogenation in a catalyst slurry system using an excess of catalyst (E4).

By analogy with these findings, it may be concluded that the rate of the Fischer-Tropsch catalyst slurry process is also limited by diffusion of hydrogen in the liquid phase, but in this case a finite rate of the chemical reaction must also be considered since the catalyst activity is low compared with Raney nickel. A considerable simplification of the problem is brought about by the experimental observation that the over-all rate of this reaction is first order and consequently one may write down the following rate equation for the steady state.

$$k_0 C_g H dV = k_L a(C_L^* - C_L)(1-H)\,dV = k_c Q_D C_s\,dV \qquad (248)$$

in which k_0 = the observed over-all first-order rate constant,

k_c = the reaction velocity constant of the catalytic reaction,

H = the volumetric fractional gas hold-up,

C_g = the concentration of hydrogen in the gas phase,

C_L = the concentration of hydrogen in the liquid phase,

C_L^* = the concentration of hydrogen in the liquid at the gas-liquid interface,

C_s = the concentration of hydrogen adsorbed on the catalyst in moles per unit mass,

Q_D = the mass of catalyst per unit volume of dispersion, and

V = the volume of dispersion or reactor space.

Assuming equilibrium at the gas-liquid and liquid-solid interface,

$$C_g = \frac{mM_sC_L^*}{RT_r} \tag{249}$$

and

$$C_L = BC_s \tag{250}$$

where m = the Henry constant,

B = the adsorption equilibrium constant of hydrogen on the catalyst,

T_r = the reaction temperature.

Thus, from Eq. (248),

$$\left(\frac{RT_r}{mM_s} \frac{1-H}{H}\right)\frac{1}{k_0} = \frac{1}{k_La} + \frac{B}{k_cQ_D'} \tag{251}$$

(where Q_D' is the weight of catalyst per unit volume of liquid) which may be interpreted to mean that the over-all "resistance" to the rate is the sum of the physical and chemical "resistances." If experiments are performed in which Q_D' and a are varied, it may be seen from Eq. (251), that a plot of $[RT_r(1-H)Q_D']/[mM_sHk_0]$ against Q_D'/a should give a straight line of slope $1/k_L$ (the physical resistance) and of intercept B/k_c (the chemical resistance). The gas-liquid interfacial area was measured by light transmission under reaction conditions but in the absence of the solid catalyst and the gas hold-up by observing the expansion of the liquid on aeration. The over-all first-order rate constant was determined from the generalized form of Eq. (242), which for a first-order reaction of the stoichiometric type, 1 (reactants) $\rightarrow \gamma$ (products), takes the form

$$(1 - \gamma)x + \gamma n \ln\left(\frac{n}{n-x}\right) = \frac{k_0P_rVH}{RT_rF} \tag{252}$$

where P_r and T_r = the reaction pressure and temperature,

F = the mass flow rate of feed to the reactor,

n = the number of moles per unit mass of feed, and

x = the number of moles converted to products per unit mass of feed.

Or in terms of the fractional conversion, $X = x/n$,

$$(1 - \gamma)X + \gamma \ln\left(\frac{1}{1 - X}\right) = \frac{k_0 V H}{F_{VI}} \qquad (253)$$

Figure 43 shows the data plotted according to Eq. (251). The slope of the line gives a value of k_L in close agreement with that predicted by Eq. (159) for the dissolution of small hydrogen bubbles in the liquid, while the intercept enables the chemical resistance to be evaluated. Figure 43 also shows that the gas-liquid contacting deteriorates as the height of the reactor is increased because of bubble coalescence which results in a decrease in the gas-liquid interfacial area with height and produces a negative deviation from the mean interfacial areas which were measured in short columns and are plotted as abscissa in Fig. 43.

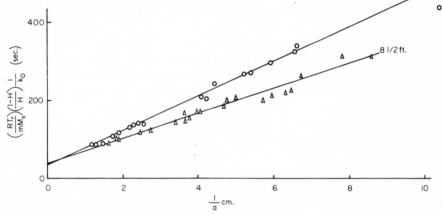

FIG. 43. Test of Eq. (251) for Fischer-Tropsch synthesis in a catalyst slurry system for conversion of 50% and two different slurry heights (E4).

C. THE PROBLEM OF SCALING-UP

In chemical process development work, the conditions for an optimum yield of the desired product are determined in a laboratory or pilot scale reactor which is frequently a mixing vessel. It then becomes necessary to reproduce the optimum agitation characteristics found on the small scale in a large commercial size. It is common practice to scale up mixing vessels on the basis of the same agitator power per unit volume of liquid whatever the scale of operation. This practice would appear to be justified when different phases are being contacted since it has been shown previously that dispersion efficiency is determined by the power dissipation per unit volume of liquid. If the agitator is operating in a fully baffled tank in the turbulent region of constant power number,

$$N_P = \frac{P}{\rho_c N^3 D^5} = \left(\frac{P}{V}\right)\left(\frac{V}{D^3}\right)\left(\frac{1}{\rho_c N^3 D^2}\right) \qquad (254)$$

If N_P is constant (turbulent agitation), (V/D^3) is constant (scale-up with geometric similarity preserved), and (P/V) is constant (scale-up at constant power per unit volume), then

$$N \propto (1/D)^{2/3} \qquad (255)$$

Thus the impeller Reynolds number $\propto D^2N \propto D^{4/3}$ and the impeller Froude number $\propto N^2D \propto D^{-1/3}$ and scale-up at constant power per unit volume implies an increasing Reynolds number and a decreasing Froude number as the scale is increased, while the impeller tip speed increases as the one-third power of the scale.

Metzner (M5, M6) has shown that the power required for a given mixing time (quality of mixing), in geometrically similar mixing vessels, increases with the scale to the power 4.1 to 5.75, depending on the type of agitator used. Since scaling up at constant power per unit volume gives a power dissipation which increases as the cube of the scale, it is evident that some loss of mixing effectiveness is involved in scaling up in this manner. Other factors influencing the scale-up of aerated mixing vessels have already been mentioned. Thus from Fig. 40 it is seen that at constant P/V, impeller "flooding" with gas is determined by the superficial gas velocity and so the maximum superficial air velocity will be the same whatever the scale of operation. At constant P/V and v_s, the gas hold-up, gas-liquid interfacial area, and mean bubble size will be the same whatever the scale of operation.

From Fig. 41, the ratio of the power drawn by the agitator in unaerated and aerated liquids, for geometrically similar mixers is determined by v_s/ND and if v_s is held constant, $P/P_0 = f(ND)$ and since, as has been shown, $N \propto D^{-2/3}$ for geometrically similar scale up at constant P/V, $P/P_0 = f(D)^{1/3}$. Moreover, the form of the relationship shown in Fig. 41 indicates that, at gas loadings approaching flooding, P/P_0 changes but little with v_s/ND and so the dependence of P/P_0 on the scale of operations at the same superficial gas velocity, if this is near flooding, is likely to be very slight.

The weight of evidence is therefore in favor of scaling up geometrically similar fermenters at constant P/V and v_s. However, three factors which are influenced by the scale of operation are worth restating at this point since they are not in accord with the above conclusion.

(1) As fermenters are scaled up with geometric similarity and constant superficial gas velocity, the bubble residence time increases and oxygen depletion of the air will eventually become apparent so that in the limit, the rate of oxygen transfer to the liquid will be determined by the rate of flow of air alone. This argument suggests that the height to diameter ratio of fermenters should be progressively decreased for large scale-up factors since from Eq. (237), $\ln(P_i/P_0)$ is directly proportional to the height of the fermenter. A logical design procedure to follow would be to decide on the gas utilization per pass (P_i/P_0) on

economic grounds and then from Eq. (237) to determine h/v_s. The value of v_s might then be set just below the impeller flooding point when h, the height of the vessel, is known. The diameter of the vessel may then be determined from the desired rate of production or holding time.

(2) Figure 16 indicates that at constant v_s, surface aeration is a rapidly increasing function of (ND), so that at constant P/V where $ND \propto D^{1/3}$ if geometric similarity is preserved, surface aeration will become increasingly important as the scale of operation is increased.

(3) Following Metzner (M5, M6), scale-up at constant P/V for geometrically similar mixing vessels will result in some loss in mixing effectiveness.

List of Symbols

a area per unit volume
A area [Eqs. (121), (149)]
A separation factor [Eq. (96)]
C concentration (mass)/(volume)
C_p heat capacity (heat)/(mass) (temp.)
c a number [Eq. (9)]
$c_{1,2,3}$ constants [Eqs. (15), (21), (41)]
d distance
d_p pipe diameter
d_0 orifice diameter
D impeller diameter
$D_{A,C,D,L}$ diffusivity (length)^2/(time)
D_p particle diameter
$D_{S.M.}$ $6H/a$, Sauter mean particle diameter
E efficiency [Eqs. (97), (98), (102), (103), (199)]
e_c eccentricity [Eq. (141)]
f friction factor [Eq. (50)]
f coalescence factor [Eq. (64)]
f_d, f_H hold-up factors [Eqs. (19), (34)]
F flow rate (mass)/(time)
F F-factor [Eq. (176)]
g gravitational constant (length)/(time)^2
G flow rate (mass)/(time) or (volume)/(time)
h height
h_c heat transfer coefficient (heat)/(time) (area) (temp.)
H hold-up, volume fraction of dispersed phase
I, I_0 light intensity [Eq. (2)]
k light-scattering factor (Fig. 4)
k, k_1 first-order reaction rate constant (time)$^{-1}$
k_1 mass-transfer coefficient [Eq. (148)] (length)/(time)
k_L liquid-phase mass-transfer coefficient (length)/(time)
K over-all mass-transfer coefficient [Eq. (74)] (length)/(time)
K_m Michaelis constant [Eq. (230)] (mass)/(volume)

l length [Eqs. (2), (148)]
L length [Eq. (9)]
L flow rate [Eq. (87)] (mass)/(time)
m Henry constant [Eq. (240)] (atm.) (volume)/(mole)
n number of moles transferred [Eq. (74)] (mass)
n concentration [Eqs. (222), (230)] (number)/(volume)
n a number [Eqs. (1), (9), (197)]
n_c refractive index of continuous phase
N rotational speed $(time)^{-1}$
N_l, N_x mole fraction [Eqs. (240), (243)]
P power (force) (length)/(time)
P pressure (atm.)
q reaction rate [Eq. (214)] (moles)/(volume) (time)
r rate (moles)/(time)
R radius [Eq. (169)], (Fig. 38)
R gas constant [Eqs. (156), (223)]
R intensity of reflected light [Eq. (5)]
S substrate concentration [Eq. (230)] (moles)/(volume)
t time
T temperature
T vessel diameter
$\sqrt{\overline{u^2}}$ root-mean-square velocity fluctuation [Eq. (14)] (length)/(time)
u tangential velocity (Fig. 26)
U velocity [Eq. (127)] (Fig. 26)
v velocity (Fig. 26)
v_s superficial gas velocity (gas flow rate)/(vessel cross-sectional area)
V volume
X conversion [Eq. (253)]
x, y distance
x, y mole fraction [Eqs. (88), (92)]
Y_s yield constant [Eq. (234)]

GREEK LETTERS

α mass obsorption coefficient [Eq. (10)] $(length)^2/(mass)$
β, β', β'' constants [Eqs. (4), (5), (6)]
γ ratio of product moles to reactant moles [Eq. (252)]
δ boundary layer thickness [Eq. (128)]
η y/δ, dimensionless distance [Eq. (128)]
θ angle (Fig. 26)
λ wavelength (Fig. 4)
λ thermal conductivity (heat)/(length)(time)(temp.)
μ viscosity (mass)/(length)(time)
ν kinematic viscosity $(length)^2/(time)$
ρ density (mass*)/(volume)
σ surface tension (force)/(length)
τ shear stress (force)/(area)
ϕ $\sqrt{kD_A/k_1}$ [Eq. (148)], dimensionless
* Slugs, grams.

DIMENSIONLESS GROUPS

N number of transfer units [Eqs. (79), (80), (88)]
N_{Gr} $D^3 g \rho_c \Delta \rho / \mu^2$, Grashoff number
N_{Nu} hD/λ, Nusselt number
N_{Pe} dv/D_L, Peclet number
N_{Pr} $C_p \mu / \lambda$, Prandtl number
N_P $P/\rho_c N^3 D^5$, power number
N_{Re} $dv\rho/\mu$, Reynolds number
N_{Re} $D^2 N \rho / \mu$, impeller Reynolds number
N_{Ra} $D^3 \Delta \rho g / D_L \mu$, Rayleigh number
N_{Sc} $\mu / D_L \rho$, Schmidt number
N_{Sh} $k_L d / D_L$, Sherwood number
N_{We} $N^2 D^3 \rho / \sigma$ or $\tau D_p / \sigma$, Weber number
N_{vi} $\mu / \sqrt{\rho D \sigma}$, impeller viscosity number

References

(A1) Aksel'Rud, G. A., *Zhur. Fiz. Khim.* (1953). 27, 1445.
(A2) Allen, H. S., *Phil. Mag.* 50, 323 (1900).
(A3) Amundsen, N. R., and Monro, W. D., *Ind. Eng. Chem.* 42, 1481 (1950).
(B1) Baender, A., and Kiesse, M., *Arch. exptl. Pathol. Pharmakol. Naunyn-Schmiedeberg's* 224, 312 (1955).
(B2) Baird, M. H. I., Ph.D. Thesis, Cambridge University, England, 1960.
(B3) Baird, M. H. I., and Hamielec, A. E., *Can. J. Chem. Eng.* 40, 119 (1962).
(B4) Baranayev, M. K., *Doklady Akad. Nauk SSSR* 66, No. 5 (1949).
(B5) Barker, J. J., and Treybal, R. E., *A.I.Ch.E. Journal* 6, 289 (1960).
(B6) Batchelor, G. K., *Proc. Cambridge Phil. Soc.* 47, 359 (1951).
(B7) Bates, R. L. (Chemineer, Inc., Dayton, Ohio), private communication (1961).
(B8) Benton, D. P., and Elton, G. A. H., *Discussions Faraday Soc.* No. 30, 68 (1960).
(B9) Boltze, E., Thesis, Gottingen University, Germany, 1908.
(B10) Bowman, C. W., Ward, D. M., Johnson, A. I., and Trass, O., *Can. J. Chem. Eng.* 39, 9 (1961).
(B11) Brown, R. W., Scott, R., and Toyne, C., *Trans. Inst. Chem. Engrs.* (*London*) 25, 181 (1947).
(C1) Cadle, R. D., and Robbins, R. C., *Discussions Faraday Soc.* No. 30, 155 (1960).
(C2) Calderbank, P. H., *Trans. Inst. Chem. Engrs.* (*London*) 36, 443 (1958).
(C3) Calderbank, P. H., *Trans. Inst. Chem. Engrs.* (*London*) 37, 173 (1959).
(C4) Calderbank, P. H., and Evans, F., *Intern. Symp. Distillation, Inst. Chem. Engrs., Brighton, England,* (*1960*), p. 51. Engrs., 1960.
(C5) Calderbank, P. H., and Evans, F., Internal Report RR/CE/14. Warren Spring Laboratory, Stevenage, England, 1961.
(C6) Calderbank, P. H., and Jones, S. J. R., *Trans. Inst. Chem. Engrs.* (*London*) 39, 363 (1961).
(C7) Calderbank, P. H., and Korchinski, I. J. O., *Chem. Eng. Sci.* 6, 65 (1956).
(C8) Calderbank, P. H., and Moo-Young, M. B., *Chem. Eng. Sci.* 16, 39 (1961).
(C9) Calderbank, P. H., and Moo-Young, M. B., *Trans. Inst. Chem. Engrs.* (*London*) 39, 337 (1961).

(C10) Calderbank, P. H., and Moo-Young, M. B., Symposium on Drops and Bubbles. Cambridge University, England, 1962 (unpublished).

(C11) Calderbank, P. H., and Rennie, J., *Trans. Inst. Chem. Engrs.* (*London*) **40**, 3 (1962).

(C12) Cameron, J. F., Intern. Conference on Radio Isotopes in Scientific Research. Paper No. NS/RIC/195 (1955).

(C13) Cauchy, A., *Compt. rend. acad. sci.* **13**, 1060 (1841).

(C14) Chalkley, H. W., Cornfield, J., and Park, H., *Science* **110**, 295 (1949).

(C15) Chao, B. T., *Phys. Fluids* **5**, 69 (1962).

(C16) Chilton, T. H., Drew, T. B., and Jebens, R. H., *Ind. Eng. Chem.* **36**, 510 (1944).

(C17) Churchill, S. W., *Discussions Faraday Soc.* No. 30, 226 (1960).

(C18) Clay, P. H., *Koninkl. Ned. Akad. Wetenschap. Proc.* **43**, 852 (1940).

(C19) Coppock, P. D., and Meiklejohn, G. T., *Trans. Inst. Chem. Engrs.* (*London*) **29**, 75 (1951).

(C20) Crofton, W., "Encyclopaedia Britannica," 9th ed. (Article on "Probability").

(C21) Crozier, R. D., Ph.D. Thesis, Chem. Eng., University of Michigan, 1956.

(D1) Danchwerts, P. V., *Ind. Eng. Chem.* **43**, 1460 (1951).

(D2) Danckwerts, P. V., *Trans. Faraday Soc.* **47**, 1014 (1951).

(D3) Davidson, J. F., and Cullen, E. J., *Trans. Faraday Soc.* **53**, 113 (1957).

(D4) Davidson, J. F., and Cullen, E. J., *Trans. Inst. Chem. Engrs.* (*London*) **35**, 51 (1957).

(D5) Davies, R. M., and Taylor, G. I., *Proc. Roy. Soc.* **A200**, 375 (1950).

(D6) Deindoerfer, F. H., and Humphrey, A. E., *Ind. Eng. Chem.* **53**, 755 (1961).

(D7) Downing, A. L., and Bayley, R. W., *Trans. Inst. Chem. Engrs.* (*London*) **39**, A53 (1961).

(E1) Elsworth, R., Williams, V., and Harris-Smith, R., *J. Appl. Chem.* **7**, 269 (1957).

(E2) Endoh, K., and Oyama, Y., *Inst. Phys. Chem. Research Sci. Papers* (*Tokyo*) **52**, 131 (1958).

(E3) Essenhigh, R. H., and Fells, I., *Discussions Faraday Soc.* No. 30, 208 (1960).

(E4) Evans, F., Calderbank, P. H., Farley, R., and Poll, A., "Symposium on Catalysis in Practice," p. 59. Institution of Chemical Engineers, London, 1963.

(F1) Flynn, A. W., and Treybal, R. E., *A.I.Ch.E. Journal* **1**, 324 (1955).

(F2) Foss, A. S., Gerster, J. A., and Pigford, R. L., *A.I.Ch.E. Journal* **4**, 231 (1958).

(F3) Friedlander, S. K., *A.I.Ch.E. Journal* **3**, 43 (1957).

(F4) Friedlander, S. K., *A.I.Ch.E. Journal* **7**, 347 (1961).

(F5) Froessling, N., *Beitr. Geophys.* **32**, 170 (1938).

(F6) Froessling, N., Lunds Univ. Arsskr. N.F. AVD. **36**, No. 4 (1940).

(G1) Garner, F. H., and Hammerton, D., "Symposium on Gas Absorption." *Trans. Inst. Chem. Engrs.* (*London*) **32**, Suppl. No. 1, 518 (1954).

(G2) Garner, F. H., and Keey, R. B., *Chem. Eng. Sci.* **9**, 119 (1958).

(G3) Garner, F. H., and Porter, K. E., "Intern. Symposium on Distillation," p. 27. Institution of Chemical Engineers, London, 1960.

(G4) Garner, F. H., and Skelland, A. H. P., *Ind. Eng. Chem.* **46**, 591 (1954).

(G5) Garwin, L., and Smith, B. P., *Chem. Eng. Progr.* **49**, 591 (1953).

(G6) Gautreaux, M. F., and O'Connell, H. E., *Chem. Eng. Progr.* **51**, 232 (1955).

(G7) Gerster, J. A., A.I.Ch.E. Tray Efficiency Report, University of Delaware, Newark, Delaware, 1958.

(G8) Griffith, R. M., *Chem. Eng. Sci.* **12**, 198 (1960).

(G9) Grober, H., *Ze. Ver. Deut. Ing.* **69**, 705 (1925).

(H1) Haberman, W. L., and Morton, R. K., Report 802, The David Taylor Model Basin, U.S. Navy Dept., Washington, D.C., 1953.

(H2) Hadamard, J., *Compt. rend. acad. sci.* **152**, 1735 (1911).

(H3) Hammerton, D., Ph.D. Thesis, University of Birmingham, 1953.

(H4) Hammerton, D., and Garner, F. H., *Trans. Inst. Chem. Engrs.* (*London*) S17, 32 (1954).

(H5) Harriott, P., *A.I.Ch.E. Journal* 8, 93 (1962).

(H6) Heertjes, P. M., Holve, W. A., and Talsma, H., *Chem. Eng. Sci.* 3, 122 (1954).

(H7) Helsby, F. W., and Birt, D. C. P., *J. Appl. Chem.* 5, 347 (1955).

(H8) Higbie, R., *Trans. Am. Inst. Chem. Engrs.* 31, 365 (1935).

(H9) Hinze, J. O., *A.I.Ch.E. Journal* 5, 289 (1955).

(H10) Hixson, A. W., and Baum, S. J., *Ind. Eng. Chem.* 36, 528 (1944).

(H11) Hu, S., and Kintner, R. C., *A.I.Ch.E. Journal* 1, 42 (1955).

(H12) Hughes, R. R., and Gilliland, E. R., *Chem. Eng. Progr. Symposium* No. 16, 51, 101 (1955).

(H13) Humphrey, D. W., and van Ness, H. C., *A.I.Ch.E. Journal* 3, 283 (1957).

(J1) Johnson, A. I., and Huang, C.-J., *A.I.Ch.E. Journal* 2, 412 (1956).

(J2) Johnson, A. I., and Hamielec, A. E., *A.I.Ch.E. Journal* 6, 145 (1960).

(J3) Johnstone, H. F., and Williams, G. C., *Ind. Eng. Chem.* 38, 993 (1939).

(K1) Kinzer, G. D., and Gunn, R., *J. Meteorol.* 8, 71 (1951).

(K2) Kneule, F., *Chem.-Ing.-Tech.* 28, 221 (1956).

(K3) Kolar, V., *Collection Chzechoslov, Chem. Communs.* 24, 3309 (1959).

(K4) Kolmogoroff, A. N., *Compt. rend. acad. sci. U.S.S.R.* 30, 301 (1941).

(K5) Kronig, R., and Brink, J. C., *Appl. Sci. Research* A-2, 142 (1950).

(L1) Langmuir, I., *Phys. Rev.* 12 (2), 368 (1918).

(L2) Lebeis, E. H., *Ind. Eng. Chem.* 53, 349 (1961).

(L3) Leibson, I., Holcomb, E. G., Cacoso, A. G., and Jacmic, J. J., *A.I.Ch.E. Journal* 2, 296 (1956).

(L4) Lewis, W. K., Jr., *Ind. Eng. Chem.* 28, 399 (1936).

(L5) Linton, M., and Sherwood, J. K., *Chem. Eng. Progr.* 46, 258 (1950).

(L6) Linton, M., and Sutherland, K. L., *Chem. Eng. Sci.* 12, 214 (1960).

(L7) Lochiel, A. C., Ph.D. Thesis, University of Edinburgh, Scotland, 1963.

(L8) Longmuir, I. S., *Biochem. J.* 57, 84 (1954).

(L9) Longmuir, I. S., *Biochem. J.* 65, 378 (1957).

(M1) Mathers, W. G., Madden, A. J., and Piret, E. L., *Ind. Eng. Chem.* 49, 961 (1957).

(M2) McCrea, D., Ph.D. Thesis, University of Edinburgh, Scotland, 1956.

(M3) McDonough, J. A., Tomme, W. J., and Holland, C. D., *A.I.Ch.E. Journal* 6, 615 (1960).

(M4) McDowell, R. V., and Myers, I. E., A.I.Ch.E. Symposium on Mechanics of Bubbles and Drops, Detroit, Michigan, 1955. (Unpublished work).

(M5) Metzner, A. B., and Norwood, K. W., *A.I.Ch.E. Journal* 6, 433 (1960).

(M6) Metzner, A. B., Feehs, R. H., Ramos, H. L., Otto, R. E., and Tuthill, J. D., *A.I.Ch.E. Journal* 7, 3 (1961).

(M7) Meyer, P., *Trans. Inst. Chem. Engrs.* (*London*) 15, 127 (1937).

(M8) Mie, P., *Ann. Physik.* 33, 1275 (1910).

(M9) Moo-Young, M. B., External Ph.D. Thesis, London University, 1961.

(M10) Monod, J., "La croissance des cultures bacteriennes," p. 37. Herman, Paris, 1942.

(M11) Morgan, G. E., and Warner, W. H., *Aeronau. Sci.* 23, 937 (1956).

(N1) Norman, W. S., *Trans. Inst. Chem. Engrs.* (*London*) 26, 81 (1948).

(O1) Oyama, Y., *Chem. Eng.* (*Tokyo*) 20, 575 (1956).

(O2) Oyama, Y., and Endoh, K., *Chem. Eng.* (*Tokyo*) 20, 666 (1956).

(P1) Pansing, W. F., Ph.D. Thesis, University of Cincinnati, Ohio, 1951.

(P2) Piret, E. L., Mattern, R. U., and Bilous, O., *A.I.Ch.E. Journal* 3, 497 (1957).

(P3) Pratt, H. R. C., *Trans. Inst. Chem. Engrs.* (*London*) 29, 126 (1951).

(R1) Roger, W. A., Trice, V. G., and Rushton, J. H., *Chem. Eng. Progr.* 52, 515 (1956).

(R2) Rose, H. E., and Lloyd, H. B., *J. Soc. Chem. Ind. (London)* **65**, 52, 65 (1946).
(R3) Rose, W. D., and Wyllie, M. R. J., *Bull. Amer. Assoc. Petrol Geol.* **34**, 1748 (1950).
(R4) Rosenberg, B., Rept. No. 727, The David Taylor Model Basin, U.S. Navy Dept., Washington, D.C., 1950.
(R5) Rumford, F., and Bain, J., *Trans. Inst. Chem. Engrs. (London)* **38**, 11 (1960).
(R6) Rushton, J. H., Foust, H. C., and Mack, D. E., *Ind. Eng. Chem.* **36**, 517 (1944).
(R7) Rushton, J. H., Lichtmann, R. S., and Mahony, L. H., *Ind. Eng. Chem.* **40**, 1082 (1948).
(R8) Rushton, J. H., and Roy, P. H., Symposium on Mechanics of Bubbles and Drops. A.I.Ch.E. Annual Meeting, Detroit, Michigan, 1955. (Unpublished work).
(S1) Sauter, J., *Forsch. Arb. Geb. Ing.* **2–8**, 312 (1928).
(S2) Schmidt, E., *Chem.-Inge.-Tech.* **28**, 175 (1956).
(S3) Schotland, R. M., *Discussions Faraday Soc.* No. **30**, 72 (1960).
(S4) Sherwood, T. K., and Pigford, R. L., "Absorption and Extraction." McGraw-Hill, New York, 1952.
(S5) Shinnar, R., and Church, J. M., *Ind. Eng. Chem.* **52**, 253 (1960).
(S6) Sloan, C. K., Wortz, C. G., and Arrington, C. H., "Angular-Dependence Light Scattering." E. I. du Pont de Nemours and Co., Wilmington, Delaware (unpublished communication).
(S7) Stein, R. S., and Keane, J. J., *J. Polymer Sci.* **17**, 21 (1955).
(S8) Stephens, E. J., and Morris, G. A., *Chem. Eng. Progr.* **47**, 232 (1951).
(S9) Stokes, G. G., e.g. H. Lamb, "Hydrodynamics," p. 598. Dover, New York, 1945.
(S10) Sullivan, D. M., and Lindsey, E. E., *Ind. Eng. Chem. Fundamentals* **1**, 87 (1962).
✗ (T1) Takadi, T., and Maeda, S. *Chem. Eng. (Tokyo)* **25**, 254 (1961).
(T2) Taylor, G. I., *Proc. Roy. Soc.* **A146**, 501 (1934).
(T3) Tomotika, S., A.R.C. Rept. and Memo. No. 1678 (1935).
(T4) Toor, H. D., and Marchello, J. M., *A.I.Ch.E. Journal* **4**, 97 (1958).
(T5) Treybal, R. E., private communication (1961).
(T6) Treybal, R. E., *A.I.Ch.E. Journal* **4**, 202 (1958).
(V1) Valentin, F. H. H., and Preen, B. V., *Chem.-Ing-Tech.* **34**, 194 (1962).
(V2) Vanderveen, J. H., and Vermeulen, T., Report UCRL-8733. Univ. of California, 1960.
(V3) Vermeulen, T., *Ind. Eng. Chem.* **45**, 1664 (1953).
(V4) Vermeulen, T., Williams, G. M., and Langlois, G. E., *Chem. Eng. Progr.* **51**, 85 (1955).
(W1) Walton, W. H., "Symposium on Particle Size Analysis," p. 141. Institution of Chemical Engineers, London, 1957.
(W2) Ward, D. M., Trass, O., and Johnson, A. I., *Can. J. Chem. Eng.* **40**, 164 (1962).
(W3) West, F. B. *et al., Ind. Eng. Chem.* **43**, 234 (1951).
(W4) Westerterp, K. R., Thesis, Technische Hogeschool, Delft, Netherlands, 1962.
(W5) Whitman, W. G., *Chem. Met. Eng.* **29**, 147 (1923).
(W6) Wilhelm, R. H., Conklin, L. H., and Savic, T. C., *Ind. Eng. Chem.* **33**, 453 (1941).
(W7) Wilke, C. R., and Pin Chang, *A.I.Ch.E. Journal* **1**, 265 (1955).
(W8) Williams, E. E., Ship Division Rept No. 7, National Physical Laboratory, Teddington, England, 1959.
(Y1) Yoshida, F., Ikeda, A., Imakawa, S., and Miura, Y., *Ind. Eng. Chem.* **52**, 435 (1960).
(Z1) Zahm, A. F., *Natl. Advisory Comm. Aeronau., Rept.* No. **253**, 517 (1926).
(Z2) Zdanovskii, P., and Karashanov, N. A., *J. Appl. Chem. (U.S.S.R.)* **32**, 129 (1959).
(Z3) Zwietering, Th. N., *Chem. Eng. Sci.* **8**, 244 (1958).

CHAPTER 7

Mixing and Chemical Reactions

Jon H. Olson

University of Delaware, Newark, Delaware

Lawrence E. Stout, Jr.

Monsanto Company, St. Louis, Missouri

I. Introduction

Chemical reactors are constructed in a great variety of forms, stirred tanks, pipelines, packed beds, fluidized beds, or circulation loops. Consideration of mixing, defined as the interlarding and diffusion of the reaction components, is important in nearly all of these reactors. The rational selection of the reactor and eventual optimization of the reactor design depend upon the ability to estimate the conversion in a reactor with some precision. Thus one needs to know how mixing will affect the performance of a reactor in order to evaluate and design the unit.

The general requirements for calculating the conversion in a chemical reactor are knowledge of the temperature and concentration fields in the reactor and the kinetics and transport parameters of the reaction. The total production of the reactor is found by integration of the point rates of reaction throughout the system. Since the point reaction rates seldom can be calculated, other methods for estimating reactor performance have been developed.

115

It is useful at the outset to indicate why mixing is important in chemical reactions. In general, two or more substances need to be mixed to make products, and this mixing must be sufficiently fine-grained that molecular transport becomes effective. Mixing is used also to provide uniformity of temperature and concentrations in chemical reactors; this uniformity is particularly important when the selectivity of the reactions is to be maximized or when side reactions are to be minimized. On the other hand, mixing in a chemical reactor is not always advantageous. For example, axial mixing in a pipeline reactor usually reduces the yield of the reaction. It is therefore important to know both the positive and negative contributions which mixing may contribute to reactor performance.

The subject of mixing and chemical reactions may be divided into subtopics in more than one way, and hence the choice made here is arbitrary. An important distinction may be made between homogeneous and heterogeneous reactions. The literature has been primarily focused upon homogeneous reactions even though one suspects that heterogeneous reactions are vastly more important (see Chapter 6 in this volume). Ideal reactor models provide a second classification. An ideal reactor is a mathematically simple model of an actual reactor. Thus the simple models of batch reactors, continuous-flow stirred-tank reactors, and piston-flow reactors provide a framework from which the more complex models of reactors may be developed.

The purpose of this chapter is to collect and interpret useful forms of experimental and theoretical information upon mixing and chemical reaction according to the classifications denoted above.

II. Homogeneous Reactions

A. CONTINUOUS-FLOW STIRRED-TANK REACTORS

This section is divided into three parts in which the interactions of mixing with reaction kinetics increase in complexity in the order given. Ideal tank reactors are discussed in Section I. An ideal mixer is one in which there is no variation of physical properties throughout the reactor; therefore, the interaction of mixing with chemical reaction is solved by definition. The least complex form of nonideal mixing is described in Section A, 2, where nonidealities of the reactor model are assumed to be represented entirely by the residence-time distribution. Finally, in Section A, 3, the mixing within the reactor is approximated by subdividing the reactor into a number of interconnected sections.

1. Ideal Tank Reactors

The effects of mixing upon chemical reactions may be introduced by considering the design of a continuous-flow, stirred-tank reactor (TR for short). The mixing in such a reactor is assumed to be ideal; that is, the concentrations and temperature are uniform throughout the reactor volume. Such mixing performance clearly is a limiting case of the effectiveness of mixing in an actual reactor. The objective in design of a TR is to select the most profitable combination of operating variables. For example, the throughput, temperature, feed composition, and reactor size are to be specified. The performance of the reactor is found from steady state heat and material balances around the reactor.

Component mass balance:

$$fw_i^{N-1} = fw_i^N - V^N r_i^N(w_j, \rho, T) \tag{1}$$

Total heat balance:

$$Q^N + fC_p^{N-1}T^{N-1} = fC_p^N T^N + V^N \sum_{i=1} r_i^N(w_j, \rho, T)\, \Delta h_i^N(w_j, \rho, T) \tag{2}$$

The superscripts $N - 1$ and N are used to denote variables of the $N - 1$ and N tanks. This notation implies that a series of tank reactors may be considered with these equations. The feed variables to the set of reactors is specified when N is zero. The subscript i denotes the ith component of the system, and the subscript j is used to denote all of the composition variables which are included in the rate expression for the ith component. Equations (1) and (2) are written in terms of mass fractions because density changes with reaction are neatly included in a mass fraction formulation of the material balance equations. The other symbols have the following definitions:

C_p = specific heat of fluid (the specific heat may be a function of temperature and composition),

f = mass flow rate,

$\Delta h_i(w_j, \rho, T)$ = the thermodynamic (exothermic is minus) heat of formation for production of one mass unit of the ith component at the conditions of the Nth reactor,

Q^N = rate of heat input to the reactor by heat transfer,

$r_i^N(w_j, \rho, T)$ = mass rate of production of the ith component per unit volume at the conditions of the Nth reactor,

V^N = volume of Nth reactor,

w_i^N = mass fraction of ith component in the Nth tank,

ρ = density of the fluid, (the density of fluid may be a function of temperature and composition).

Equations (1) and (2) are often converted to molal concentration units when the reaction rate equation is so written. Examples of the various forms which Eqs. (1) and (2) may take are discussed extensively in textbooks on chemical engineering reaction kinetics (A5, K3, S4, W1).

Although Eqs. (1) and (2) are complicated in the most general form, most

practical cases yield far simpler forms. In particular, if the reaction is isothermal, first order, irreversible, and constant density, then for $A \to P$:

$$r_a^1 = -k\rho w_a^1; \qquad w_a^1 = w_a^0(1 + K)^{-1} \tag{1a}$$

$$\frac{Q^1}{f} = -\frac{Kw_a^0 \,\Delta H_p}{(1 + K)M_A} \tag{2b}$$

where ΔH_p = thermodynamic heat of reaction per mole of P produced,

M_A = molecular weight of A,

K = dimensionless reaction parameter, $V\rho k/f$.

The major point for emphasis about Eqs. (1) and (2) is that the contribution
of mixing has been specified by the form of the equations. An observer attached
to a fluid particle would note an abrupt change in the concentrations and
temperatures upon "entering" the tank and thereafter no changes would be
observed. The design engineer must judge if this discontinuous approximation
yields a sufficiently accurate estimate of the equipment performance; this
point will be discussed below. The other problem the design engineer must
solve is to decide if the TR is the best form of reactor. Factors which may be
included in finding the "best choice" for reactor type go beyond the scope of
this chapter; availability, construction time, control stability, flexibility, and
costs are all important considerations upon the ultimate reactor choice.

Nevertheless, it is worthwhile noticing that a TR may often offer selectivity
advantages. If one has a reaction sequence of the type

$$A + B \to I \tag{3a}$$

$$I + B \to P \quad \text{desired} \tag{3b}$$

$$I + I \to X \quad \text{waste} \tag{3c}$$

then a TR with large excess of B will suppress the unwanted side reaction.
Examples of this type of selectivity advantage for the TR are given by van der
Vusse (V2) and Kramers (K3).

The corresponding thermal stability problem is obtained when $I + I \to P$
is highly exothermic and $I + B \to P$ is less exothermic. Reduction of the I
level in the reactor decreases the potential instability of the processes. The
generalization of the cases illustrated by the equations above is that tank
reactors permit one to hold the concentrations of reaction intermediates at
low (and final) values.

2. Nonideal, Continuous-Flow Stirred-Tank Reactors: Residence-Time Distri-bution Effects

The least complex method for representing imperfect mixing within a TR is
by altering the residence-time distribution. There are two ways for estimating

the contribution of the residence-time distribution to the conversion of the TR, and the selection of the method used depends upon the apparent order of the kinetic rate expression. Linear reaction kinetics provides a special case, and in Section A, 2a, it is shown that the conversion of the reactor is uniquely determined by the residence-time distribution and the linear rate expression. However, when the reaction kinetics are nonlinear, the conversion of the reactor is not uniquely specified by the residence-time distribution and the kinetic rate expression. Nevertheless, the residence-time distribution does provide useful information for reactors in which nonlinear kinetics are occurring. In Section A, 2, b, Zweitering's (Z1) criteria of complete segregation and maximum mixedness are developed as useful estimates of the importance of mixing within these systems.

a. Linear Kinetics. Observation of any TR demonstrates that the instantaneous mixing implied by the ideal model is not achieved. The engineer must then determine what difference nonideal mixing can make upon the process. In Chapter 4 of Volume I, Gray reviewed some of the models for nonideal mixing and illustrated calculation of the residence-time frequency distribution, $f(t)$, from these models. For the important special case of first-order kinetics knowledge of the reaction rate expression and $f(t)$ are sufficient to determine the conversion in the reactor. First-order linear reaction kinetics are defined by kinetic rate equations of the form

$$r_i^N(w_j, \rho, T) = -\rho^N k^N(w_i^N - w_i^{Ne})$$ (4)

The w_i^{Ne} term is the weight fraction of the ith reactant component at equilibrium. The rate constant, k, usually is given as an empirical Arrhenius function of temperature

$$k^N = Ae^{-\frac{\Delta H_{act}}{RT^N}}$$ (5)

When one is considering nonideal tank reactors in which the temperature in the reactor is uniform, the linearity of the rate expression Eq. (4) permits the conversion in the reactor to be calculated as

$$w_i^N = w_i^{Ne} + (w_i^{N-1} - w_i^{Ne})\int_0^\infty f^N(t)e^{-k^N t}dt$$ (6)

where $f(t)dt$ = fraction of the mass flow through the reactor with a residence time on the period $t, t + dt$.

For linear or pseudo-first-order kinetics, Eq. (6) yields the same result as Eq. (1a) for an ideal TR but is obviously more complex to use. Equation (6) also implies that the conversion in pseudo-first-order isothermal reactions in nonideal tank reactors is controlled by the eventual form of the residence-time distribution and not by the details of the mixing process. Consequently, one does not need very much additional information to specify the performance of a nonideal tank reactor when the reaction kinetics are linear. In addition, one can give a rule of thumb from which the applicability of the ideal TR may be estimated. In the models of Marr and Johnson (M4), van der Vusse (V2), Dickens (D4), Conover (C7), and Larson (L1), the residence-time frequency distribution is "very close" to ideal if the pump circulation rate is greater than five times the input flow rate. This condition is rather easy to achieve in practice except for extremely viscous materials, and consequently design estimates are usually made using ideal TR models.

   b. *Nonlinear Kinetics.* Additional information is required when one is considering nonlinear reaction kinetics or temperature effects in nonideal tank reactors. Zweitering (Z1) and others (C5, C7, D5) have demonstrated that it is easy to construct models of processes in which the residence-time frequency distributions are identical but in which the conversions for second-order isothermal kinetics $(r = -k_2(w\rho)^2)$ are not equal. It is apparent that additional information is required before the conversion of the reactor can be determined uniquely for reactors in which second-order reactions are occurring. Zweitering showed, however, that useful limits upon the performance of a reactor can be obtained from the residence-time distribution function and the nonlinear, kinetic rate equation. Zweitering's approach is to define two types of mixing within the reactor; these two types of mixing are defined uniquely from the residence-time distribution function. Zweitering assumed that the two cases represent the least and the most mixing which can occur in the process vessel. Consequently, calculation of reaction conversion for these two mixing models sets limits upon the extent to which mixing affects the conversion of the reactor. Before indicating how the reaction conversions are to be calculated, it is instructive to develop a physical understanding of the two mixing models.

   The least amount of mixing occurs in what Zweitering calls a completely segregated reactor. The term "complete segregation" or "macromixing" means that the flow process produces fluid elements of considerable size and that these fluid elements do not mix or interact with one another in the reactor. The term "complete segregation" is an attempt to describe the absence of interaction of the large fluid elements; in turn the concept of mixing implies some form of intermingling of these large fluid clumps. The conversion in a completely segregated reactor is found by treating each of the fluid elements as a batch reactor and then integrating over the residence-time distribution. This procedure is presented by the following equation:

$$w_i^N = w_i^{N-1} + \int_0^\infty f^N(t) \int_0^t \frac{r_i^N}{\rho}(w_j, \rho, T)dt' \, dt \qquad (7)$$

in which w has dimensions of concentration gm./cm.^{3}. (The temperature, T, is assumed constant in Zweitering's paper.)

   The greatest amount of mixing occurs in what Zweitering calls the "maximum mixedness reactor." Maximum mixedness is used to denote the largest amount of interaction between the fluid elements which is consistent with the residence-time distribution. In an ideal TR each fluid element has a uniform probability of mixing with any other element; Zweitering showed that this uniform interaction probability is consistent with the residence-time distribution of an ideal TR. When one considers an arbitary residence-time distribution, however, it is no longer possible for all fluid lumps to interact with one another uniformly. Zweitering established that the conversion in a maximum mixedness reactor is given by the solution of the differential equation

$$\frac{dw_i^N}{dh} = \frac{r_i^N}{\rho}(w_j, \rho, T) + (w_i^{N-1} - w_i^N)\frac{f^N(-h)}{\int\limits_{-h}^{\infty} f^N(t)dt} \tag{8}$$

The parameter, h, of Eq. (8) is a dummy variable in the integration but is chosen symbolically as the mixing-history parameter. The boundary condition and the limits for integration of Eq. (8) require justification and depend upon the behavior of the argument

$$g^N(h) \equiv \frac{f^N(-h)}{\int\limits_{-h}^{\infty} f(t)dt} \tag{8a}$$

as h approaches $-\infty$. There are three cases to be considered which are described below.

1. $g^N(h)$ equals a positive constant as h approaches $-\infty$. This case occurs whenever the residence-time frequency function exponentially approaches zero for large values of the time parameter. The boundary condition for this case is established by selecting a small value of h and specifying w_i^w by the approximate algebraic equation

$$O = \frac{r_i^N}{\rho}(w_j, \rho, T) + (w_i^{N-1} - w_i^N)g^N(h) \tag{8b}$$

2. $g^N(h)$ approaches zero as h approaches $-\infty$. This case occurs whenever the residence-time frequency function approaches zero as t^{-n}, n greater than 1, for large values of the time parameter. The boundary condition for this case is found by specifying $w_i^N = w_i^{N-1}$ at some small value of h.
3. $g^N(h)$ approaches $+\infty$ at some value of h. This boundary condition occurs when the residence-time frequency function approaches zero by a step function at some finite value of the time parameter. The boundary condition for this case is found by specifying $w_i^N = w_i^{N-1}$ for $-h$ equal to the value of the time parameter at which the step function occurs.

Thus the initial condition for Eq. (8) is established for some arbitrary and small value of h (cases 1 and 2) or for some particular value of h (case 3). The conversion of the maximum mixedness reactor is found by integrating Eq. (8) from the initial value of h to $h = 0$. The conversion of the reactor is given by the value of w_i^N at the final limit of the integration.

It has not been established if Eqs. (7) and (8) always give a limiting performance of a reactor, but Zweitering did suggest that useful estimates are obtained. However, numerical experience with Eqs. (7) and (8) indicates that there is not much difference in the effluent concentrations for simple kinetic schemes under isothermal conditions. Hence, when one is dealing with continuous tank reactors under isothermal conditions, evaluation of Eqs. (7) and (8) may suggest that no further consideration of the mixing within the tank need be made. On the other hand, if the kinetic scheme is complex (for example, a polymerization reaction), then further examination of the mixing process may be justified.

3. Nonideal Tank Reactors—Signal-Flow Models

In this section the effects of nonideal mixing within a reactor on conversion will be considered. This section differs from the previous one in that several forms of mixing are assumed to occur within a single reactor. The tank reactor is replaced by a more complex and hopefully more realistic model in which various forms of mixing are connected together. This technique is popular with users of analog computers; the name "signal-flow models" comes from the signal-flow diagrams used as an intermediate step in analog computation (C8, M3, M5).

Several examples of signal-flow analysis are given in this section. The general attack is to identify the types of mixing which occur and then to make quantitative estimates of the more important factors. The first example is the most complicated case given; the purpose of this example is to illustrate most of the intermediate mixing models which are used in this technique. The second and third examples are presented as being useful reductions of the complexities of the first model.

Qualitatively when one observes a tank reactor, it is noted that the only zone which really appears to be well mixed is that surrounding the impeller. Larger zones of much lower intensity of mixing appear in the remainder of the tank. If some inert tracer particle is introduced into the tank, the particle is observed to make circulation loops out from and into the zone of high shear surrounding the impeller. Figure 1a depicts the circulation patterns of a tank reactor. Other qualitative features of the mixing within a TR are short-circuiting of feed to the tank effluent, stagnant zones, and circulation loops which do not pass through the impeller region. These qualitative features do not appear in the model shown in Fig. 1a, but the techniques for including these possibilities are no different than those illustrated. The signal-flow model for this mixer is shown in Fig. 1b. Some general comments upon the model are given prior to the detailed description of the example.

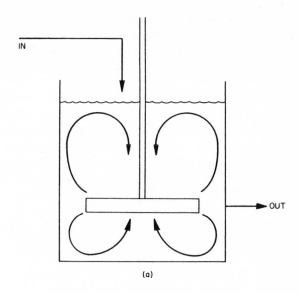

(a)

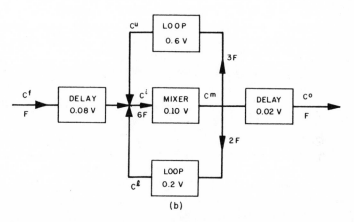

(b)

FIG. 1. (a) Flow model for a tank reactor—the curved lines represent the flow patterns developed by an impeller mixer. (b) Signal-flow diagram for a tank reactor—the numbers inside each of the boxes represent the fraction of the total tank volume which is assigned to the portion of the reactor represented by the box. The numbers beside the arrowheads outside the boxes represent the ratio of the volumetric feed rate in the flow path to the volumetric feed rate to the reactor. The two delay sections are assumed to be piston-flow reactors, the two loops are assumed to be segregated tank reactors, and the mixer is assumed to be an ideal tank reactor.

The performance of the reactor can be estimated if the residence-time distributions and the mixing in each of the sections of the reactor can be estimated. One purpose of making a signal-flow model is to gain a representation of a tank reactor which more adequately represents the nonidealities of the unit. An advantage of the signal-flow technique is the possibility of incorporating some analytical solutions of the equation of motion into parts of the model. For example, Larson (L1) calculated the flow patterns of the circulation loops from a potential flow model of the tank. These analytical models apparently are more accurate in design scale-up than extrapolations of empirical data. Thus by using a more complicated model for the tank reactor in the pilot-plant scale, one hopes to be able to make a better estimate of reactor performance in plant-scale operation.

The use of these signal-flow models for nonideal tank reactor systems is best indicated by illustration. Assume that an irreversible second-order reaction takes place isothermally and without volume change in the tank reactor shown in Fig. 1a. The sequence of steps in setting up a model is then to assume appropriate forms of reaction rate equations for each of these sections and then combine these equations to obtain steady state conditions. The algebra of these computations is straightforward but often cumbersome. The following assumptions have been made in constructing the signal-flow diagram shown in Fig. 1b.

1. The feed to the tank reactor is not mixed with any other material in the reactor until the feed arrives at the mixer impeller. In addition the feed enters the impeller after a simple piston-flow delay following injection into the tank. The delay time between injection into the tank and arrival at the mixer depends markedly upon the location of the feed pipe. The delay time may be estimated from the potential flow model for fluid circulation in tanks (L1) or may be guessed as some fraction of the mean circulation time of the upper loop.

2. The fluid mixing zone surrounding the impeller is a zone of sufficiently high shear that the perfect mixer model is appropriate (M2). The volume of this zone is taken as the volume swept by the impeller.

3. The fluid discharge from the impeller is split into two fractions which flow into the upper and lower portions of the tank and into a fraction which exits from the reactor. The total discharge rate of the impeller can be estimated from correlations (Chapter 4, M5) whereas the split between the upper and lower loops can be evaluated from other correlations (D4, L1).

4. The flow in the upper and lower circulation loops is "completely segregated" and has a residence-time distribution equivalent to a TR. Several authors (D4, H3, L1, M3, V2) have attempted to measure this residence-time distribution. Holmes et al. (H3) and van der Vusse (V2)

show that the residence-time distribution for the circulation loops is similar to that for 4 or 5 TR in series, whereas Dickens found that the residence-time distribution for the loops was similar to 1–2 TR in series. The estimation of the residence-time distribution in the circulation loops is difficult, and there is very little experimental data from which estimates may be made.

5. The effluent from the reactor is obtained after a short delay from the effluent of the impeller. This assumption is as good (or bad) as that for the inlet feed to the reactor.

Assuming that the volumes of the various sections of the tank reactor have been assigned and the discharge rates from the impeller to the various sections are known, the mean residence time in each section of the vessel may be easily computed. As an example consider flows and volumes shown in Fig. 1b. Further assume an irreversible, second-order reaction, $2A \rightarrow P$ takes place in the reactor. The concentrations leaving the various sections are computed as follows:

Input to Mixer: The effluent concentration of the inlet delay section is calculated by assuming piston flow of the inlet stream. The result is

$$C^1 = \frac{C^f}{1 + K\phi_1 C^f} \tag{9}$$

where A^f = feed concentration;
A^r = reference concentration;
$C^f = A^f/A^r$, dimensionless feed concentration;
C^1 = dimensionless concentration leaving first delay section;
F = volumetric feed rate to reactor;
$K = A^r k_2 V/F$;
V = volume of the tank;
V_1 = volume of inlet delay section;
$\phi_1 = V_1/V$.

Output of Upper Loop: Using Eq. (7) for second-order kinetics and the residence-time frequency distribution of a TR, one finds the concentration leaving the upper loop to be

$$C^u = \int_0^\infty \frac{C^m e^{-\tau/\phi_u}}{1 + KC^m\tau} d\left(\frac{\tau}{\phi_u}\right) \tag{10a}$$

$$C^u = C^m y e^{+y} \int_y^\infty \frac{e^{-\eta}}{\eta} d\eta \tag{10b}$$

$$C^u = C^m y e^{+y} E_1(y) \tag{10c}$$

where C^m = dimensionless output concentration of the mixer,
   C^u = dimensionless output concentration of the upper circulation loop,
   $E_1(y)$ = a tabulated exponential integral,
   $y = (KC^m\phi_u)^{-1}$,
   $\phi_u = V_u F/VF_u$,
   τ = dimensionless tank residence time, Vt/F.

Output of Lower Loop: The output concentration for the lower loop is found in the same way as Eq. (10c) and is

$$C^\ell = C^m z e^{+z} \int_z^\infty \frac{e^{-\eta}}{\eta}\, d\eta \tag{11}$$

where C^ℓ = dimensionless output concentration of the lower loop,
   $z = (KC^m\phi_\ell)^{-1}$,
   $\phi_\ell = V_\ell F/VF_\ell$.

Output Mixer: The effluent concentration of the mixer is found from a material balance upon an ideal stirred tank reactor. The result is

$$C^m = \frac{-1 + \sqrt{1 + 4K\phi_m C^i}}{2K\phi_m} \tag{12}$$

where $\phi_m = \dfrac{V_m F}{V(F + F_\ell + F_u)}$.

Input Mixer: The input concentration to the mixer, C_i, is the weighted average of the concentration of the three streams which flow into the mixer. Thus one finds

$$C^i = \frac{FC^1 + F_u C^u + F_\ell C^\ell}{F + F_u + F_\ell} \tag{13}$$

where F = feed rate to the tank reactor,
   F_u, F_ℓ = feed rates to the upper and lower loops.

Output of System: The effluent to the entire system is found by calculating the conversion in the output delay section which is fed from the mixer.

$$C^0 = \frac{C^m}{1 + K\phi_2 C^m} \tag{14}$$

where V_2 = volume of outlet delay section,
   $\phi_2 = V_2/V$.

   The use of Eqs. (9) to (14) is illustrated by assuming that $C^f = 1.0$, $K = 1.0$, and the holding time parameters, ϕ_1, ϕ_ℓ, ϕ_u, ϕ_m, and ϕ_2, are those taken from

Fig. 1b, and then performing the necessary algebra. The final effluent concentration for this example is $C^0 = 0.596$. This value is equal to the effluent concentration of a completely segregated TR (Section II, A, 2, b) to the third significant figure. One may attempt to understand this surprising result by investigating the effect of changing the values of the parameters in the signal-flow model and by altering the order in which the various sections are interconnected. It is found that the effluent concentration is very strongly influenced by the models chosen for the inlet and exit flows. In fact if the outlet delay section is taken directly from the impeller, the exit concentration is 0.603. This value is intermediate to the value for an ideal TR, 0.608, and the value for a completely segregated TR, 0.596. Thus one of the disadvantages of the signal-flow model is that the conversion of the reactor is most sensitive to parameters which are hard to estimate. A second objection to this model is that the differences in the conversion for moderate changes in K are too small to be important. In fact, it is only when the effluent concentration C^0 is very low that any significant difference is observed in these models.

A logical extension of the signal-flow model developed above was presented by Manning (M3). The two cases are shown in Figs. 2a and 2b. It has been noted that the location of the feed and exit ports is one of the more significant variables in the signal-flow representation of imperfect mixing in tank reactors. Accordingly, the distinction between the two cases of Fig. 2 is made on the way in which feed is injected or removed from the tanks. The feed is injected directly into the mixing zone in the micro-macro mixer model shown in Fig. 2a. This model corresponds to locating the feed port adjacent to the impeller or turbine and the exit port at some position well removed from the inlet. The symmetric opposite case is made in the macro-micro mixer model in Fig. 2b; the effluent is collected near the discharge of the impeller and the feed enters into a circulation loop. The volume of the perfect mixer section in Fig. 1b is reduced to zero or "very small" in the Manning model. In addition the circulation loops are assumed to be perfectly segregated systems with exponential residence-time distributions. The time constants of these distributions are determined from the pumping rate of the turbine. The advantage of these models is their simplicity.

To illustrate how these models are used, assume a second-order irreversible reaction occurs in a micro-macro mixer. The dimensionless effluent concentration of the unit, C^{ia}, is found from a material balance upon the loop and upon the mixer. The analysis is similar to the development of Eq. (10a) for the loop and Eq. (13) for the mixer. The result is

$$C^{ia} = C^m Z e^{+z} E_1(z) \tag{15a}$$

$$C^m = \frac{FC^f + PC^{ia}}{F + P} \tag{15b}$$

where C^{ia} = dimensionless output concentration of the micro-macro model;
C^f = dimensionless feed concentration to the reactor;
C^m = dimensionless output concentration of the mixer;
F = volumetric feed rate;
$K = k_2 A^r V / F$, the second-order dimensionless rate parameter;
P = volumetric pumping rate of the impeller;
$Z = F/[(F + P)KC^m]$.

Equations (15a, b) and (15b) may be solved together by trial and error.

The effluent concentration of the macro-micro mixer is also found from material-balance computations, but the result is somewhat more complex than the result for the micro-macro system. The volumes of the feed loop and the circulation loop are adjusted to make the mean residence time (volume/feed rate) of the two loops identical. Thus one finds that a fraction $\phi = F/(F + P)$ of the tank volume is assigned to the feed loop and the remainder $(1 - \phi) = P/(F + P)$ is assigned to the circulation loop. One finds the concentration

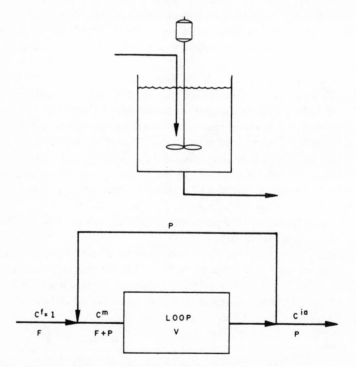

FIG. 2a. Micro-macro mixer—the feed to this reactor is injected directly into the impeller and there is mixed with the fluid of the circulation loop. The circulation loop is represented as a segregated tank reactor.

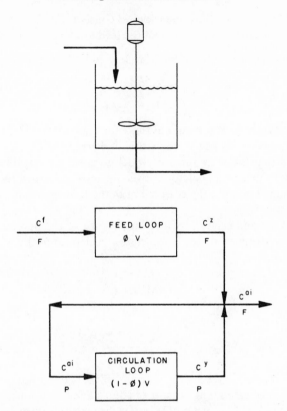

FIG. 2b. Macro-micro mixer—the feed to this reactor is injected some distance away from the impeller and flows to the impeller through the segregated feed loop. The fluid leaving the reactor is taken directly from the mixer. The major portion of the fluid in the tank passes through the reactor to the circulation loop.

of the reactant leaving the macro-micro reactor model by first evaluating the conversions in the feed loop and the circulation loop and then mixing these two streams to find the produce concentration. The results of these computations are as follows:

Output of the macro-micro mixer:

$$C^{ai} = \phi C^y + (1 - \phi)C^z \tag{16a}$$

Output of the circulation loop:

$$C^i = C^{ai} Y e^Y E_1(Y) \tag{16b}$$

Output of the feed loop:

$$C^z = C^f Z e^z E_1(Z) \tag{16c}$$

where C^{ai}, C^y, and C^z are the dimensionless concentrations leaving the macro-micro mixer, the circulation loop, and the feed loop. In Eqs. (16a) to (16c)

$$Y = C^{ai}K(1 - \phi)$$

$$Z = C^f K \phi$$

$$\phi = F/(F + P)$$

and the other symbols are the same as those for Eqs. (15a) and (15b). Equations (16a) to (16c) again may be solved by trial and error.

It is found that the macro-micro model gives slightly greater conversions than the micro-macro model for irreversible, second-order reactions. This result is obtained because the conversion in the feed loop is higher than in a circulation loop.

To gain some appreciation of the importance of the pumping rate upon conversion, a mixing efficiency parameter, η, is defined for the two models as Macro-micro:

$$\eta^{ai} = \frac{C^{ai} - C^{sf}}{C^{tr} - C^{sf}} \tag{17a}$$

Micro-macro:

$$\eta^{ia} = \frac{C^{ia} - C^{sf}}{C^{tr} - C^{sf}} \tag{17b}$$

where $C^{sf} = Xe^x E_1(X)$ and is the reactant effluent concentration for an ideal segregated tank reactor of equivalent dimensions,

$C^{tr} = -1 + (1 + 4K)^{1/2}/2K$ and is the reactant effluent concentration for a perfectly mixed tank reactor of equivalent size,

$K = k_2 A^r V/F$,

$X = (C^f K)^{-1}$.

The mixing efficiency parameters is a measure of the degree to which the conversion of the unit approaches that of a perfectly stirred tank reactor. The denominator is chosen to emphasize the relatively small difference in conversion between the two limiting models. Figure 3 shows the effect pumping has on the mixing efficiency parameter when the dimensionless rate parameter, K, equals five. One finds that the macro-micro mixer has substantially lower mixing efficiency than the micro-macro model. This result emphasizes the importance of the feed circulation loop.

It is somewhat surprising to find that the conversion of the reactor drops as mixing efficiency increases. The conversion is lower for the micro-macro model than for the macro-micro model, yet the mixing efficiency shown in Fig. 3 is lower for the latter. This result stems from the particular kinetics chosen in the examples. There are very few chemical reactions which are of the type $2A \rightarrow P$, that is, true irreversible, second-order, but for these reactions mixing does

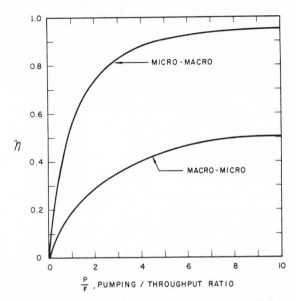

η

$\dfrac{P}{F}$, PUMPING / THROUGHPUT RATIO

FIG. 3. Effect of throughput upon mixing efficiency—the mixing efficiency, η, defined as the approach of conversion of the reactor to the conversion of an ideal tank reactor, increases as the pumping rate in the tank increases. The efficiency of a macro-micro unit does not reach unity.

decrease conversion. For the more common case of a reaction between two different reactants to produce a product, the irreversible, second-order kinetics can apply only when the two reactants are mixed in stoichiometric balance on a molecular scale upon entry to the reactor. Such mixing might be approximated by having a highly efficient mixing-tee at the inlet port of the reactor vessel. Under these circumstances the signal-flow analysis of a reactor describes how the performance of the reactor is affected by "large" scale mixing, for clearly the reactants are already mixed upon a "small" or "local-molecular" scale. The computations illustrate that large-scale mixing decreases the conversion of a second-order reaction.

This section has illustrated the techniques for obtaining the conversion in a stirred-tank reactor when the mixing is described by signal-flow models. The major advantage of these signal-flow models is that the mixing is well-defined and hence the conversion of the reactor may be calculated. The chief disadvantage of these models for mixing is that the sizes and types of the components in the signal-flow diagram are difficult to extrapolate from laboratory and pilot-plant scale to plant sizes. This extrapolation is uncertain because there is so little quality data for large units; clearly there is a need to publish more information upon the performance of large equipment (S5).

Throughout this section accurate knowledge of the reaction kinetics has been assumed. However, the types of reactions carried out in tank reactors usually are quite complex chemical mechanisms; consequently, the kinetics of these reactions are only approximated by the simple forms used for illustration. Very often the chemical mechanisms of a commercial process are not known; the development of such processes demands pilot-plant experiments. Although determination of the reaction mechanisms of a process is often very difficult, the results of mechanism studies are of extraordinary importance in the rational design of reactors. A small change in a reaction rate parameter often makes a much larger contribution to the performance of the reactor than rather large changes in the type of mixing. The reactor designer should never lose sight of the importance of chemistry in the process.

B. PIPELINE AND TUBULAR REACTORS

A pipeline or tubular reactor is the second major classification of continuous reactor types. In idealized form, the pipeline reactor is a continuous analog of a well-stirred batch reactor; consequently, the pipe reactor is a good choice when one wants to keep reaction intermediates at a high level. Other advantages of the pipeline reactor is that it permits control of the temperature history of the reactants, it is easy to design and operate, and it is inexpensive to construct.

This section is divided into three parts. The performance of an ideal, piston-flow reactor is described in the first part. The sources of deviation from the ideal, piston-flow model are classified and evaluated in the second part. An intermediate mixing model is described in the third part. Thus order of the discussion of pipe reactors parallels the previous section on tank reactors.

1. Ideal, Piston-Flow Reactors

This section describes an idealized mixing model for tubular reactors. The effect of mixing upon conversion is "solved" by definition; there is no mixing in an ideal piston-flow reactor. The conversion of the reactor is found from integration of the mass and heat conservation equations.

A justification for the piston-flow reactor model is developed by examination of the velocity profile in a pipe. The velocity profile for the turbulent flow of Newtonian fluids in pipes is quite flat; the deviation of point velocities from the mean velocity is quite small. A first approximation of the behavior of a pipeline reactor then is to ignore the variations in velocity and assume that the fluid passes through the pipe in the same way a piston passes through a cylinder. Consequently, this model for the reactor is called "piston flow." One-dimensional flow approximations equivalent to the piston-flow model often are used in engineering; for example, thermodynamic analysis of processes usually are made assuming uniform properties in entrance and exit streams.

Piston-flow velocity profiles are closely approximated in packed beds (C1). Thus the fluid mechanics of a packed-bed reactor universally are described by the piston-flow assumption. However, one notes that a packed-bed reactor is a heterogeneous system, for the bed packing usually is a catalyst. Nonetheless when the catalyst particles are small and the catalyst pellets have nearly the same temperature as the fluid surrounding them, the heat and mass balance equations have essentially the same form as the homogeneous reaction case. Sometimes the bed packing is chemically inert; the function of the packing is to reduce mixing in the axial direction and to improve temperature uniformity within the reactor. These remarks are made to justify consideration of packed-bed reactors along with tubular reactors.

The design of an ideal, piston-flow reactor depends upon the solution of first-order ordinary differential equations of the type
Component mass balance:

$$\nabla \cdot (\rho w_i \mathbf{v}) - r_i(w_j, \rho, T) = 0 \qquad (18a)$$

Heat balance:

$$\nabla \cdot (\rho C_p T \mathbf{v}) + \sum_i \Delta h_i(w_j, \rho, T) r_i(w_j, \rho, T) = 0 \qquad (18b)$$

where $\mathbf{v}$, the velocity vector, is at most a function of axial position.

Equations (18a) and (18b) are written in Gibbs vector notation to allow any geometry in which $\mathbf{v}$ is uniform across the cross section (necessary for the piston-flow model) and to allow for temperature and concentration variations in the transverse direction. The solution of Eqs. (18a) and (18b) with a vast assortment of boundary conditions is a topic of much of the chemical engineering literature (A5, K2, L4, S4, W2). To illustrate a most simple case, consider an isothermal first-order reaction which occurs without change in fluid density:

$$\rho \mathbf{v} \frac{dw_1}{dz} + \rho k w_1 = 0 \qquad (19a)$$

$$w_1(z) = w_1(0)e^{-kz/\mathbf{v}} \qquad (19b)$$

where z = distance from reactor entrance; i.e., the reactor performance is identical to that of a batch reactor in which $(z/\mathbf{v})$ is used as the measure of time.

The more sophisticated solutions of Eqs. (18a) and (18b) yield some very interesting problems; for example, if the reaction is exothermic and the rate of heat removal is too small, explosions can occur. Barkelew (B1) showed that the conditions for explosion are very sensitive to the temperature profile within the bed; consequently, execution of an exothermic reaction may be difficult to control. When the reaction kinetics are more complex than first-order irreversible, Eqs. (18a) and (18b) yield challenging optimization problems (A5), a topic of considerable current interest to chemical engineers. Since the purpose of this chapter is to describe the contribution of mixing to chemical reactors, the reactor-design features of Eqs. (19a) and (19b) are ignored. Note however that the ideal, piston-flow model assumes that no mixing occurs within the reactor; hence, the ideal, piston-flow model is a bench mark to which other mixing models may be compared.

2. Modified Piston-Flow Reactors

This section is divided into three parts; in the first, the methods for describing the departure from ideal piston flow are presented, the effects of such mixing upon reactions with linear kinetics are evaluated in the second, and finally the computational techniques for estimating the effect of mixing upon nonlinear chemical reactions are described in the third.

a. Deviations from Ideal Flow. It has already been noted that turbulent flow in a pipe is described by a velocity profile which deviates slightly from the piston-flow model. The contribution of the real velocity profile to the average flow of material in pipes has attracted considerable attention (D2, D6, L2, N1). The average flow of mass in pipes is assumed to obey the equation

$$\mathcal{D}_{ax} \frac{\partial^2 \langle A \rangle}{\partial z^2} = \langle v \rangle \frac{\partial \langle A \rangle}{\partial z} + \frac{\partial \langle A \rangle}{\partial t} \tag{20}$$

where $\langle A \rangle = \int\limits_0^1 2b\mathbf{v}(b, z)A(b, z)db/\langle v \rangle,$

$\langle v \rangle = \int\limits_0^1 2b\mathbf{v}(b, z)db,$

$\mathcal{D}_{ax}$ = effective axial diffusivity,
$b = r/R$ dimensionless radius,
z = axial length,
t = time,
$A(b, z)$ = concentration profile in the pipe,
$\mathbf{v}(b, z)$ = velocity profile in the pipe.

Equation (20) represents a diffusive dispersion process superimposed upon piston flow. The processes which cause deviations from plug flow are axial turbulent diffusion, usually small, and the variations of the velocity profile, the so-called Taylor diffusion (A5, C9, T1). Barkelew (B1) found Eq. (20) to represent flow in long tubular systems quite well. Sjenitzer (S1) has correlated a significant body of experimental data by the empirical expression

$$\frac{\mathcal{D}_{ax}}{\langle v \rangle d} = 1.57 \times 10^7 f_f^{3.6} \left(\frac{\ell}{d}\right)^{0.14} \tag{21}$$

where d = pipe diameter,
f_f = Fanning friction factor,
ℓ = system length.

Mixing in laminar flow also has been analyzed as Taylor diffusion, and Aris (A4) found that Eq. (20) holds for "long" pipes if the estimate of $\mathscr{D}_{ax}$ is given as

$$\frac{\mathscr{D}_{ax}}{\langle v \rangle d} = \frac{1}{192} \frac{\langle v \rangle d}{\mathscr{D}_m} + \frac{\mathscr{D}_m}{\langle v \rangle d} \tag{22a}$$

where $\mathscr{D}_m$ = molecular diffusivity. Farrell and Leonard (F1) solved the coupling of radial diffusion with axial convection for laminar flow through moment analysis to quantitatively establish that Eq. (22a) is useful for pipes longer than

$$\ell > \frac{0.2 \langle v \rangle d^2}{\mathscr{D}_m} \tag{22b}$$

The analysis of Farrell and Leonard indicate that Eq. (20) is a poor representation of mixing in tubes for pipes shorter than demanded by Eq. (22b). Under these conditions accurate evaluation of the conversion in the reactor demands solution of the mass balance equation [Eq. (23) below].

The axial dispersion of mass in packed beds has been extensively investigated, and the experimental results are correlated using Eq. (20) to define the effective axial diffusivity. The Peclet group $\langle v \rangle d_p / \mathscr{D}_{ax}$ has been correlated with Reynolds number and the Schmidt group (B3), but the simplest correlation relates the effective axial diffusivity as a dimensional quantity to the product of the particle diameter and the interstitial velocity (C1). Figure 4 depicts this simple

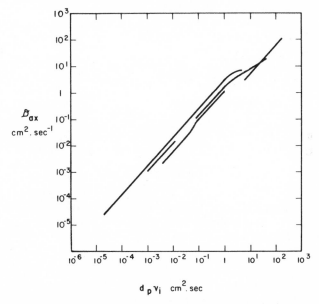

FIG. 4. Axial dispersion in packed beds—the lines show the results of several investigations and serve to indicate the uncertainty of axial dispersion estimates.

correlation. It is found that the apparent axial diffusivity increases when the tube-to-particle-diameter ratio, (d/d_p), is less than ten; this increase is explained as Taylor diffusion (C9) at the walls of the tube. Consequently, the data of Fig. 4 should not be used for beds in which the tube-to-particle-diameter ratio is less than six (C2).

The purpose of using Eq. (20) as the mass balance equation in tubular reactors is to convert the general mass balance equation, Eq. (23),

$$\nabla \cdot (\mathscr{D} \nabla \rho w_i) = \nabla \cdot (\rho w_i \mathbf{v}) - r_i(w_j, \rho, T) - \frac{\partial \rho w_i}{\partial t} \tag{23}$$

(where $\mathscr{D}$ = a general diffusivity) into an ordinary differential equation. This approximation is justified whenever the transverse variation of $r_i(w_j, \rho, T)$ may be neglected or the reactor is sufficiently long. Since the axial mixing data in tubular reactors is not very precise ($\pm 20\%$), the simplest boundary conditions should be chosen in the solution of Eq. (20).

b. *Linear Kinetics.* A method for incorporating axial mixing data in the design of reactors is illustrated in this section. The axial mixing parameter serves to alter the residence-time distribution of the reactor and also to decrease the magnitude of the concentration gradient through the reactor.

The design method is illustrated by considering an isothermal, first-order reaction. The contribution of mixing to the conversion of the reactor is found from the solution of the equation

$$\frac{\mathscr{D}_{ax}}{\langle v \rangle \ell} \frac{d^2 C}{dX^2} = \frac{dC}{dX} + KC \tag{24}$$

The general solution of Eq. (24) is

$$C(X) = A_1 e^{r_1 x} + A_2 e^{r_2 x} \tag{25}$$

$$r_1 = \frac{1 + \sqrt{1 + 4Kg}}{2g} \qquad r_2 = \frac{1 - \sqrt{1 + 4Kg}}{2g}$$

$$C = A/A^r \qquad K = k\ell/\langle \mathbf{v} \rangle \qquad g = \mathscr{D}_{ax}/\langle \mathbf{v} \rangle \ell \qquad X = z/\ell$$

The constants A_1 and A_2 depend upon the boundary conditions assumed for the reactor. The selection of the correct boundary conditions is described extensively in the literature [van der Laan (V1); Wehner and Wilhelm (W2)]. From this discussion the following two boundary conditions are the most important for reactor design:

$$1 = C(0) - \frac{\mathscr{D}_{ax}}{\langle \mathbf{v} \rangle \ell} \frac{dC}{dX} \bigg|_0 \tag{26a}$$

$$\frac{dC}{dX} \bigg|_1 = 0 \tag{26b}$$

and

$$C(0) = 1 \tag{26c}$$

$$\lim_{X \to \infty} C(X) = 0 \tag{26d}$$

Equation (26a) indicates that there is no diffusive transport process outside of the reactor at the entrance, and Eq. (26b) states that the gradient at the end of the reactor is zero. Equations (26a) and (26b) are used when the inverse Peclet group, $(\mathscr{D}_{ax}/\langle \mathbf{v} \rangle \ell)$, is greater than 0.05, i.e., for short pipeline reactors. Equations (26c) and (26d) are the limiting forms of Eqs. (26a) and (26b) when the diffusive flux at the bed entrance is ignored and the reactor is assumed to be a finite section of an infinite reactor. Equations (26c) and (26d) apply when the inverse Peclet number is less than 0.05. Solutions of Eqs. (24), (26a), and (26b) are given in graphical form (C2, L5), and the solution of Eq. (24) with boundary conditions Eqs. (26c) and (26d) is simple but informative:

$$C = Me^{-K} \tag{27a}$$

$$M = \exp\left[+Kg'(1 - 2g' + 5g'^2) \cdots \right] \tag{27b}$$

where

$$g' = K\left(\frac{\mathscr{D}_{ax}}{\langle \mathbf{v} \rangle \ell}\right)$$

Equations (27a) and (27b) indicate that mixing reduces the conversion of a tubular reactor since as $\mathscr{D}_{ax}$ goes to zero, g also approaches zero. It may be noted that K can represent any linear process within a tubular reactor. As a result of the superposition theorem for linear systems the conversion of the reactor also may be calculated (z1) from the equation

$$C = \int_0^\infty C(t)f(t)dt \tag{7a}$$

where $C(t) = \exp(-Kt\langle \mathbf{v} \rangle/\ell) = e^{-kt}$ = concentration in a batch reactor at
time, t;

$f(t)$ = the residence-time frequency distribution of the tubular reactor.

Equation (7a) is the Laplace transform of the residence-time distribution; one finds that Eqs. (25) and (7a) are equivalent when k is regarded as the Laplace transform variable.

Thus the contribution of mixing to reactor performance is easy to calculate for first-order isothermal reactions. The boundary conditions for the problem depend upon the magnitude of the Peclet number. It should be noted that the performance of packed-bed reactors fifty or more particle diameters in length closely approaches the performance of a piston-flow reactor. Further, the piston-flow model for empty tubular reactors is valid only for long reactors in which the residence-time distribution again is very near to piston flow.

c. *Nonlinear Kinetics.* The piston-flow model including axial mixing is extremely useful for reactor design in which the process is nonlinear. The over-all reaction rate of a chemical reactor can be calculated if the concentration is known at every point in the reactor. The axial dispersion piston-flow model permits determination of the point concentration distribution throughout the system, and consequently, the yield can be found.

Equation (20) is combined with a similar energy balance equation to give the nonlinear set of ordinary differential equations that represent the system. The boundary conditions of Eqs. (26c) and (26d) are still used. These equations usually are evaluated numerically (A1, C3, D3, F2, L5, T1). Figure 5 shows the results of such computational efforts. It is noted that mixing makes only a secondary contribution to the performance of reactors for which Eq. (20) is an adequate model. However, Tichacek (T1) notes that some selectivity gains can be obtained by considering axial mixing.

3. Piston-Flow Reactors—More Complex Mixing Models

The dispersion model for mixing in tubular reactors can be made more elegant and harder to use by including consideration of transverse fluxes. The resulting set of partial differential equations has been attacked analytically and numerically.

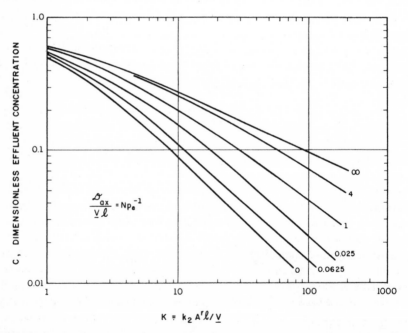

FIG. 5. Tubular reaction conversion—the curves show the fraction, C, of reactant remaining for a second-order irreversible reaction executed in a tubular reactor. Axial mixing is represented by the inverse Peclet group parameter.

If piston flow is assumed, for example, in a packed bed, but transverse temperature and concentration gradients are assumed, then the partial differential equation, Eq. (23), can be separated and analytic techniques used. Amundson (A3) developed a complete solution to the linear packed-bed problem. If laminar flow is assumed and the bulk reaction is restricted to first order, then the partial differential equation system can be expanded as a nonorthogonal, doubly infinite series (D2, L2, N1). The major difficulty with all of these analytic solutions is that numerical effort of considerable scope is required before any useful results can be obtained. Hence direct numerical solution seems attractive.

The direct numerical solution of Eqs. (18a) and (18b) requires a digital machine with a large memory because concentration and temperature gradients upon the entire transverse and axial grid normally are stored (split boundary value problem). This memory demand can be reduced if the axial mixing is neglected or if the bed is regarded as a fraction of a bed of infinite extent (C4, D1, M7). The other requirement for numerical solution of Eqs. (18a) and (18b) is radial mixing parameter data. The radial mixing coefficients historically were measured before the axial mixing values, and the radial coefficients are available from the literature (L5). When a highly exothermic reaction occurs in a packed bed, the performance is critically dependent upon the tube diameter and external tube heat transfer coefficient. The contribution of radial mixing processes to a particular tubular reactor is illustrated by Fig. 6. Gray (G2) has compared some two-dimensional tubular reactor solutions with the corresponding one-dimensional cases. Some of these results are displayed in Fig. 6. From examination of similar cases, Gray concludes that a two-dimensional solution is warranted for reactions which are 20 kcal. or greater exothermically.

Initial design estimates for tubular chemical reactors justifiably are made using the ideal piston-flow model. The contribution of axial mixing to reactor output for "long" reactors can be learned by integration of the set of ordinary material and heat balance equations. The transverse mixing is important in short tubular reactors or any tubular reactor in which the exothermic reaction heat is greater than $+20$ kcal. For these cases numerical solution of the partial differential equation is required for analytic design. One notes that tubular reactors are easy to operate at the pilot-plant scale, and consequently experimental determination of mixing and other factors which contribute to the reactor yield are often useful.

C. Batch Kinetics

Although the chemical and petroleum industry has achieved great improvements in productivity and product control by operating in a continuous manner, there remains a sizable portion of industrial chemicals which are most economically produced in batch lots. Consequently, it is of some importance to judge quantitatively the contribution of mixing to reactor performance in

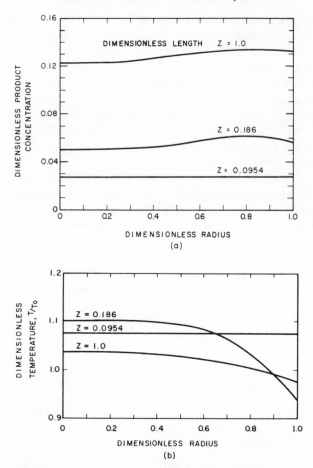

FIG. 6. (a) Concentration profiles in a packed-bed reactor—this plot shows the dimension-less concentration of product as a function of radius at various downstream distances. The temperature profile along the wall has been adjusted to give the maximum conversion in the reactor. (b) Temperature profiles in a packed-bed reactor—this plot shows the dimensionless temperature profiles corresponding to the concentration profiles of (a). The exothermic reaction used in this example requires cooling by heat transfer through the reactor-tube wall.

such systems. Some guide-line comments may be useful. First, batch kinetic systems are often heterogeneous; these reactions are discussed in Section III. Second, the typical problem in operating a batch reactor is to duplicate bench-scale chemistry in pilot-plant or semi-works equipment. Consequently, the chemical kinetics of the reaction often are not known because the chemistry is not understood or known. In addition, the reactant concentration may be far from optimum, for a chemist often controls his batch reaction by using excess solvent ("mild conditions"). Thus the mixing problem in running such a chemical reaction may be deciding how to introduce a relatively small amount of one reactant into a sea of the dilute second reactant. In such batch systems only a very small portion of the reactor may be making product at any time; in engineering terms the space velocity (feed flow rate in reactor volumes per unit time) may be extraordinarily low. Finally, there are some batch reaction systems which are dominated by mixing phenomena. Such processes may involve extremely viscous materials; one immediately thinks of a batch bulk-polymerization reaction. Batch viscous reactions are dominated by the rate at which new surface is generated in the system since the penetration depth of molecular diffusion is usually quite small. Thus one finds that only with extremely fast reactions or with very viscous systems does mixing affect the operation of batch reactors.

This discussion of batch homogeneous reactors is divided into two parts: stirred-tank reactors and piston-flow reactors. Batch tank reactors present all of the complexities of their continuous cousins plus the complications of transient variables. Consequently, the operation of such reactors is not developed extensively in the literature. Piston-flow reactors can operate in a batch manner when the catalyst life is very short; for example, when the function of the bed packing is to transfer sensible heat to an endothermic, gas-phase reaction. The linearized version of transient or batch operation of a packed bed has been extensively developed; the interest in the problem undoubtedly stems from the interesting mathematical characteristics of the problem. The mixing processes in packed-bed models are represented by several forms of diffusional processes.

The batch operation of stirred tanks is usually attacked by extension of the signal-flow models described previously. The simplest model is obtained when the mixing in the reactor is considered perfect and the reaction is isothermal; under these conditions an estimate of the holding time required for some desired production is obtained by integrating the rate equation. Selectivity gains sometimes can be obtained by adding one reactant to an excess of another; such possibilities are beyond the scope of this chapter and are discussed extensively in chemical engineering kinetics textbooks. The next level of difficulty comprises the nonisothermal operation of a batch reactor.

The question of mixing in a batch tank reactor is avoided by the assumption of ideal mixing. A useful rule of thumb is as follows: If the reaction half-time is five times greater than the average loop circulation time, then the perfect mixing assumption is justified for usual engineering precision. Thus for very fast reactions it may be necessary to consider the detailed performance of the reactor. The detailed description of the reactor is particularly important when one is considering polymerization reactions or when the desired extent of reaction is high. For these cases reactor yields are most easily obtained by analog simulation of the process. Analog simulation not only provides a detailed description of the expected operation of the stirred-tank reactor but also suggests appropriate control schemes and operator instructions.

As an example of the effect of mixing upon batch reactor operation, consider the condensation polymerization of a bifunctional polymer which is described by the reaction sequence

$$B_n + B_m \underset{k_r}{\overset{k_f}{\rightleftharpoons}} B_{m+n} + H_2O \tag{28a}$$

$$B_n + M_n \underset{k_r}{\overset{k_f}{\rightleftharpoons}} M_{m+n} + H_2O \tag{28b}$$

Following Kilkson (K2), the reaction rates for this system can be written as

$$\frac{dB_n}{k_f dt} = \sum_{j=1}^{n-1}(B_j B_{n-j}) - 2B_n \hat{B} - B_n \hat{M}$$

$$+ \frac{k_r}{k_f}(H_2O)\left[2(\hat{B} - \sum_{j=1}^{n}(B_j)) + (\hat{M} - \sum_{j=1}^{n}(M_j)) - (n-1)B_n\right] \tag{28c}$$

$$\frac{dM_n}{k_f dt} = \sum_{j=1}^{n-1}(B_j M_{n-j}) - M_n B + \frac{k_r}{k_f}(H_2O)\left[(\hat{M} - \sum_{j=1}^{n}(M_j)) - (n-1)M_n\right] \tag{28d}$$

where $\hat{B} = \sum_{j=1}^{\infty} B_j,$

$$\hat{M} = \sum_{j=1}^{\infty} M_j.$$

Kilkson showed that Eqs. (28c) and (28d) can be solved in terms of the molecular-weight distribution moments by application of the zeta transform. This technique was applied to a batch polymerization by Woods (W5).

Figure 7 shows the simple stirred tank in which the reaction is assumed to occur. The turbine impeller yields circulation loops of the polymer; the viscosity of the polymerizing mass is so high that material in the circulation loops remains segregated. One notes that the "water of condensation" is removed by diffusive transport to the surface of the polymerizing mass. The contribution of this diffusive process to the rate of reaction is approximated by assuming that the water concentration is proportional to the circulation time of the segregated fluid element. This arbitrary assumption is based upon the observation that the interior circulation loops have longer circulation times than the outer circulation loops, and that the exterior circulation loops have lower average water content than the interior circulation loops. The effect of mixing upon the product molecular weight distribution is found from computing the ratio of the second moment of the product number distribution to the square of the first moment of the product number distribution to the similar ratio for simple batch polymerization at the same degree of completion. Computation of this ratio indicates that the assumed mixing pattern significantly alters the product molecular weight distribution. Although the water concentration has

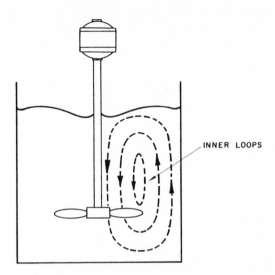

Fig. 7. Circulation loops in batch polymerization. Inner loops have higher water concentration owing to the greater diffusion path to the surface.

been assumed uniform in the circulation loops, in fact the water concentration should be higher at the bottom of the tank. Thus if the circulation time is low at the end of the polymerization, the polymer may be expected to have poor uniformity.

This example is intended to illustrate that the detailed fluid mechanics can have an important effect upon the performance of a reactor. It is also intended to show that the prediction of the contribution of mixing to yield is not easily estimated; one must study a transient "signal-flow model" problem.

D. MIXED-MODEL SYSTEMS

Sometimes the interaction of mixing with yield and selectivity in chemical reactors is estimated by constructing hypothetical models for the reactor and then computing the progress of the reaction in the hypothetical model. This approach is called the "mixed-model method" because the hypothetical model usually consists of stirred-tanks and piston-flow reactors. If one can create a mixed model which is both simple and also represents the fluid mechanics of the reactor reasonably well, the method can be successful; otherwise the mixed-model approach to reactor design is little more than a guessing game. In a sense the mixed-model approach already has been illustrated in the discussion of signal-flow models for stirred tanks. A major point which should be made is that such models are highly arbitrary. The mixed-model method is developed by reviewing some examples.

Perhaps the simplest example of the mixed-model approach is that shown on Fig. 8a, the mixer-tube (MT) and tube-mixer (TM) models of Cholette and Blanchet (C5) and Cholette and Cloutier (C6). The real reactor is assumed to be made up of a perfect mixer followed by a tubular reactor (MT) or a tubular reactor which feeds into a mixer (TM). The adjustable parameter in the model is the fraction of the total volume which is a perfect mixer. The potential contribution of mixing upon reactor yield is estimated by comparing the MT and TM cases. Figure 8b shows how the conversion of the two models varies with the ratio of holding time V/F, to a characteristic reaction time for equal volumes in the perfect mixer and tubular reactor sections. One notes that the mixer-tube arrangement is poorer than the tube-mixer reactor; the relative difference between the two forms increases as the conversion in the reactor increases.

A second example of the mixed-model technique is shown in Fig. 9. Adler and Hovorka (A2) developed the model to find a way for approximating the residence-time distributions of short tubular vessels. It was found that a wide variety of residence-time distributions are represented by the model when the adjustable parameters are appropriately specified. It may be seen that the model consists of a section of piston flow which feeds into a stirred tank. The stirred-tank interchanges material with a closed container. The parameters to be specified are the number of units of the form shown in Fig. 9 which are in the entire system, the fraction of the volume of the unit which is piston flow, the fraction of the volume of the unit which is stirred tank, and the interchange rate between the stirred tank and the closed mixer. Although one can fit a great variety of residence-time distributions with this model, it is extremely difficult to relate the parameters of one model to another. The lack of generality of the model occurs because the model is only remotely connected to the actual fluid mechanics of the reactor. It is apparent, however, that the reactor conversion can be specified by evaluating the conversions in each of the separate elements. With reference to Fig. 9, the conversion for a first-order reaction in one section of the Adler and Hovorka model is given by the following equation:

$$\frac{C^0}{C^f} = \frac{e^{-K\phi_D}}{\left(K\phi_T + 1 + \frac{F_e}{F} - \frac{1}{1 + (F/F_c)\phi_E K}\right)} \tag{29}$$

where $K = kV/F$;

ϕ_D, ϕ_T, ϕ_E = volume fraction of section in delay, stirred-tank, and exchange stirred-tank forms;

F_e/F = feed-rate ratio to the exchange stirred tank.

The complete model consists of N units of the type shown connected in series. Although additional examples of the mixed-model technique are found in

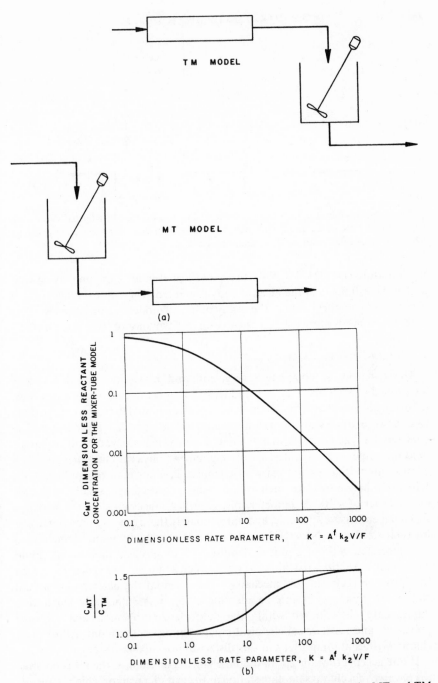

FIG. 8. (a) The MT and TM mixed models. (b) Fraction remaining in the MT and TM mixed models—the upper figure shows the decrease in reactant concentration with increasing reactor holding time, and the lower figure indicates that the tube-mixer configuration always gives higher conversion to product than the mixer-tube arrangement.

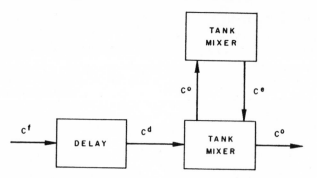

FIG. 9. Adler's mixed model—the parameters of this model may be adjusted to fit nearly any residence-time distribution. The upper tank mixer represents dead space in the reactor which exchanges fluid with the flow regions at low rates.

the literature (D5, H1, L3, W4), these additional models are only extensions and embellishments of the ideas already presented. There is a singular lack of experimental confirmation of the utility of mixed models; in order to test these models one needs more data on the mixing performance of large reactors (S5).

E. MIXING-HISTORY MODELS

The concepts of complete segregation and maximum mixedness were mentioned in Sections II, A, 2, b and II, B, 2, c; it is the purpose of this section to review some of the methods which have been proposed for cases intermediate to the two extremes of maximum mixedness and complete segregation. The concept of a fluid clump is an aid to defining mixing history. A fluid clump (or fluid molecule) is the smallest set of neighboring physical molecules to which continuum mechanics and classical chemical kinetics may be applied. J. G. Kirkwood has shown that such a fluid clump is exceedingly small—in liquids of the order of a 100 Å diameter sphere. Other suggestions for the size of a fluid element have been made, and unfortunately there is no clear demarcation for their size limits. Nevertheless, the concept of a fluid volume element may be interpreted without a precise definition of its physical dimensions. Fluid clumps are the entities to which one refers when a "point" in a space is specified by a position vector in fluid mechanics. The essential characteristics of a fluid clump are that: (a) it can be identified, (b) it is sufficiently small that "uniformity" is achieved within the fluid element by molecular diffusive processes, and (c) classical chemical kinetics apply to the point. All of these characteristics are important in this discussion of mixing-history models.

If one accepts the concept of a fluid elements or clumps, then it is possible to conceive of a clump sampling station at the exit of a reactor which is operating at steady state. One then asks: "How did this particular fluid clump arrive at its final state?" The answer to this question is the mixing history of this fluid element; a mixing-history model is a construction or an equation from which the mixing history is evaluated.

It is useful to describe first the mixing history for a completely segregated and a maximum-mixed reactor in terms of the mixing-history analysis. The approach is then extended to systems which fall between the two limits.

Figure 10 shows a series of five constructions which represent several levels

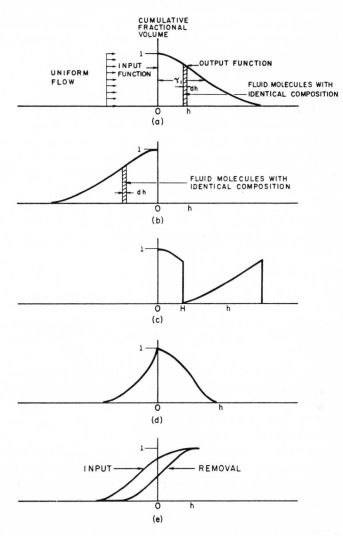

FIG. 10. Mixing-history models—these models provide ways for evaluating the contribution of mixing for reactors with any residence-time distribution. (a) Complete segregation, (b) maximum mixedness, (c) Weinstein's model, (d) Wood's model, and (e) generalized model.

of complexity of the mixing-history analysis. Figure 10a shows a model which may be used to visualize the mixing history of a completely segregated reactor. The diagram consists of two functions, the input function, $I(h)$, which is to the left of $h = 0$, and the output function, $O(h)$, which is to the right of $h = 0$. In Fig. 10a the input function is a unity step function at $h = 0$ and the output curve is the smooth curve which decreases toward the positive abscissa. The piano-shaped area under the $I(h)$ and $O(h)$ curves represents the reactor. Fluid is assumed to flow horizontally and uniformly from left to right through the reactor. A fluid clump is described as a differential area, $dOdh$, and the residence time for a fluid clump is proportional to the horizontal distance drawn between the input and output curves. The $O(h)$ curve shown in Fig. 10a is constructed with the relation $O(h) = \int_{h}^{\infty} f(t)dt$; this definition yields the desired residence-time distribution function for the reactor.

If one follows a number of clumps through the reactor, it may be seen that all elements with the same value of the history parameter, h, have the same composition. The steady state effluent of the reactor consists of clumps with all possible values of the parameter, h; these elements are then mixed at the reactor exit to yield the total production of the reactor. The term "complete segregation" is applied to this reactor because none of the fluid elements within the reactor interact with other clumps. The general features of the complete segregation model which is retained in the other constructions are the following:

1. Fluid flows uniformly from left to right and reaction occurs only within the reactor.
2. The effluent concentration of the reactor is found by mixing the concentrations along the output curve.
3. Fluid elements with the same value of the h parameter have identical concentration.
4. A fluid clump is modeled as a differential area.

The maximum-mixedness model described by Zweitering (Z1) is illustrated in Fig. 10b. Figure 10b is obtained by rotating Fig. 10a about the ordinate; the input function, $I(h)$, is the smoothly increasing curve which terminates at $h = 0$; the output function, $O(h)$, is a unit step decrease at $h = 0$. The mixing is described by specifying that the feed material which enters the reactor along $I(h)$ is mixed with all clumps of the same h value in the reactor. This fluid is represented by the area $I(h)\,dh$; Eq. (8) is obtained from a material balance around $I\,dh$ including a chemical reaction. Zweitering showed that the mixing described by this model is the most interaction of fluid clumps consistent with the residence-time distribution.

Thus the effluent of the maximum-mixedness reactor is a set of fluid clumps with identical composition, and the physical molecules within these clumps have a distributed residence time. In contrast, the complete segregation model yields exit fluid clumps with distributed concentrations but uniform residence time within the clumps. Thus the two models are complementary to one another.

The function of the $I(h)$ curve in the mixing-history analysis is to define the schedule under which new material is added to and mixed with the fluid in the reactor. Since the $I(h)$ function may be defined independently from the residence-time distribution, the process of mixing is separated in the mixing history from the other factors which produce the particular residence-time distribution of the reactor. Thus "mixing" and changes in the residence-time distribution are not equivalent.

Although it has been noted that there is very little real difference in reaction yields for simple, isothermal, homogeneous reactions for the "maximum-mixedness" and "complete segregation" models, one finds that adiabatic reactions or polymerization reactions are affected substantially by the mixing history. Accordingly, it is useful to find ways for calculating reaction conversion for mixing levels intermediate between the limiting cases of complete segregation and maximum mixing. Figure 10c shows a mixing-history model for intermediate mixing developed by Weinstein (W3). Reactants entering the system are considered initially to be completely segregated and then enter into a zone of maximum mixedness. A time, H, is specified at which the reactants cease to be completely segregated; the material remaining in the reactor at that point is fed into the maximum-mixedness section. In Fig. 10c the maximum-mixedness section is modeled by reflecting the "tail" of the distribution shown in Fig. 10a. Weinstein found that the over-all reactor yield is dependent significantly upon the choice of H when the reaction is considered adiabatic rather than isothermal. The advantage of Weinstein's model for mixing history is that only one additional parameter needs to be specified to be able to compute the conversion of any adiabatic reaction. On the other hand, as a result of higher thermal than molecular diffusion, reactants which enter a stirred tank tend to achieve the exit temperature of the vessel before the reactant fluid molecule attains uniform composition.

A fourth model is used to describe polymerization reactions. In polymerization reactions, one finds that the bulk viscosity of the reactants increases as the reaction progresses; thus it is likely that the reactants are initially well mixed and become segregated as the reaction proceeds. Figure 10d shows an intermediate mixing model which may represent mixing history in such polymerization reactions. The well-mixed section of the reactor is represented by the section to the left of $h = 0$ and the segregated portion is represented by the section to the right. An additional function, namely, the input function, is required to calculate the conversion in the reactor. The moments[1] of condensation polymerization reactions have been found to be affected by the

[1] The Nth moment of B_i is defined as $(\sum_{j=1}^{\infty} j^N B_j)/(\sum_{j=1}^{\infty} B_j)$.

form of the input curve, $I(h)$. In fact, it may be possible to estimate the mixing history of a reactor by measuring the residence-time distribution and the molecular weight distribution of a polymer product (W5).

The calculation of the concentration of material leaving the fourth mixing-history model illustrates the techniques which apply to the other three models as well. The concentration in the reactor for h less than zero is found from the solution to the analog of Eq. (8), namely,

$$\frac{dw_i^N}{dh} = \frac{r_i^N}{\rho} (w_j^N, \rho, T) + \frac{(w_i^{N-1} - w_i^N)}{I(h)} \frac{dI(h)}{dh} \tag{30a}$$

The boundary condition for Eq. (30a) is selected using the same rules developed for Eq. (8). The concentration in the reactor for h greater than zero is found from the equation for batch reactions

$$w_c^N(h) - w_c^N(O) = \int_0^h \frac{r_i^N}{\rho} (w_j^N, \rho, T)dh' \tag{30b}$$

The effluent concentration from the reactor is calculated from the analog of Eq. (7), namely,

$$w_i^N = - \int_0^\infty w_i^N(h) \frac{dO(h)}{dh} dh \tag{30c}$$

Equations (30a) to (30c) are equivalent to describing the mixing in the early portion of the reactor as maximum-mixedness and the later portion as complete-segregation.

The mixing-history models can be extended in a number of obvious ways. For example, the reactor system may consist of a number of reaction zones in series; this approach changes the number of peaks in the reactor model. Alternatively, the reactor may be represented by the model shown in Fig. 10e; in this model material may be added and removed from the reactor history element simultaneously. The disadvantage of these extensions is that the residence-time distribution must be calculated separately; there is no simple *a priori* way by which the residence-time distribution may be related to the form of the curves.

All of the mixing-history models shown in Fig. 10 suffer from the disadvantage of only being remotely connected to the fluid mechanics of the reactor. Thus the same criticisms given for the mixed-model technique apply to these mixing-history models as well. In addition, it is likely that the history of fluid clumps or elements is more complicated than can be represented by the models; for example, the fluid clumps or fluid molecules actually interchange mass with one another, but this interchange is not represented by any of the models developed.

Other mixing-history models attempt to reflect more realistically the actual fluid mechanics of mixing. Rosenhouse (R2), following Batchelor (B2), developed a "surface-extension" model for the mixing process of liquids. The action of turbulence or energy dissipation in water is assumed to increase the surface area between two "eddies." Figure 11 aids in showing how the model is developed. At time one, two eddies, already fairly small, are touching one another along the x-axis. The action of a turbulent velocity fluctuation is to

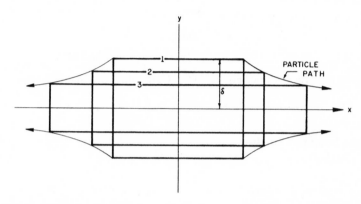

FIG. 11. Eddy extension model—this model provides a way whereby the conversion in an eddy of any size may be computed.

stretch the eddies along the x-axis. It is assumed that the fluctuating velocity components are uniform within the eddy; this assumption limits the analysis to fairly small or high wave-number eddies. Considering a constant-volume slab of incompressible fluid which has thickness, δ, and cross-sectional area, A, then incompressibility of the fluid requires that

$$A \frac{d\delta}{dt} + \delta \frac{dA}{dt} = 0 \tag{31a}$$

The rate of change of δ is also the velocity in the y direction and thus

$$v_y = -y \frac{d(\ln A)}{dt} \tag{31b}$$

It is assumed that the relative rate of growth of surface area between the slabs is a measurable constant, thus

$$v_y = -yg$$

where $g = d(\ln A)/(dt)$, the relative growth rate of eddy surface. In the material balance equation for component C_a one has

$$\frac{\partial C_a}{\partial t} = \mathscr{D}_m \frac{\partial^2 C_a}{\partial y^2} + gy \frac{\partial C_a}{\partial y} + \frac{R_a(C_a)}{A^r} \tag{32}$$

in which it is assumed that the concentration gradients parallel to the growing interface are small. The material balance equation applies most accurately for systems in which large concentration gradients exist normal to the interfaces between eddies, i.e., systems in which fast reactions occur. If one introduces the change of variables

$$\zeta = \left(\frac{g}{\mathscr{D}_m}\right)^{1/2} y e^{-gt}$$

$$\tau = \frac{1}{2}(e^{-2gt} - 1)$$

$$\zeta_0 = \zeta(y_0)$$

$$y_0 = \text{original eddy half-thickness}$$

Then the material balance equation for C_a in gm. mole/liter becomes

$$\frac{\partial C_a}{\partial \tau} = \frac{\partial^2 C_a}{\partial \zeta^2} + \frac{R_a(C_a)}{A^r g} \tag{33}$$

with the boundary conditions

$$C_a(\zeta, 0) = 0 \qquad -\zeta_0 \leq \zeta \leq 0 \tag{33a}$$

$$C_a(\zeta, 0) = 1 \qquad \zeta_0 \geq \zeta > 0 \tag{33b}$$

$$\left.\frac{\partial C_a}{\partial \zeta}\right|_{\zeta_0} = 0 \qquad \left.\frac{\partial C_a}{\partial \zeta}\right|_{-\zeta_0} = 0 \tag{33c, d}$$

Equation (33) and the boundary conditions are typical for absorption with chemical reaction and thus a considerable body of solutions exist for various choices of the rate expression, $R_a(C_a)/g$. The change of variables is made to switch from the laboratory coordinate system to a simplified material coordinate system for the decay of a typical eddy. The equations are useful only if one has measurements of g and can specify reasonable values for ζ_0, the original size of the eddy, and τ_f, the total period of the decay of the eddy. Some measurements of the area growth rate have been made, and the period of the eddy decay is taken as the residence time in the velocity field.

The area extension model for mixing and chemical reaction has been used in a somewhat different form to analyze the performance of mixing around an impeller in a stirred tank. Rice et al. (R1) concluded that the original shape of the eddies was best represented as cylindrical; the original cylindrical size of the eddies was of the order of 1 to 10 μ. Similarly, O'Hern and Rush (O1) experimentally investigated the size of barium sulfate precipitates formed in a tank reactor and in a Hartridge-Roughton mixer. This investigation showed that the Hartridge-Roughton mixer yielded much smaller precipitates than the stirred tank; the nucleation rates were enhanced by high concentrations of reactants under conditions of very high shear. It is interesting to note that the mean particle size in the tank experiment was of order of 8 μ whereas the rapid-mixer yielded particles in the range of 0.02 to 2 μ. These particles give an order of magnitude estimate of the eddy size in both systems. Thus there is some experimental evidence consistent with the area extension model of chemical reactions.

F. STATISTICAL MODELS

Statistical theories of turbulence have been used with limited but encouraging success to explain mixing phenomena. Some of these theories have been discussed in Chapter 2 of Volume I. In this section the application of these statistical theories to reaction kinetics is considered. It shall be found that these statistical formulations do not yield direct information upon the connection of mixing with reactor performance.

The continuity equation for a single reacting component in a constant density system is written as

$$\frac{\partial w_i}{\partial t} + \nabla \cdot (v w_i) = -\frac{1}{\rho} \nabla \cdot \mathbf{j}_i + r_i(w_j, \rho, T)/\rho \tag{34}$$

where $\mathbf{j}_i$ = mass diffusive flux of the ith species,

v = mass average velocity,

w_i = mass fraction of ith species,

ρ = mass density.

The standard approach in turbulence theory is to replace each variable by a time-averaged component and a fluctuating component. Thus, in the continuity equation, the concentration and velocity are represented as

$$w_i = \bar{w}_i + w_i'$$
$$v = \bar{v} + v'$$

where the overbar represents a time-averaged variable. It is necessary to specify the period in which the averaging is to be done; this period may be defined by instrument limitations. It is generally assumed that the turbulence is isotropic (independent of the orientation and parity of the coordinate system) for the "high frequency" or physically "small" fluctuations even though the turbulence which produces these fluctuations may not be isotropic.

If one applies Eq. (34) to a stirred-tank reactor or to nearly uniform, time-averaged flows in which $\bar{v}$ and $\nabla \cdot \mathbf{j}_i$ are zero for a proper choice of the coordinate system, then the equation becomes

$$\frac{\partial w_i}{\partial t} + \nabla \cdot (v' w_i) = -\frac{1}{\rho} \nabla \cdot \mathbf{j}_i' + \frac{r_i}{\rho}(w_j, \rho, T) \tag{34a}$$

If the turbulence is isotropic for the eddy sizes of interest, then upon averaging, Eq. (34a) yields

$$\frac{\partial \bar{w}_i}{\partial t} = \frac{r_i}{\rho}(\bar{w}_j, \rho, T) \tag{35}$$

For a first-order irreversible reaction $(A \rightarrow P)$ one finds

$$- \frac{d\overline{w}_a}{dt} = +k\overline{w}_a \tag{36}$$

Thus the time-averaged equation of a true first-order reaction in an idealized turbulent system has the same form as a nonturbulent system. If, however, the reaction is only pseudo-first order, then the time-averaged rate expressions are more complex and are illustrated by the case

$$A + B \rightarrow P \quad \text{Isothermally in excess } B,$$

$$\frac{d\overline{w}_a}{dt} = - \frac{k_2 \rho}{M_b} (\overline{w}_a \overline{w}_b + \overline{w'_b w'_a}) \tag{37}$$

The second term on the right-hand side of Eq. (37) represents the contribution of turbulence to reactor performance. If in addition small temperature fluctuations also occur in the reactor,

$$\frac{d\overline{w}_a}{dt} = - \frac{k_2 \rho}{M_b} [\overline{w_a w_b} + \overline{w'_a w'_b} + k_3(\overline{w_a w'_b T'} + \overline{w_b w'_a T'} + \overline{w'_a w'_b T'})] \tag{38}$$

where $k_2 =$ a second-order rate constant,

$$k_3 = \frac{\Delta H_{act}}{RT^2}.$$

Again the last four terms represent the contribution of fluctuating variables to the reaction rate. It is obvious that additional equations will be required to evaluate the missing variables. Corrsin (C10) presented a preliminary attack upon the problem; the reaction considered was $2A \rightarrow P$ irreversibly and isothermally. The time-averaged rate expression then is

$$\frac{d\overline{w}_a}{dt} = - \frac{k_2 \rho}{M_a} (\overline{w^2} + \overline{w'^2_a}) \tag{39}$$

A suitable rate expression for $\overline{w'^2_a}$ is obtained by multiplying Eq. (34a) by w'_a and then time averaging. The result is

$$\frac{d\overline{w'^2_a}}{dt} = 2\mathscr{D}_m \overline{w'_a \nabla^2 w'} - \frac{2k_2 \rho}{M_a} (2\overline{w'^2_a \, w_a} + \overline{w'^3_a}) \tag{40}$$

The diffusional term can be represented as a function of the diffusive microscale, ℓ_{aa}.

$$\overline{w'_a \frac{\partial^2 w'_a}{\partial x^2}} = - \frac{2}{\ell^2_{aa}} \overline{w'^2_a} \tag{41}$$

Thus

$$\frac{d\overline{w_a'^2}}{dt} = -2\left(\frac{6\mathcal{D}_m}{\ell_{aa}^2} + \frac{2k_2\rho}{M_a}\,\overline{w_a}\right)\overline{w_a'^2} - \frac{2k_2\rho}{M_a}\,\overline{w_a'^3} \qquad (42)$$

In Eq. (42) the diffusive microscale, ℓ_{aa}, is assumed to be independent of orientation, an assumption consistent with isotropic turbulence. As is often found in statistical mechanics or turbulence, the number of unknowns is always greater than the number of equations; specifically, the equation for $\overline{w_a'^2}$ gives rise to the variable $\overline{w_a'^3}$ and the undetermined parameter ℓ_{aa}. One may continue the process of generating equations for correlation variables by multiplying Eq. (34a) by $w_a'^2$, time averaging, subtracting the steady state rate, and expressing the diffusive flux term $\mathscr{m}_{a^2a}$ in a similar manner to Eq. (41) to obtain

$$\frac{d\overline{w_a'^3}}{dt} = -3\left(\frac{6\mathcal{D}_m}{(\mathscr{m}_{a^2a})^2} + \frac{2k_2\rho}{M_a}\,\overline{w_a}\right)\overline{w_a'^2} - \frac{3k_2\rho}{M_a}\,(\overline{w_a'^4} - (\overline{w_a'^2})^2) \qquad (43)$$

Thus development of an equation for $\overline{w_a'^3}$ has introduced two new variables, the correlation length parameter, $\mathscr{m}_{a^2a}$, and the fourth-order concentration fluctuation term, $\overline{w_a'^4}$.

To obtain some experience with the quantities involved in the turbulent reaction rate equations, Pigford (P2) assumed a stationary turbulent system, used reasonable values for ℓ_{aa}, $\mathscr{m}_{a^2a}$, and $\overline{w_a'(0)^2}$, and further assumed that $\overline{w'^4} = 3(\overline{w'^2})^2$ (Gaussian kurtosis); Pigford then integrated Eqs. (39), (42), and (43) for a range of values of the adjustable parameters. The reaction rate for turbulent systems was essentially the same as the batch rate $-k_2\rho\overline{w_a^2}/(M_a)$ for all values of the parameters considered. It was concluded that concentration fluctuations are not likely to be sufficiently intense to alter $2A \rightarrow P$ kinetics.

The contribution of turbulent mixing (Chapter 2 in Volume I) to reaction rate is not very realistically evaluated through examination of the $2A \rightarrow P$ reaction. A more rewarding problem is the mixing of A and B together when A and B are dissimilar; the reaction zone is the portion of the reactor which contain both A and B at any instant. The $A + B \rightarrow P$ differs drastically from the $2A \rightarrow P$ in that no reaction can take place until A and B are mixed together. In a batch reaction the reaction rate initially will increase as the surface area between eddies of "mostly" A and "mostly" B increases.

The average reaction rate for the $A + B \rightarrow P$ case is given by Eq. (37). For fast reactions and at the beginning of most reactions the $\overline{w_a'w_b'}$ fluctuation term is likely to be negative; the negative value means that on the average reactants A and B are not found in the same place. However, the number of unknowns in the set of equations analogous to Eqs. (39), (42), and (43) telescopes even more rapidly for the $A + B$ case than for the $2A$ problem. In particular, equations for the microscale lengths must be derived, for these parameters are likely to change drastically during the course of the reaction.

Corrsin has developed an approach whereby the required microscale lengths may be estimated for certain reacting systems. For mixing in isotropic turbulent systems, Corrsin (C12) has shown that the diffusion microscale length can be evaluated from a relatively crude theoretical development of the concentration spectral distribution. As discussed in Chapter 2 of Volume I, the concentration spectral distribution states the probability with which a given eddy will be in a specified size range. The concentration spectral distribution is affected by the properties of the fluid medium and the reaction kinetics; also, Corrsin (C11) has developed the theory whereby the spectral function may be computed for simple reaction kinetics. Pao (P1) has shown that the spectral functions may be computed for a connected set of first-order kinetic equations. Thus for a fairly inclusive range of kinetics it is possible to estimate microscale lengths from theoretical models. Such computations reduce the number of arbitrary assumptions needed for solution of Eq. (42) or Eq. (43).

There is a special limiting type of kinetics for which the statistical theory of mixing with reaction is much less complex. When the chemical reaction is extremely fast, then the reaction of $A + B$ to yield products occurs upon a plane, for by definition A and B cannot coexist. This idealization of infinitely rapid chemical kinetics is approximated adequately by neutralization reactions in solution or free radical reactions in turbulent flames. Toor (T2) has shown that a relatively simple transformation will eliminate the reaction kinetic terms in Eq. (43) and that consequently the problem of mixing plus reaction for fast reaction kinetics is equivalent to the problem of mixing alone. Keeler *et al.* (K1) have verified Toor's analysis experimentally; consequently, one can predict the extent of reaction in these systems from measurements of the "unmixedness" in the reactor.

The statistical theory of turbulence has produced a much more profound understanding of mixing. When chemical reaction is coupled with mixing, however, the theory becomes more complex and less supported by critical experimentation. For the special case of extremely rapid reactions, the statistical theory for mixing plus reactions becomes equivalent to the theory for mixing alone; thus with the statistical theory one can convert measurements of unmixedness into estimates of the extent of reaction.

III. Heterogeneous Systems

The interplay of mixing and reaction yield in heterogeneous systems is both poorly formulated and poorly understood. The typical attack upon a heterogeneous reactor problem is to construct a suitably simplified model to represent the system, collect a body of pilot-plant data, and then hope that the model can be scaled up for plant design.

The organization of this section is similar to the discussion of homogeneous systems. Thus heterogeneous reactions in tank reactors, piston-flow systems, and mixed reactors will be discussed. The complexity of heterogeneous systems demands a less general presentation than given for homogeneous systems.

A. Tank Reactors

Tank reactors are often used to execute heterogeneous reactions. Thus liquid-gas reactions are often conducted in a stirred-tank reactor; Calderbank (Chapter 6 in this volume) has illustrated solutions when these reactions are mass transfer limited. Dissolution and precipitation reactions are examples of liquid-solid reaction systems often done in a stirred tank. Hydrogenation reactions with solid catalysts illustrate three-phase reactions typically done in a tank autoclave. Thus heterogeneous-phase tank reactors are common in the chemical industry.

The interaction of mixing with reactor performance in the examples given usually is to reduce the mass transfer resistance between the various phases. Harriott (H2) has reviewed this problem extensively; this chapter is limited to more direct effects of mixing upon reaction rates. The specific effect of mixing upon reactor performance in tank reactors is illustrated by a liquid-liquid system in which the reaction occurs only in the dispersed or bubble phase. It will be shown that the rates at which bubbles coalesce has a profound effect upon the reactor operation. It is useful to note, however, that distinctions of the nature of the mixing are not made in discussions of heterogeneous systems. The model for mixing in a liquid-liquid tank reactor is at best a first-order approximation.

When two dispersed-phase liquid drops collide in a stirred-tank reactor, the drops often coalesce. The union of the two drops is assumed by Curl (C13) to mix the contents of the two drops "perfectly." It is experimentally verified that the size distribution of drops in a liquid-liquid tank reactor is quite narrow; simply stated, there is a high probability that a "coalesced" drop will divide into two drops of the original size containing the average concentration of the two original drops. Following Curl, a material balance is made upon the drop concentration distribution:

$$\frac{\partial P}{\partial \tau} = P_0 - P + M \left\{ 4 \int_0^C P(C - \alpha)P(C + \alpha)d\alpha - P \right\} + K \frac{\partial PC^N}{\partial C} \quad (44)$$

where A^r = reference concentration;

C = dimensionless concentration, A/A^r;

F = volumetric feed rate;

K = dimensionless reaction rate, $(V\phi k_N(A^r)^{N-1})/(F\phi_f)$;

k_D = drop mixing rate, the fraction of the dispersed phase which coalesces per unit time;

k_N = Nth order irreversible reaction rate constant;

M = dimensionless drop mixing rate, $(k_D V \phi)/(F \phi_f)$;
N = reaction order;
P = dimensionless concentration probability density function, i.e., the probability that the concentration in the dispersed phase will be in the range C to $C + dC$;
P_0 = dimensionless inlet concentration probability density function;
V = volume of tank;
τ = dimensionless time, $(t \phi_f F)/(V \phi)$;
ϕ, ϕ_f = volume fraction dispersed phase in the tank and inlet.

The mixing of the dispersed-phase drops is described by the right-hand term in braces in Eq. (44). The integral represents the rate at which drops of a specified concentration are formed by coalescence of two drops of the same average concentration. The second term in braces represents the rate at which drops of the specified concentration disappear by mixing with other drops in the tank. The rate at which the coalescence occurs appears in the dimensionless parameter, M. Madden (M1) and Curl (C13) have measured the coalescence rate as a function of stirring rate and dispersed-phase fraction. Howarth (H4) notes that the measured dispersion rate is consistent with

$$k_D = C_1 \phi^{0.5} N^{2.2} \exp(-C_2/N^2) \qquad (45)$$

where N = stirring rate, r.p.m. Here, C_1, C_2 are constants for the particular system and geometry. Equation (45) does not permit one to predict a coalescence rate for any given system; it is necessary to find C_1 and C_2 experimentally.

Curl solved Eq. (44) numerically for several representative cases; Fig. 12

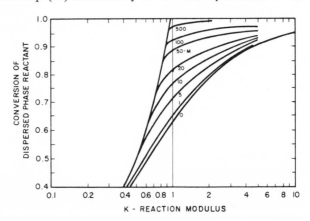

FIG. 12. Conversion in dispersed-phase reactions—the mixing parameter, M, defines the rate at which dispersed-phase bubbles coalesce. The intrinsic reaction rate for a zero-order reaction is included in the reaction parameter, K. The plot indicates that the conversion is increased substantially by mixing of the bubbles. [Curl, *A.I.Ch.E. J.* **9**, 175 (1963)].

shows the effect of coalescence upon reactor yield when the tank reactor operates at steady state with a zero-order reaction. It may be seen that the mixing by coalescence has a profound effect upon the conversion of the reactor. Thus coalescence represents a mixing process in a chemical reactor with which significant changes in the reaction rate or selectivity can be obtained. There is need for further experimental work upon the coalescence process in agitated media.

B. Two-Phase Piston-Flow Models

The extension of the piston-flow models for reacting systems to two-phase flow is represented by the performance of a liquid-liquid extraction column in which a reaction occurs in the dispersed phase. This problem may be considered an extension of the work of Sleicher (S2) and Miyauchi (M8) and Miyauchi and Vermeulen (M9) upon axial mixing in continuous liquid-liquid extraction columns. On the other hand, extractive reaction has been considered by Piret *et al.* (P3) and Trambouze and Piret (T3) and staged mixer-settler operation has been analyzed by Sleicher (S3). Deans and Lapidus (D1) showed that it is often useful to represent axial mixing in a continuous system by approximating the continuous system as a series of tank vessels. Thus one may use the extractive reactor analysis for staged vessels as an approximation to the continuous column if the number of stages is properly chosen.

For a countercurrent system in which a first-order reaction occurs in the dispersed or "bubble" phase,

$$\left(\frac{1}{N_{Pe}}\right)_B \frac{d^2 C_B}{dX^2} = \frac{dC_B}{dX} + M(C_B - C_C) + K_B C_B \tag{46a}$$

$$\left(\frac{1}{N_{Pe}}\right)_C \frac{d^2 C_C}{dX^2} = -\frac{dC_C}{dX} + M\phi(C_C - C_B) \tag{46b}$$

$$C_B(0) = 1 \qquad \frac{dC_B}{dX}\bigg|_1 = 0$$

$$C_C(1) = C_C^1 \qquad \frac{dC_C}{dX}\bigg|_0 = 0$$

where A_B, A_C = concentration of reactant in the bubble and continuous phases;

b = bubble radius;

$C_B = A_B/A_B(0)$ dimensionless bubble or dispersed-phase concentration;

$C_C = K_D A_C/A_B(0)$ dimensionless continuous-phase concentration;

f_B, f_C = volume fraction in the bubble and continuous phases;

$K_B = k\ell/v_B$ dimensionless dispersed-phase reaction rate;

K_D = dimensionless equilibrium distribution ratio;

K_L = over-all mass transfer coefficient based upon dispersed-phase concentration;

k = dispersed phase first-order reaction rate;

$M = K_L 3\ell/f_B v_B b$ = dimensionless mass transfer coefficient;

$(N_{Pe})_B = v_B \ell/\mathscr{D}_{ax}$ Peclet number of the bubble phase (the Peclet group for the continuous phase is defined analogously);

v_B, v_C = interstitial velocities of the dispersed and continuous phases;

$X = z/\ell$ dimensionless length;

$\phi = v_B f_B/v_C f_C$.

The algebra of the solution to Eqs. (46a) and (46b) is lengthy (M9) but is identical to the nonreactive case when $K_D C_D$ terms are replaced with $(K_D + K_R)C_D$ terms. The mathematics of the problem are nearly identical to May's (M6) bubble interchange model for fluidized beds. Numerical solutions for Eqs. (43a) and (43b) without chemical reaction have been prepared by Sleicher (S2) and Miyauchi (M8).

Comparison of experimental data with the numerical results for Eqs. (43a) and (43b) is disturbing, for it is found that the boundary conditions are far more important than the Peclet numbers of the system. The boundary conditions chosen are approximations, however, for these boundary specifications do not adequately represent the drop formation process. Hence the simple extension of the concept of axial mixing to two-phase systems is not particularly rewarding for this problem.

C. HETEROGENEOUS-PHASE MIXED-MODEL SYSTEMS

The use of mixed models for estimating the performance of homogeneous reactions was illustrated in Section II, D. The purpose of this section is to extend the mixed-model technique to heterogeneous reactions. Fluidization certainly is one of the more important reactor types and is an example of a heterogeneous reaction which has been analyzed successfully through mixed models.

The flow of gas through a fluidized bed is a complex process. As the gas flow rate through the bed increases, the bed of solid particles expands. After the bed is about 11% expanded, bubble-shaped zones free of solids form at the distributor plate and rise through the system. The motion of the bubbles through the bed is responsible for extremely efficient mixing of the solids in the bed. Motion picture studies have shown that wake of the rising bubbles produces turbulent agitation of the solids and that the solids which flow into the wake are supplied from zones above the particles. The gas velocity through the bubble must be larger than through the solids in order to preserve the "interface" between the bubble and the remainder of the system. This velocity difference produces a convective mixing in the gas phase. Thus an operating fluidized bed comprises two "phases" of solid density, a "bubble" phase in which the solid particle density is low and a "dense" phase. The density of particles in the "dense" phase does not change with total throughput to the bed; as the throughput to the bed increases more bubbles form. Hence in superficial behavior, the gas-solids fluidized bed resembles a liquid-liquid extraction in which the dispersed phase is stagnant.

Two mixed-model configurations for a fluidized bed are shown in Fig. 13. May (M6) assumed that the dense phase is similar to a piston-flow reactor in which mixing occurs by axial dispersion. The bubble phase is assumed to pass through the bed without mixing. Reactants are supplied to the dense phase from the bubble phase, and the rate of this mass transfer is characterized by the "crossflow ratio," an empirical parameter which is strongly dependent upon bed diameter. Reaction occurs only in the dense phase. Material balance equations for the May model are similar to liquid-liquid extraction equations and are

$$(N_{Pe})_C^{-1} \frac{d^2C_C}{dX^2} = \frac{dC_C}{dX} - M\phi(C_B - C_C) + K_C C_C \qquad (47a)$$

$$0 = \frac{dC_B}{dX} + M(C_B - C_C) \qquad (47b)$$

$$1 = C_C(0) - (N_{Pe})_C^{-1} \frac{dC_C}{dX}\bigg|_0 \qquad (47c)$$

$$\lim_{X \to \infty} C_C(X) = 0 \qquad (47d)$$

$$C_B(0) = 1 \qquad (47e)$$

The nomenclature for Eqs. (47a) and (47b) is identical to Eqs. (46a) and (46b) except for the following modifications:

A_i = area/volume of the nonspherical bubbles,
$K_C = k\ell/v_C$ continuous-phase dimensionless rate parameter,
$k_g A_i$ = equivalent bubble mass transfer coefficient (May denotes this parameter as the crossflow ratio),
$M = k_g A_i \ell/f_B v_B$ dimensionless mass transfer coefficient.

The solution of Eqs. (47a) and (47b) again is algebraically complex, but experimental data can be adequately represented by a suitable choice of the crossflow ratio. Thus Eqs. (47a) and (47b) represent an adequate design model.

Orcutt et al. (O2) attempted to find a more fundamental model for the crossflow process, and the results of his analysis yield the following estimate for the crossflow ratio

$$M = 2.44 \frac{\ell_m \mathscr{D}^{1/2}}{g^{1/4} b^{7/4}} + 3.16 \frac{\ell_m u_m}{g^{1/2} b^{1/2}} \qquad (48)$$

$$b = 0.252 \frac{(u - u_m)^2}{g\left(\dfrac{\ell}{\ell_m} - 1\right)} \qquad (49)$$

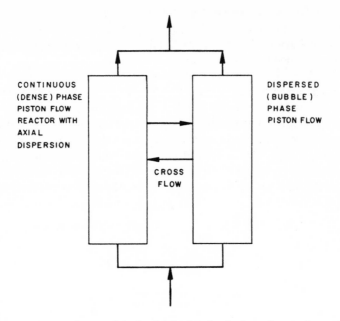

FIG. 13a. May's crossflow model of a fluidized bed—the interchange of mass between the bubble and dense phases of the fluidized bed is specified by the crossflow ratio, a dimensionless mass transfer coefficient.

where b = bubble radius,

$\quad \mathscr{D}_m$ = gas phase diffusivity,

$\quad g$ = gravitational acceleration,

$\quad \ell, \ell_m$ = bed height at operating and minimum fluidization conditions,

$\quad u, u_m$ = superficial velocity in bed at operating and incipient bubble formation conditions.

Figure 13b shows one of Gray's (G1) models for the fluidized bed. The fluidized bed is viewed as a series of tank reactors connected by flow in the bubble phase. The number of units to be used is estimated by dividing the bed length by the Davidson bubble diameter, defined with Eq. (49) as $2b$. The crossflow parameter is estimated from Eq. (48). Gray's model replaces the axial dispersion parameter by the number of tanks in the system. A series of difference equations is written for the system:

$$C_B^{N-1} = C_B^N + M(C_B^N - C_C^N) \qquad (50a)$$

$$C_C^{N-1} = C_C^N + \phi M(C_C^N - C_B^N) + K_C C_C^N \qquad (50b)$$

$$C_B^0 = C_C^0 = 1 \qquad (50c)$$

where the superscript denotes the bubble number. The exit concentration of

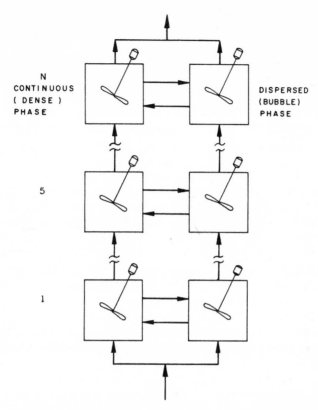

N
CONTINUOUS
(DENSE)
PHASE

5

1

DISPERSED
(BUBBLE)
PHASE

FIG. 13b. Gray's bubble-flow model of a fluidized bed—axial dispersion in the fluidized bed is represented by flow through a series of tank reactors. The rate of interchange between the bubble and the continuous phase is found from Davidson's [Orcutt *et al.* (O2)] theoretical calculation.

the reactor is found from the finite difference solution to Eqs. (50a) and (50b) and is

$$C = \frac{1}{r_1 - r_2}\left\{ r_2 r_1^N - r_1 r_2^N + (r_2^N - r_1^N)\frac{(1 + \phi)^{-1} + \dfrac{M}{N} + \dfrac{K_C}{N}}{(1 + \phi)^{-1} + \dfrac{M}{N} + \left(1 + \dfrac{M}{N}\right)\dfrac{K_C}{N}} \right\} \quad (51)$$

where N = the total number of bubble reactors in the fluidized bed,

r_1 and r_2 = the roots to the equation

$$\left((1 + \phi)^{-1} + \frac{M}{N} + \left(1 + \frac{M}{N}\right)\frac{K}{N}\right)r^2$$

$$- \left(2(1 + \phi)^{-1} + \frac{M}{N} + \frac{K}{N}\right)r + (1 + \phi)^{-1} = 0 \quad (52)$$

Comparison of Gray's model with experimental data shows that the model is adequate for design estimates. The advantage of Gray's model is that there are fewer arbitrary parameters than the May model.

The fluid mechanics of a fluidized bed are extremely complicated, and it is clear that a useful understanding of the fluid motion has not yet been achieved. Thus one can safely predict the development of other models for the operation of a fluidized bed reactor.

IV. Summary

The contribution of mixing to reactor performance occurs in two ways. First, mixing affects the residence-time distribution of the reactor; for example, changes in impeller speed, type, and position are known to alter the residence-time distribution of the tank. The form of the residence-time distribution has an important effect upon yields of reactions of all orders, and consequently many authors have incorrectly equated mixing entirely with contributions from the residence-time distribution. For reactors in which first-order isothermal chemical kinetics occur, it is found that only the residence-time distribution is needed to estimate conversion. Thus the contribution of mixing to reactor performance is often a secondary factor.

A number of authors attempt to identify a second conceptualization of mixing which is separate from the residence-time distribution. It is argued that the residence-time distribution owes its form to the gross velocity contours within the reactor, but the mixing processes are the mechanisms whereby the molecules within fluid velocity contours interact with each other. The contribution of this latter type of mixing to over-all reactor performance is much more difficult to estimate. The history models are far too arbitrary, and the statistical models introduce unknowns more rapidly than one can generate equations. Additional attention should be devoted to discovering what happens in mixing of small fluid eddies (micromixing), for if these mixing processes were better understood, then a more rational attack for mixing design could be established. Some encouraging results already have been developed for the special case of very fast reactions in isotropic turbulence.

Mixing, both residence-time effects and micromixing, are much better understood for homogeneous reactions than for heterogeneous systems. In single phase systems it is nearly always possible to select some model or analysis which will yield useful design estimates upon investment of a modest amount of work. In multiple phase systems, one finds that the techniques for design estimation are not well established and that critical comparison of design models to operating data are extremely scarce. Thus reactions in multiphase equipment is a research area in which nearly any reasonable experiment provides useful information.

Although this chapter is primarily devoted to descriptions of models for estimating the effect of mixing upon chemical reactions, it is not likely that the important research of the future will be developed along these lines. It is more likely that as the fundamental fluid mechanics of the mixing processes are understood, the less necessary mixing models will become. Mixing design needs to be made less an art and more a science.

List of Symbols

A pre-exponential constant of an Arrhenius rate equation, sec.$^{-1}$
A eddy surface area, cm.^{2}
A concentration at point specified by the superscript, gm. mole/liter
A^f feed concentration, gm. mole/liter
A_i interfacial area/volume of fluidized bubbles, cm.$^{-1}$
A^r reference concentration, gm. mole/liter
B_n molal concentration of bifunctional polymer made from n-monomer units, gm. mole/liter
b r/R dimensionless radius
b bubble radius, cm.
$C = A/A^r$ dimensionless concentration at point specified by superscript
C_p specific heat of stream, gm. cal./(gm. °C.)
d pipe diameter, cm.
d_p particle diameter, cm.
$\mathscr{D}$ general diffusivity parameter, cm.^{2}/sec.
$\mathscr{D}_{ax}$ effective axial diffusivity, cm.^{2}/sec.
$\mathscr{D}_m$ molecular diffusivity, cm.^{2}/sec.
$E_1(y) \equiv \int_y^\infty (e^{-\eta}/\eta)d\eta$ a tabulated experimental integral
F steady state volumetric flow rate, cm.^{3}/sec.
f steady state mass flow rate, gm./sec.
f fractional yield of product
$f(t)$ residence-time distribution, sec.$^{-1}$
f_B, f_C volume fraction in bubble and continuous phase
f_f Fanning friction factor
g eddy area growth rate, sec.$^{-1}$
g gravitational acceleration, cm./sec.^{2}
$g(h)$ parameter defined by Eq. (8a), sec.$^{-1}$
ΔH_{act} apparent activation energy, gm. cal./gm. mole
ΔH_a thermodynamic heat of reaction for the production of one unit mole of a at the conditions which exist in the tank reactor, gm. cal./(gm. mole of a)
h time parameter in mixing-history models, sec.
$\Delta h_i(w_j, p, T)$ thermodynamic heat of formation for the production of one unit mass of i at the conditions which exist in reactor, gm. cal./(gm. of i)
$I(h)$ cumulative input function, dimensionless
$\mathbf{j}_i$ diffusive flux, grams i/(sec. cm.^{2})
K dimensionless rate parameter defined by the kinetics of the system
$K_B = k\ell/v_B$ dimensionless bubble phase rate constant
$K_C = k\ell/v_C$ dimensionless continuous-phase rate constant
K_D liquid-liquid phase distribution ratio
K_L over-all mass transfer coefficient, cm./sec.
k first-order reaction rate constant, sec.$^{-1}$

k_2 second-order reaction rate constant, liter/(mole sec.)

k_D drop mixing rate, sec.$^{-1}$

k_f forward second-order reaction rate constant, liter/(mole sec.)

k_r reverse second-order reaction rate constant, liter/(mole sec.)

ℓ system length, cm.

ℓ_{aa} two-point double diffusive microscale, cm.

M dimensionless drop mixing rate

M dimensionless mass transfer coefficient in Eq. (46)

M_i molecular weight of the ith species, gm./(gm. mole)

M_n concentration of monofunctional polymer made from n-monomer units, gm. mole/liter

$\prime\prime\prime_{a^2a}$ two-point triple diffusive microscale, cm.

N tank number

N reaction order

N stirring rate, r.p.m.

$(N_{Pe})_B, (N_{Pe})_C$ Peclet number of bubble and continuous phase, dimensionless

$O(h)$ the output function

P volumetric pumping rate of the impeller, cm.^{3}/sec.

P dimensionless concentration probability density function

Q^N rate of heat input to reactor by heat transfer, gm. cal./sec.

R gas constant, gm. cal./(gm. mole °K.)

$R_a(A)$ molal rate of production of component A, gm. mole. (liter sec.)

r radial distance, cm.

R radius, cm.

$r_i(w_j, \rho, T)$ mass rate of production of the ith component on the Nth tank per unit volume, gm./(cm.^{3} sec.)

r_1, r_2 dimensionless roots in Eqs. (25) and (52)

T temperature, °K.

T_0 inlet temperature to fixed-bed reactor, °K.

t time, sec.

$u_1 u_m$ superficial velocity in a packed bed, cm./sec.

V volume of tank, cm.^{3}

v mass average fluid velocity, cm./sec.

v_B interstitial velocity in the bubble phase of a fluidized bed, cm/sec.

v_C interstitial velocity in the continuous phase of a fluidized bed, cm/sec..

v_i interstitial velocity in a packed bed, cm./sec.

v_y fluid velocity in the y direction, cm./sec.

w_i mass fraction of the ith component, dimensionless

X dimensionless distance, z/ℓ, x/ℓ

y_0 original eddy half-thickness, cm.

z axial distance parameter, cm.

δ eddy thickness, cm.

η^{ai} fractional macro-micro reactor model effectiveness

ξ dimensionless eddy material coordinate [Eq. (33)]

ρ mass density, gm./cm.^{3}

$\tau = (Vt)/F$ dimensionless tank residence time

$\tau = (t\phi_0 F)/\phi V$ dimensionless time parameter in a two-phase tank reactor

τ dimensionless eddy life [Eq. (33)]

ϕ dimensionless average residence time of the portion of the reactor indicated by the subscript

ϕ fraction of reactor volume in feed loop of the macro-micro model

ϕ ratio of volumetric throughput of the bubble to the continuous phases.

References

(A1) Adler, J., and Vortmeyer, D., *Chem. Eng. Sci.* **18**, 99 (1963); **19**, 413 (1964).
(A2) Adler, R. J., and Hovorka, R. B., "A finite-stage model for highly asymmetric residence time distributions," Case Inst. of Technology, Cleveland, Ohio, 1961.
(A3) Amundson, N. R., *Ind. Eng. Chem.* **48**, 26 (1956).
(A4) Aris, R. R., *Proc. Roy. Soc. (London)* **A235**, 67 (1956).
(A5) Aris, R. R., "Introduction to the Analysis of Chemical Reactors." Prentice-Hall, Englewood Cliffs, New Jersey, 1965.
(B1) Barkelew, C. H., *Chem. Eng. Progr. Sym. Ser.* **55**, No. 25, 37 (1959).
(B2) Batchelor, G. K., *J. Fluid Mech.* **5**, 113 (1959).
(B3) Bischoff, K. B., and Levenspiel, O., *Chem. Eng. Sci.* **17**, 245, 257 (1962).
(C1) Cairns, E. J., and Prausnitz, J. M., *Ind. Eng. Chem.* **51**, 1441 (1959).
(C2) Cairns, E. J., and Prausnitz, J. M., *Chem. Eng. Sci.* **12**, 20 (1960).
(C3) Carberry, J. J., *Can. J. Chem. Eng.* **36**, 207 (1958).
(C4) Carberry, J. J., and Wendel, M. M., *A.I.Ch.E. J.* **9**, 129 (1963).
(C5) Cholette, A., and Blanchet, J., *Can. J. Chem. Eng.* **39**, 192 (1961).
(C6) Cholette, A., and Cloutier, L., *Can. J. Chem. Eng.* **37**, 105 (1959).
(C7) Cholette, A., Blanchet, J., and Cloutier, L., *Can. J. Chem. Eng.* **38**, 1 (1960).
(C8) Conover, J. A., M. S. thesis, St. Louis Univ., St. Louis, Missouri, 1960.
(C9) Converse, A. O., *A.I.Ch.E. J.* **6**, 344 (1960).
(C10) Corrsin, S., *Phys. Fluids* **1**, 42 (1958).
(C11) Corrsin, S., *Phys. Fluids* **7**, 1156 (1964).
(C12) Corrsin, S., *A.I.Ch.E. J.* **10**, 870 (1964).
(C13) Curl, R. L., *A.I.Ch.E. J.* **9**, 175 (1963).
(D1) Deans, H. A., and Lapidus, L., *A.I.Ch.E. J.* **6**, 656, 663 (1960).
(D2) DeBaun, R. M., and Katz, S., *Chem. Eng. Sci.* **16**, 97 (1961).
(D3) DeMaria, J., Longfield, J. E., and Butler, G., *Ind. Eng. Chem.* **53**, 259 (1961).
(D4) Dickens, E. S., Ph.D. thesis, Univ. of Delaware, 1964.
(D5) Douglas, J. M., *Chem. Eng. Progr. Symp. Ser.* **60**, No. 48, 1 (1964).
(D6) Dranoff, J. S., *Math. Computation* **15**, 403 (1961).
(F1) Farrell, M. A., and Leonard, E. F., *A.I.Ch.E. J.* **9**, 160 (1962).
(F2) Froment, G. F., *Chem. Eng. Sci.* **17**, 849 (1962).
(G1) Gray, R. D., M.S. thesis, Univ. of Delaware, 1963.
(G2) Gray, R. D., Ph.D. thesis, Univ. of Delaware, 1965.
(H1) Haddad, A., Wolf, D., and Resnick, W., *Can. J. Chem. Eng.* **42**, 216 (1964).
(H2) Harriott, P., *Can. J. Chem. Eng.* **42**, 60 (1962).
(H3) Holmes, D. B., Voncken, R. M., and Decker, J. A., *Chem. Eng. Sci.* **19**, 201 (1964).
(H4) Howarth, W. J., *Chem. Eng. Sci.* **18**, 47 (1963).
(K1) Keeler, R. N., Petersen, E. E., and Prausnitz, J. M., *A.I.Ch.E. J.* **11**, 221 (1965).
(K2) Kilkson, H., *Ind. Eng. Chem. Fundamentals* **3**, 281 (1964).
(K3) Kramers, H., and Westerterp, K. R., "Chemical Reactor Design and Operations." Academic Press, New York, 1963.
(L1) Larson, P. Z., Jr., M. S. thesis, Univ. of Delaware, 1963.
(L2) Lauwerier, H. A., *Appl. Sci. Res. Sect.* **A8**, 366 (1959).
(L3) Levenspiel, O., *Can. J. Chem. Eng.* **40**, 135 (1962).
(L4) Levenspiel, O., "Chemical Reaction Engineering." Wiley, New York, 1962.
(L5) Levenspiel, O., and Bischoff, K. B., *Advances in Chem. Eng.* **4**, 95 (1963).
(M1) Madden, A. J., and Damerell, G. L., *A.I.Ch.E. J.* **8**, 233 (1962).
(M2) Manning, F. S., and Wilhelm, R. H., *A.I.Ch.E. J.* **9**, 12 (1963).
(M3) Manning, F. S., Wolf, D., and Keairns, D. L., *A.I.Ch.E. J.* **11**, 723 (1965).

(M4) Marr, G. R., Jr., and Johnson, E. F., *Chem. Eng. Progr. Symp. Ser.* **57**, No. 36, 109 (1961).

(M5) Marr, G. R., Jr., and Johnson, E. F., *A.I.Ch.E. J.* **9**, 383 (1963).

(M6) May, W. G., *Chem. Eng. Progr.* **55**, 49 (1959).

(M7) Mickley, H. S., and Letts, R. W. M., *Can. J. Chem. Eng.* **41**, 273 (1963); **42**, 21 (1964).

(M8) Miyauchi, T., U.S. Atomic Energy Commission, Univ. of Calif. Radiation Lab.-9311, 1957.

(M9) Miyauchi, T., and Vermeulen, T., *Ind. Eng. Chem. Fundamentals* **2**, 113, 304 (1963).

(N1) Nichols, D. G., and Lamb, D. E., "A general solution for chemical reaction with axial and radial diffusion in laminar tube flow," submitted to *A.I.Ch.E. J.*, in revision.

(O1) O'Hern, H. A., and Rush, F. E., Jr., *Ind. Eng. Chem. Fundamentals* **2**, 267 (1963).

(O2) Orcutt, J. C., Davidson, J. F., and Pigford, R. L., *Chem. Eng. Progr. Symp. Ser.* **58**, No. 38, 1 (1962).

(P1) Pao, Yih-Ho, *Am. Inst. Aero. Astro. J.* **2**, 1550 (1964).

(P2) Pigford, R. L., unpublished research, 1960.

(P3) Piret, E. L., Penney, W. H., and Trambouze, P. J., *A.I.Ch.E. J.* **6**, 394 (1960).

(R1) Rice, A. W., Toor, H. W., and Manning, F. S., *A.I.Ch.E. J.* **10**, 91 (1964).

(R2) Rosenhouse, H., Ph.D. thesis, Univ. of Delaware, 1964.

(S1) Sjenitzer, F., *Petrol. Engr.* **30**, No. 12, D13 (1958).

(S2) Sleicher, C. A., Jr., *A.I.Ch.E. J.* **5**, 145 (1959).

(S3) Sleicher, C. A., Jr., *A.I.Ch.E. J.* **6**, 529 (1960).

(S4) Smith, J. M., "Chemical Engineering Kinetics." McGraw-Hill, New York, 1956.

(S5) Stout, L. E., Jr., and Otto, R. E., *Chem. Eng. Progr. Symp. Ser.* **57**, No. 36, 69 (1961).

(T1) Tichacek, L. J., "Selectivity in experimental reactors," A.I.Ch.E. San Francisco Meeting, 1959.

(T2) Toor, H. L., *A.I.Ch..E. J.*, **8**, 70 (1962).

(T3) Trambouze, P. J., and Piret, E. L., *A.I.Ch.E. J.* **6**, 574 (1960).

(V1) van der Laan, E. T., *Chem. Eng. Sci.* **7**, 187 (1958).

(V2) van der Vusse, J. G., *Chem. Eng. Sci.* **17**, 507 (1962).

(W1) Walas, S. M., "Reaction Kinetics for Chemical Engineers." McGraw-Hill, New York, 1959.

(W2) Wehner, J. F., and Wilheim, R. H., *Chem. Eng. Sci.* **6**, 89 (1956); **8**, 309 (1958).

(W3) Weinstein, H., Ph.D. thesis, Case Inst. of Technology, Cleveland, Ohio, 1963.

(W4) Wolf, D., and Resnick, W., *Ind. Eng. Chem. Fundamentals* **2**, 287 (1963).

(W5) Woods, C. R., Jr., B.S.E. thesis, Univ. of Delaware, 1965.

(Z1) Zweitering, T., *Chem. Eng. Sci.* **11**, 1 (1959).

CHAPTER 8

Mixing of High Viscosity Materials

H. F. Irving

Baker Perkins Company
Saginaw, Michigan

R. L. Saxton

E. I. du Pont de Nemours & Company
Wilmington, Delaware

I. Introduction

This chapter presents a theoretical treatment of mixing of highly viscous materials and describes the devices used in industry for this task. Highly viscous materials will be defined as those having viscosities in excess of one thousand poises, and include, at the lower viscosity level, thermoplastic resins such as nylon, and, at the upper, concentrates of carbon black in rubber. Asphalts, putties, doughs, rubber, and molten thermoplastics all come under this definition.

169

II. Fundamental Concepts[1]

To introduce an engineering approach to mixing high viscosity materials it is helpful to recall what a *mixture* is and what a *mixing process* can be expected to accomplish (M4).

A. PROPERTIES OF MIXTURES

The term "mixture" is defined as "a complex of two or more ingredients which do not bear a fixed proportion to one another and which, however thoroughly commingled, are conceived as retaining a separate existence." In this discussion a "mixing" process is considered one in which a mixture is produced by the subdivision and distribution of the components throughout the entire volume of the system without the ultimate elements of the components undergoing any intrinsic change. If intrinsic changes such as dissolution, molecular weight reduction, or deagglomeration occur, then processes other than mixing are involved.

An inevitable problem is evaluation of the quality of mixtures, or the "goodness of mixing." Two properties that are useful in characterizing mixtures are the "scale of segregation," and the "intensity of segregation" (D1) (see also Chapter 2). The scale of segregation is a measure of the size of undistributed portions of the components, while the intensity refers to the difference between the compositions of undistributed portions and the desired mean composition. The natures of these properties are illustrated in Fig. 1; scale increases to the right, and intensity increases upward. In the figure an increase in intensity of segregation is shown by a greater difference between adjacent regions in number of dots per unit area. Increasing scale is shown by increasing size of regions of a given areal concentration of dots.

In general, reduction of the scale of segregation is accomplished by relative motion of the components imposed by moving surfaces of the mixing device. Reduction of intensity can only be brought about by diffusion. The effect of reduced intensity can, of course, be achieved by carrying the scale of segregation down to the dimensions of the ultimate units of the components, as by emulsifying immiscible liquids to a molecular scale. But ordinarily, mixing by itself cannot reduce the intensity of segregation, although if the components are mutually soluble, the rate of reduction of intensity, i.e., the rate of dissolution, will be increased due to the shortened average path length for diffusion. If there is no diffusion, the intensity of segregation remains constant; i.e., the components retain their identity and remain as separate phases. This situation is approximated in the mixing of molten plastics and other high viscosity systems.

[1] Adapted from *Ind. Eng. Chem.* (M4, M5).

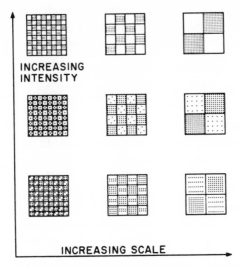

FIG. 1. Scale and intensity of segregation [reprinted from *Ind. Eng. Chem.* (M4)].

B. MIXING IN LAMINAR-FLOW SYSTEMS

In the mixing of viscous fluids, then, diffusion is virtually of no help; there is no turbulence, no eddy diffusivity. Under these conditions, the major mechanism of mixing is *shear*. Shear performs mixing by drawing out the components into thinner and thinner layers. This action reduces the size of regions occupied exclusively by one component. In Fig. 2 is depicted a system of two viscous liquids enclosed between parallel plates. Initially, the minor component (dark-colored) exists as discrete cubes, randomly distributed as in a blend. Under shear, induced by movement of the upper boundary surface, these particles would be stretched out, and eventually this view of the system would show only striations of light and dark material. If enough shear were imposed, the combined thickness of each pair of layers could be brought below the limit of resolution, and the eye would see only uniform gray.

The average combined thickness of a pair of layers, or the separation of a

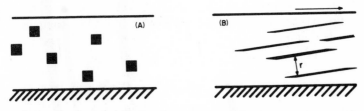

FIG. 2. Mixing action of fluid shear: (A) initial state, (B) sheared state [reprinted from *Ind. Eng. Chem.* (M4)].

pair of like interfaces, is a measure of the scale of segregation of the components, and hence of goodness of mixing. This characteristic property of mixtures prepared in laminar-flow systems is called the "striation thickness," r.

C. CALCULATION OF STRIATION THICKNESS

The striation thickness is related to the amount of interfacial area developed by the mixing operation. Spencer and Wiley (S2) have considered how this interfacial area can be calculated, and the following treatment (M4) stems from their work.

Consider a striated mixture of dark with clear liquid. If the radii of curvature of all interfacial surfaces are large relative to the striation thickness, the total volume, V, may be expressed in terms of interfacial area, S,

$$V = \frac{rS}{2} \tag{1}$$

This neglects interfacial area at ends and edges of striations of the minor component. The factor of two enters because all except two striations have two interfaces. For a system in unidirectional shear the ratio of final to initial interfacial area is given by Spencer and Wiley (S2):

$$\left(\frac{S}{S_0}\right)^2 = 1 - 2\left(\frac{\partial u_x}{\partial y}\right)\cos\alpha_x \cos\alpha_y + \left(\frac{\partial u_x}{\partial y}\right)^2 \cos^2\alpha_x \tag{2}$$

where $\cos\alpha =$ a direction cosine of the normal to the original surface at point (x, y, z),

$u_x = x$-component of the displacement vector of point (x, y, z),

$\partial u_x/\partial y =$ a constant for the case of simple shear in the x-direction; a measure of the shearing strain imposed.

In this case

$$\frac{\partial u_y}{\partial y} = \frac{\partial u_z}{\partial y} = 0$$

Equation (2) shows that the new is proportional to the original surface and depends on the orientation of the original surface to the direction of shear. This relation has been applied to the case in which the minor component exists originally as randomly distributed, discrete cubes. If the cubes are assumed oriented with edges parallel to the coordinate axes, the direction cosines of the surfaces are:

x, y-plane. $\cos\alpha_x = 0$, $\cos\alpha_y = 0$, $\cos\alpha_z = 1$

y, z-plane. $\cos\alpha_x = 1$, $\cos\alpha_y = 0$, $\cos\alpha_z = 0$

x, z-plane. $\cos\alpha_x = 0$, $\cos\alpha_y = 1$, $\cos\alpha_z = 0$

From Eq. (2) the area ratios for the three types of sides are:

$$x, y\text{-plane} \quad \left(\frac{s'}{s'_0}\right)^2 = 1$$

$$y, z\text{-plane} \quad \left(\frac{s'}{s'_0}\right)^2 = 1 + \left(\frac{\partial u_x}{\partial y}\right)^2$$

$$x, z\text{-plane} \quad \left(\frac{s'}{s'_0}\right)^2 = 1$$

and since

$$S = 2(s'_{x,y} + s'_{y,z} + s'_{x,z})$$

and $S_0 = 6s'_0$

$$\frac{S}{S_0} = \tfrac{1}{3} + \tfrac{1}{3}\sqrt{1 + (\partial u_x/\partial y)^2} + \tfrac{1}{3} \tag{3}$$

In any practical mixing operation it is apparent that

$$\partial u_x/\partial y > 10$$

so Eq. (3) can be reduced to

$$\frac{S}{S_0} = \frac{1}{3}\frac{\partial u_x}{\partial y} \tag{4}$$

The initial ratio of interfacial surface to volume is

$$\frac{S_0}{V} = \frac{6Y_b}{l_b} \tag{5}$$

Combining this with Eq. (4) gives

$$\frac{S}{S_0} = \frac{S/V}{S_0/V} = \frac{l_b}{3rY_b} = \frac{1}{3}\frac{\partial u_x}{\partial y} \tag{6}$$

if γ be substituted for $\partial u_x/\partial y$, the amount of resultant shear, the striation thickness is expressed by

$$r = \frac{l_b}{\gamma Y_b} \tag{7}$$

If the components differ in viscosity, a correction factor must be applied:

$$r = \frac{l_b}{\gamma_a Y_b} \cdot \frac{\mu_b}{\mu_a} \tag{8}$$

In the derivation of this correction it is assumed that the boundary surfaces of the system are wet by the major component. Relative motion of portions of these surfaces imposes a shear rate on this component. The shear stress induced is the same in both components. The shear rate of the minor component, however, depends on its viscosity:

$$\tau_a = \tau_b = \mu_a \dot{\gamma}_a = \mu_b \dot{\gamma}_b \tag{9}$$

Rearranging and multiplying both shear rates by the residence time give

$$\gamma_b/\gamma_a = \mu_a/\mu_b \tag{10}$$

from which

$$r = \frac{l_b}{\gamma_a Y_b} \cdot \frac{\mu_b}{\mu_a} \tag{11}$$

The ratio of viscosities is taken as unity for $(\mu_b/\mu_a) \leq 1$, because the minor component can never be deformed more rapidly than the matrix of major component. The shear strain, γ_a, is calculated as the product of the rate of shear imposed on the major component and the duration of shearing. This calculation is not simple, because in real compounding devices shear rates and residence times vary among and along the flow paths. However, in principle, in a laminar-flow system an element of fluid can be traced along its streamline. So if the velocity distributions in the mixing device can be formulated, shear rates and residence times for each part of the path can be computed and summed up to yield the corresponding shear strain. By tracing a number of flow paths in this manner, a profile of shear strain in the product is obtained.

Stated in words, Eq. (11) says that the final striation thickness equals the initial striation thickness divided by the product of the amount of shear supplied and the relative fluidity.

The assumption of orientation of the cube faces parallel to the coordinate axes should not seriously restrict application of Eq. (11), because for large amounts of shear the initial orientation of symmetrical particles should have little effect on the final striation thickness.

D. MIXING IN THE SINGLE-SCREW EXTRUDER

The single-screw extruder is widely used as a mixing device in the plastics industry. A common practice is to blend a granular thermoplastic with colorants or other additives, and extrude either the final shape—e.g., rod, sheet, or tubing—or strands to be cut again into granules. In the extruder the plastic is melted, and the other component distributed through the viscous liquid. Although diffusion of soluble additives such as dyes can assist the mixing process, this effect is not large.

A theory, which permits calculation of the goodness of mixing achieved in the single-screw extruder, was developed by combining the foregoing theory (Section C) of mixing in laminar-flow systems with that of melt extrusion (C1, C2). The equations developed permit the effects of screw geometry and average residence time to be investigated (M5). The calculated pattern of mixing in the extrudate and the effect of residence time agreed qualitatively with experiment. This laminar-flow mixing process was susceptible to analysis because the flow patterns could be described mathematically.

1. Flow Patterns in Screw Channel

In the single-screw extruder fluid flows through a channel of rectangular cross section (Fig. 5). The two sides are the leading and trailing surfaces of the flight, the bottom is the screw root, and the top is the inside surface of the barrel. Fluid is conveyed because of the relative motion of barrel and channel. The conveying force is the viscous drag transmitted to fluid in the channel from the barrel surface, and all the fluid motion derives from it.

For simplicity the channel is visualized as uncoiled and laid flat (Fig. 3). The barrel surface is considered to move across the channel at the screw helix angle, ϕ, and with velocity $U = \pi D N$.

For convenience, this velocity, U, is resolved into components V and T, parallel and transverse to the channel, respectively. Component V induces a flow, called the drag flow, $Q_{\mathscr{D}}$, having a linear velocity profile (Fig. 4A). This flow is the maximum pumping capacity of the extruder, and does not depend on the viscosity of the fluid. If flow is restricted at the outlet, however, the pressure that the screw must develop rises, and a pressure gradient is set up in the channel. The effect of this gradient is introduced as a fictitious flow called pressure flow, $Q_{\mathscr{P}}$, which has the usual parabolic velocity distribution (Fig. 4B). For a given pressure gradient this flow is inversely proportional to viscosity. A third flow occurs in the clearance between the barrel and the flight because of the pressure difference across the flight. This is called leakage flow, $Q_{\mathscr{L}}$. The discharge from an extruder is the net volumetric rate of flow across a plane perpendicular to the axis of the screw. This is calculated as the algebraic sum of drag, pressure, and leakage flow, when these are expressed as rates of flow parallel to the screw axis.

$$Q = Q_{\mathscr{D}} - Q_{\mathscr{P}} - Q_{\mathscr{L}} \tag{12}$$

Because the leakage flow is ordinarily much smaller than the others, it will not be considered further.

In the transverse plane the barrel surface drags liquid toward the leading face of the flight. In the absence of leakage, all of this liquid must be forced downward to return across the bottom of the channel in a pressure-type flow (Fig. 5). This "circulating" movement is an important factor in the mixing action of the extruder. In Fig. 6 is shown an observed flow pattern, obtained by solidifying and cross sectioning material from the channel of a 2-in. extruder.

Mathematical description of these flows is provided by simplified extrusion theory (C1, C2) based on these assumptions:

(1) The viscosity is the same at all points in the major component. This implies a Newtonian fluid and isothermal operation.
(2) The width of the channel is large compared to the depth, so that the influence of the sides of the channel on velocity distributions can be neglected.
(3) Entrance and exit effects on flow are negligible.
(4) The liquid wets all surfaces and is incompressible.

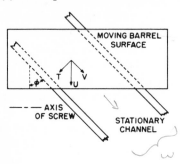

FIG. 3. Extruder channel [reprinted from *Ind. Eng. Chem.* (M5)].

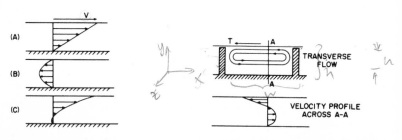

FIG. 4. Velocity profiles in parallel plane: (A) drag flow, (B) pressure flow, and (C) resultant flow [reprinted from *Ind. Eng. Chem.* (M5)].

FIG. 5. Velocity profile in transverse plane [reprinted from *Ind. Eng. Chem.* (M5)].

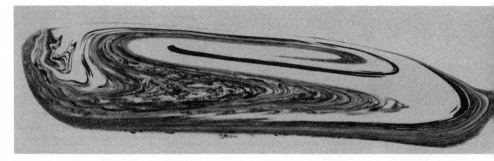

FIG. 6. Observed flow pattern in transverse plane of 2-in. extruder channel (courtesy of E. I. du Pont de Nemours & Co.).

The local velocity down the channel may be expressed (C1, C2) as

$$v = \frac{Vy}{h} + \frac{(y^2 - hy)}{2\mu}\frac{(dP)}{(dz)} \tag{13}$$

The first term on the right is the drag flow velocity; the second, the pressure flow velocity. Making the substitutions

$$H = \frac{y}{h} \tag{14}$$

and

$$a = \frac{h^2}{6V\mu}\left(\frac{dP}{dz}\right) \tag{15}$$

simplifies Eq. (13) to

$$v = V[(1 - 3a)H + 3aH^2] \tag{16}$$

The velocity profile predicted by this equation is shown in Fig. 4C. At any point along the channel h, V, μ, and dP/dz have finite values; in principle, therefore, a is a determinable quantity. It is seen to characterize the velocity profile in the plane parallel to the channel. If there is no pressure rise along the channel, $dP/dz = 0$, and $a = 0$. If there is no net flow

$$Q_V = hw\int_0^1 v\,dH = hw\int_0^1 V[(1 - 3a)H + 3aH^2]dH = \frac{whV}{2}(1 - a) = 0 \tag{17}$$

from which when $Q_V = 0$, $a = 1.0$. It appears that $a = 0$ for drag flow alone and 1.0 for no net flow (pressure flow equal to drag flow), and the flow is linear in a for intermediate conditions. It can be shown that $a = Q_\vartheta/Q_\vartheta$.

The velocity profile in the transverse plane may be written by analogy with that in the parallel plane

$$s = \frac{Ty}{h} + \frac{(y^2 - hy)}{2\mu}\frac{dP}{dx} \tag{18}$$

Defining

$$c = \frac{h^2}{6T\mu}\left(\frac{dP}{dx}\right) \tag{19}$$

simplifies this equation. If, as assumed, leakage flow is negligible, there is no net flow in the transverse plane and $c = 1.0$. Equation (18) can then be reduced to

$$s = T(3H^2 - 2H) \tag{20}$$

The velocity profile in the transverse plane is shown in Fig. 5.

The resultant path of an element of fluid is a helix of oval cross section, in which the helix angle undergoes a cyclical change around each turn. The effect of increased pressure flow is to reduce the average value of this helix angle; in its passage through the channel an element travels around its loop in the transverse plane more often the higher the ratio: $Q_{\mathscr{P}}/Q_{\mathscr{D}}$. Consequently, increasing the average residence time by restricting flow at the die has a specific effect: The amount of shear contributed by this circulation in the transverse plane is increased, while shear in the parallel plane is not changed.

2. Calculation of Shear Received by a Fluid Element

The shear strain imposed on an element of liquid in the channel was calculated as the sum of the products of shear rate and residence time for each part of the flow path. Shear contributions in the parallel and transverse planes were calculated separately. Shear received between the end of the screw and the die was approximated in the calculations by the shear over an unflighted extension of the screw having a length, L_E, equal to the screw diameter. The shear imposed by the die was calculated. All shear contributions were resolved into components parallel to the screw axis (axial shear), and perpendicular to it (circumferential shear). Finally, these components were corrected for the change in shape experienced in forming the rod-shaped extrudate; the amounts of axial shear and circumferential shear were summed, and the resultant amount of shear, γ, for the element in the extrudate was obtained by vector addition. Lastly, the striation thickness was computed by means of Eq. (11).

In principle, the final position of an element in a laminar-flow process can be predicted from the initial position and the geometry of the system. In these calculations, the starting height in the channel, H_1, determined the radial position in the extruded rod. Shear in the channel was calculated for a number of starting heights, and the results were obtained as striation thickness vs. reduced radius in the rod.

The procedure employed for calculating shear strain is shown schematically in the accompanying diagram.

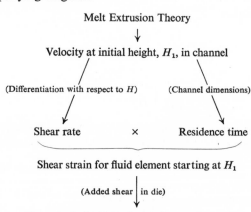

Melt Extrusion Theory

Velocity at initial height, H_1, in channel

(Differentiation with respect to H)   (Channel dimensions)

Shear rate   ×   Residence time

Shear strain for fluid element starting at H_1

(Added shear in die)

Shear strain for fluid element in round extrudate
at radius corresponding to H_1

3. Calculations

Shear in Transverse Plane. The flow path in the transverse plane was approximated by assuming the element to cross the entire width of the channel at one height, H_1, then at once to start back on the complementary streamline at H_2 (Fig. 7).

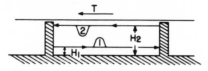

FIG. 7. Simplified picture of flow in transverse plane [reprinted from *Ind. Eng. Chem.* (M5)].

The relationship between complementary streamlines is found by equating volumetric flows in the two directions.

$$\int_0^{H_1} s\,dH = -\int_{H_2}^{H_1} s\,dH \tag{21}$$

Substituting for s from Eq. (20) and integrating give

$$H_1^2 - H_1^3 = H_2^2 - H_2^3 \tag{22}$$

The average velocity of an element is obtained from the distance traveled divided by the time to complete one cycle

$$\bar{s}_1 = \frac{2w}{\dfrac{w}{s_1} - \dfrac{w}{s_2}} = \frac{2s_1 s_2}{s_2 - s_1} \tag{23}$$

The average shear rate is then found

$$\frac{1}{h}\frac{d\bar{s}_1}{dH_1} = \frac{2\left(s_2 \dfrac{ds_1}{dH_1} + s_1 \dfrac{ds_2}{dH_2}\dfrac{dH_2}{dH_1}\right) - \bar{s}_1\left(\dfrac{ds_2}{dH_2}\dfrac{dH_2}{dH_1} - \dfrac{ds_1}{dH_1}\right)}{h(s_2 - s_1)} \tag{24}$$

The residence time in the channel of a screw of length, L, and helix angle ϕ is

$$\theta = \frac{L}{\bar{v}_1 \sin \phi} \tag{25}$$

where $\bar{v}_1$ is the average velocity down the channel of an element starting at height H_1. The amount of shear in the transverse plane is given by

$$\gamma_T = \frac{1}{h}\left(\frac{d\bar{s}_1}{dH_1}\right)\frac{L}{\bar{v}_1 \sin \phi} \tag{26}$$

It is convenient to modify this equation by making the shear rate and average velocity dimensionless. It can be shown that

$$\bar{v}_1 = \bar{v}_1|_{a=0}(Q/Q_{\mathscr{D}})$$

$$= \bar{v}_1|_{a=0}(1 - a)$$

Also,

$$T/V = \tan \phi$$

Hence,

$$\gamma_T = \frac{L}{h(1 - a)\cos\phi}\left(\frac{d\bar{s}_1}{dH_1} \times \frac{1}{T(\bar{v}_1/V)_{a=0}}\right) \tag{27}$$

For simplicity function, F_T, is defined as

$$F_T = \frac{d\bar{s}_1}{dH_1} \times \frac{1}{T} \times \frac{1}{v_1/V}\bigg|_{a=0} \tag{28}$$

which depends only upon H_1; F_T can be computed for a series of values of H_1 and the corresponding amounts of shear obtained from the equation

$$\gamma_T = \frac{L}{h(1 - a)\cos\phi} \times F_T \tag{29}$$

The amount of shear in the transverse plane is seen to increase with increasing a, the ratio of pressure to drag flow.

Shear in Parallel Plane. The average velocity down the channel for an element starting at height H_1 is computed as the distance traveled in the parallel plane during one cycle in the transverse plane divided by the time for that cycle.

$$\bar{v}_1 = \frac{v_1(w/s_1) - v_2(w/s_2)}{w/s_1 - w/s_2} = \frac{v_1 s_2 - v_2 s_1}{s_2 - s_1} \tag{30}$$

The average shear rate is obtained by differentiating with respect to H_1; the amount of shear is given by

$$\gamma_V = \frac{1}{h}\left(\frac{d\bar{v}_1}{dH_1}\right)\frac{L}{\bar{v}_1 \sin\phi} = \frac{L}{h \sin\phi}\left(\frac{d\bar{v}_1}{dH_1}\right)\left(\frac{1}{\bar{v}_1}\right) \tag{31}$$

It can be shown that the term in parentheses is independent of pressure flow; the calculations are simpler for the case of $a = 0$. In this case $v = VH$ and

$$\frac{dv_1}{dH_1} = \frac{dv_2}{dH_2} = V$$

It is convenient to define the function

$$F_V = \left(\frac{d\bar{v}_1}{dH_1} \times \frac{1}{\bar{v}_1}\right)_{a=0} = \frac{\left(s_2 + H_1 \dfrac{ds_2}{dH_2}\dfrac{dH_2}{dH_1} - s_1 \dfrac{dH_2}{dH_1} - H_2 \dfrac{ds_1}{dH_1}\right)}{H_1 s_2 - H_2 s_1}$$

$$- \frac{\dfrac{\bar{v}_1}{V}\bigg|_{a=0}\left(\dfrac{ds_2}{dH_2}\dfrac{dH_2}{dH_1} - \dfrac{ds_1}{dH_1}\right)}{H_1 s_2 - H_2 s_1} \tag{32}$$

so that the amount of shear in the parallel plane is given by

$$\gamma_V = \frac{L}{h \sin \phi} F_V \tag{33}$$

Pressure flow is seen not to affect γ_V.

Like F_T, F_V is dimensionless and depends only on H_1. Values of these functions are listed in Table I.

Table I

Values of Functions F_V and $F_T{}^a$

H_1	F_V	F_T
0.050	17.527	−21.679
0.100	7.880	−8.527
0.150	4.796	−4.171
0.200	3.309	−1.983
0.250	2.441	−0.648
0.300	1.870	0.264
0.350	1.461	0.936
0.400	1.149	1.458
0.500	0.675	2.225
0.600	0.278	2.754
0.750	−0.446	3.155

[a] Reprinted from *Ind. Eng. Chem.* (M5).

The two shear contributions of the channel are resolved into components parallel and perpendicular to the screw axis (i.e., to the direction of extrusion) as these are the logical directions to consider in the extruded rod.

$$\gamma_{\lambda,C} = \gamma_V \sin \phi - \gamma_T \cos \phi = \frac{L}{h}\left(F_V - \frac{F_T}{1-a}\right) \tag{34}$$

$$\gamma_{U,C} = \gamma_V \cos \phi + \gamma_T \sin \phi = \frac{L}{h}\left(\frac{F_V}{\tan \phi} + \frac{F_T \tan \phi}{1-a}\right) \tag{35}$$

Once F_V and F_T have been computed for the desired values of H_1, it is only necessary to substitute values of the several variables in turn to obtain the effects of helix angle, ϕ, ratio of pressure flow to drag flow, a, and thread depth, h.

Relationship between Shear in Channel and in Extrudate. The effect of the change in shape of the flow profiles in the channel to the piston-type flow in the extruded rod was evaluated by calculating multipliers for both components of shear. The annular volume elements considered are shown in Fig. 8. The radii of these elements lie on the same streamlines.

The axial shear in the extrudate is given by

$$\gamma_{\lambda,R} = \gamma_{\lambda,C} \frac{dL_R}{dL} \frac{dy_1}{dr_R} \tag{36}$$

where dL_R/dL and dy_1/dr_R express the effect on axial shear of changes in dimensions experienced by an element of volume in passing from the screw channel to the rod-shaped extrudate. In passing from channel to rod the volume of the annular element remains constant, so that

$$\pi D_C \, dL \, dy_1 = 2\pi r_R \, dL_R \, dr_R$$

$$\frac{dy_1}{dr_R} = \frac{2r_R dL_R}{D_C dL} \tag{37}$$

Between two streamlines the volumetric flow rate is constant:

$$\pi D_C \, dy_1 f = 2\pi r_R \, dr_R f_R \tag{38}$$

From Eqs. (37) and (38)

$$\frac{dL_R}{d_L} = \frac{f_R}{f} \tag{39}$$

Substitution from Eqs. (37) and (39) into (36) gives the axial multiplying factor

$$\begin{aligned}
\frac{dL_R}{dL} \times \frac{dy_1}{dr_R} &= \frac{2r_R}{D_C} \left(\frac{f_R}{f} \right)^2 \\
&= \frac{2r_R}{D_C} \left[\frac{Q}{\pi R_R^2} \middle/ \frac{6Q(H_1 - H_1^2)}{\pi Dh} \right]^2 \\
&= \frac{h^2 \rho_R D^2}{18 \, R_R^3 (H_1 - H_1^2)^2 D_C} \\
&= \frac{D^2 h^2 \rho_R}{18 \, R_R^3 (H_1 - H_1^2)^2 [D - 2h(1 - H_1)]} \tag{40} \\
&\simeq \frac{Dh^2 \rho_R}{18 \, R_R^3 (H_1 - H_1^2)^2} \tag{41}
\end{aligned}$$

for $(h/D) \leq 0.05$.

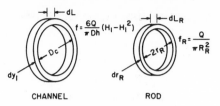

FIG. 8. Volume elements in channel and extruded rod [reprinted from *Ind. Eng. Chem.* (M5)].

The corrected axial shear is, therefore

$$\gamma_{\lambda,R} = \frac{LhD^2\rho_R}{18\ R_R^3[D - 2h(1 - H_1)](H_1 - H_1^2)^2}\left(F_V - \frac{F_T}{1 - a}\right) \quad (42)$$

$$\cong \frac{LhD\rho_R}{18\ R_R^3(H_1 - H_1^2)^2}\left(F_V - \frac{F_T}{1 - a}\right) \quad (43)$$

The amount of circumferential shear is

$$\gamma_{U,R} = \gamma_{U,C}\left(\frac{2r_R}{D_C}\right)\left(\frac{dy_1}{dr_R}\right)$$

$$= \frac{2LD\rho_R^2}{3(H_1 - H_1^2)[D - 2h(1 - H_1^2)]^2}\left(\frac{F_V}{\tan\phi} + \frac{F_T\tan\phi}{1 - a}\right) \quad (44)$$

$$\cong \left(\frac{2L}{3D}\right)\left(\frac{\rho_R^2}{(H_1 - H_1^2)}\right)\left(\frac{F_V}{\tan\phi} + \frac{F_T\tan\phi}{1 - a}\right) \quad (45)$$

If $(h/D) > 0.05$, Eqs. (42) and (44) must be used. In these calculations Eqs. (43) and (45) were used only for $h = 0.1$ in.

Relationship between Position in Channel and in Extrudate. The velocity distribution parallel to the axis is assumed to be parabolic with distance from screw root to barrel surface. The liquid leaving the screw is assumed to follow a streamline path to the final position in the extruded rod. The material near the root will then form the center of the rod, and that near the barrel the outside layer.

If the forward flow in the channel has a parabolic profile

$$f = k(H_1 - H_1^2)$$

where k is a proportionality constant that depends on screw speed and geometry.

The fraction of the total flow passing between the root and a point at height H_1 is

$$\frac{qH_1}{Q} = \frac{k\displaystyle\int_0^{H_1}(H_1 - H_1^2)dH_1}{k\displaystyle\int_0^1(H_1 - H_1^2)dH_1} = H_1^2(3 - 2H_1) \quad (46)$$

In the extrudate, piston-type flow obtains so that the fraction of the total flow that passes between the center and a circle of radius r is

$$\frac{q_r}{Q} = \frac{\pi r_R^2}{\pi R_R^2} = \rho_R^2 \tag{47}$$

Equating the two fractional flows gives the position in the extrudate, ρ_R, in terms of position in the channel, H_1:

$$\rho_R = H_1(3 - 2H_1)^{0.5} \tag{48}$$

Shear between Screw and Die. Liquid in the space between the screw and the die is sheared both axially and circumferentially. The amounts of shear imposed were approximated by shear in the annulus around an unflighted extension of the screw root of length L_E.

The forward velocity in this annulus is

$$f = \frac{6Q}{\pi Dh}(H_1 - H_1^2)$$

from which the residence time is computed

$$\theta_E = \frac{L_E}{f} = \frac{\pi DhL_E}{6Q(H_1 - H_1^2)} \tag{49}$$

The axial shear rate is

$$\frac{df}{dy} = \frac{6Q}{\pi Dh^2}(1 - 2H_1) \tag{50}$$

from which the amount of axial shear is

$$\gamma_{\lambda,E} = \frac{L_E(1 - 2H_1)}{h(H_1 - H_1^2)}$$

The shape factor is the same as for the channel [Eq. (40)]. Hence, the axial shear in the extrudate contributed by the extension is

$$\gamma_{\lambda,E,R} = \frac{L_E D^2 h\rho_R(1 - 2H_1)}{18R_R^3(H_1 - H_1^2)^3[D - 2h(1 - H_1)]} \tag{51}$$

As the circumferential shear rate is $\pi DN/h$, the amount of circumferential shear in the extrudate is given by the product of shear rate, residence time [Eq. (49)], and the shape factor

$$\gamma_{U,E,R} = \frac{2DL_E\rho_R^2}{9 \sin\phi \cos\phi (1 - a)(H_1 - H_1^2)^2[D - 2h(1 - H_1)]^2} \tag{52}$$

If $h/D < 0.05$, D_C can be considered equal to D, as in Eqs. (43) and (45). This approximation was used in the calculations for $h = 0.1$ in.

Shear in Die. The velocity distribution in the circular die is assumed to be parabolic

$$f_D = \frac{2Q(1 - \rho_D^2)}{\pi R_D^2}$$

The axial shear rate is then

$$\frac{df_D}{dr_D} = \frac{1}{R_D}\frac{df_D}{d\rho_D} = \frac{1}{R_D}\frac{2Q}{\pi R_D^2}(-2\rho_D)$$

The residence time is

$$\theta_D = \frac{L_D}{f_D} = \frac{\pi R_D^2 L_D}{2Q(1 - \rho_D^2)}$$

Hence the amount of axial shear is

$$\gamma_{\lambda,D} = -\frac{2L_D\rho_D}{R_D(1 - \rho_D^2)} \tag{53}$$

The effect of the change in shape of the volume elements in forming the extruded rod must be added as the appropriate multipliers

$$\gamma_{\lambda,D,R} = \gamma_{\lambda,D}\left(\frac{dL_R}{dL_D}\right)\left(\frac{dr_D}{dr_R}\right)$$

Even though the diameters may be equal, the positions of an element in the circular die and in the extruded rod are not the same because of the difference between the velocity profiles. The relation between the two positions can be derived as follows.

The fraction of the total flow passing between the center of the tube and a circle of radius r_D is

$$\frac{q_r}{Q} = \frac{\displaystyle\int_0^{r_D} 2\pi f_D r_D dr_D}{Q}$$

$$= \frac{R_D^2}{Q}\int_0^{\rho_D} \frac{4Q}{R_D^2}(1 - \rho_D^2)\rho_D d\rho_D$$

$$= 2\rho_D^2\left(1 - \frac{\rho_D^2}{2}\right) \tag{54}$$

Combining Eqs. (54) and (47) gives

$$\rho_D^2 = 1 - (1 - \rho_R^2)^{1/2} \tag{55}$$

Between streamlines the volumetric flow rate is constant,

$$2\pi f_D r_D \, dr_D = 2\pi f_R r_R \, dr_R \tag{56}$$

so that the volume of the annular element remains constant

$$2\pi r_D \, dr_D \, dL_D = 2\pi r_R \, dr_R \, dL_R \tag{57}$$

Combining Eqs. (56) and (57) gives

$$\frac{dL_R}{dL_D} = \frac{f_R}{f_D} \qquad (58)$$

and from Eq. (57)

$$\frac{dr_D}{dr_R} = \frac{r_R dL_R}{r_D dL_D} = \frac{R}{R_D} \frac{\rho_R f_R}{\rho_D f_D} \qquad (59)$$

The shear in the extruded rod contributed by the die is, therefore

$$\gamma_{\lambda,D,R} = \frac{-L_D R_D^2 \rho_R}{2R_R^3 (1 - \rho_D^2)^3} \qquad (60)$$

4. Results of Calculations

The equations developed permitted calculation of the amount of shear imposed on an element of liquid as a function of its initial position in the channel. They were used to investigate the effects of screw geometry and average residence time on the mixing performance of the single-screw extruder. Striation thickness in the extrudate was computed for eleven reduced radii corresponding to eleven initial heights, H_1, in the channel. The value of H_1 giving the minimum shear in the extrudate was also determined. The assumed conditions and the values of the variables investigated are listed in Tables II and III.

Table II

Dimensions Used in Computation of Extruder Mixing Performance[a]
(in inches)

Minor component	
Cube side length, l_b	0.125
Volume fraction, Y_b	0.1
Screw dimensions	
Length, L	24.0
Diameter, D	2.0
Length of extension, L_E	2.0
Die dimensions	
Land length, L_D	3.0
Radius, R_D	0.25
Extrudate radius, R_R	0.25

[a] Reprinted from *Ind. Eng. Chem.* (M5).

Table III

Values of Variables in Computation of Extruder Mixing Performance[a]

Helix angle, ϕ (deg.)	10, 17.7, 30, 45
Ratio of pressure flow to drag flow, a	0, 0.333, 0.50, 0.667, 0.80, 0.90, 0.95
Thread depth, h (in.)	0.10, 0.30
Viscosity ratio	1.0

[a] Reprinted from *Ind. Eng. Chem.* (M5).

In Fig. 9 three graphs are plotted showing typical calculated distributions of shear strain in the screw channel. It is easy to see that the shear strain is not evenly distributed: material near midchannel ($H \sim 0.5$) receives relatively little. When this profile of shear strain was translated into a corresponding profile of striation thickness in a cylindrical extrudate, the characteristic pattern shown by Fig. 10 was obtained.

In Fig. 10 are shown three typical graphs of striation thickness against reduced radius in the extruded rod. The parameter is a. Mixing was predicted to be relatively good near the center and edge of the section, poorest at approximately half the radius. The maximum striation thickness was lowered by a factor of 2.4, and the average by a factor of 5.8 by reducing throughput to 5% of the drag flow ($a = 0.95$). In Fig. 11 the calculated maximum striation thicknesses are plotted against the fraction of drag flow or the fraction of the maximum pumping capacity; the parameter is helix angle. Mixing was improved by reducing helix angle and increasing pressure flow.

At constant output per turn (output ÷ screw speed), and equal thread depth, a screw with a higher helix angle would operate at a higher ratio of pressure flow to drag flow than one of lower helix angle. This would not overcome the improvement in mixing produced by lower helix angles, as shown by the lines of constant output per turn sketched in Fig. 11.

Increased thread depth was predicted to improve mixing. For all helix angles the effect of thread depth was greatest for $0.50 \le a \le 0.667$. The largest difference in goodness of mixing was found at $\phi = 45°$, $a = 0.667$; for this condition the striation thicknesses at the point of poorest mixing (r_{max}) were 0.0298 and 0.0119 in. for $h = 0.1$ and $h = 0.3$ in. respectively. At the lesser thread depth goodness of mixing fell more rapidly with increasing helix angle. As a result, the effect of thread depth became greater the higher the helix angle (Table IV).

Table IV

Effect of Thread Depth on Maximum Striation Thickness[a]

ϕ	r_{max} (in.)[b] ($h = 0.1$ in.)	($h = 0.3$ in.)	$\left[\dfrac{r_{max}\|h = 0.1 \text{ in.}}{r_{max}\|h = 0.3 \text{ in.}}\right]$
10	0.0140	0.00895	1.57
17.7	0.0235	0.0134	1.75
30	0.0346	0.0173	2.00
45	0.0395	0.0188	2.10

[a] Reprinted from *Ind. Eng. Chem.* (M5).
[b] For $a = 0$.

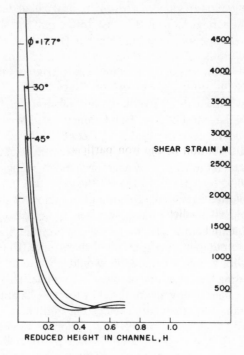

FIG. 9. Calculated distribution of shear strain in screw channel.

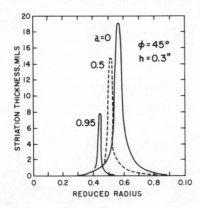

FIG. 10. Calculated patterns of goodness of mixing over cross section of cylindrical extrudate [reprinted from *Ind. Eng. Chem.* (M5)].

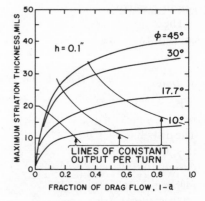

FIG. 11. Effect of helix angle and pressure flow on maximum striation thickness [reprinted from *Ind. Eng. Chem.* (M5)].

To test the theory, a few mixing experiments were carried out; the calculated and observed behaviors were in qualitative agreement. The feed material in all experiments was a blend of $\frac{1}{8}$ in. cubes of natural polyethylene with similar cubes of the minor component, a concentrate of "tracer" particles of iron (average particle size one micron) in polyethylene. The concentrate was 4.35% by volume of the blend. The viscosity of the concentrate was varied by using base polyethylenes of different viscosity. The feed blend was extruded by a 2-in. diam. extruder into $\frac{1}{8}$ in. rod. Cross sections of the rod, 20 μ thick, were cut on a microtome, mounted on slides, and photomicrographed. The uniformity of distribution of the iron particles over the area of the sample section gave a qualitative measure of goodness of mixing of the two polyethylene components.

The pattern of mixing over the cross section of rod extruded at a low value of a agreed with that predicted (Fig. 12). The preservation of this pattern of relatively poor mixing about the mid-radius requires that the liquid follow a smooth course between the end of the screw and the die. The material near the screw root should form the center of the rod; that near the barrel, the outside; and the least sheared material, from mid-channel, should fill the space about the mid-radius. This pattern is not often clearly observed, because flow between the end of the screw and the die is usually distorted by passing through filters containing fine-mesh wire screens.

The action of pressure flow is beneficial, as predicted. Figure 13 shows the improvement in mixing obtained by increasing a from 0.17 to 0.30.

The calculations were carried out for components of equal viscosity, but the effect of differing viscosities could be found by multiplying any calculated striation thickness by the viscosity ratio: μ_b/μ_a [see Eq. (11)]. This ratio significantly influences the goodness of mixing achieved for a given set of conditions (compare Figs. 12A and 12B).

These qualitative agreements between theory and experiment reinforce the concept that shear is the primary mechanism of mixing, and also give some indication that these methods of calculation are applicable to real machines.

5. Improvement of Mixing by Extruders

Both experiment and theory teach that the single-screw extruder is a relatively inefficient mixing device. The foregoing theory suggests the reason: The distribution of shear strain in the screw channel is nonuniform. To correct this, it is probably best to add a section designed especially for mixing, and design the rest of the screw for the other functions of solids conveying, plasticating (melting), and conveying melt against pressure. The mixing section must break the regular helical streamlines established in the screw channel and force material from the center of the main channel to pass near a surface. Mixing extensions of several types have been found effective; some of these will be described later on.

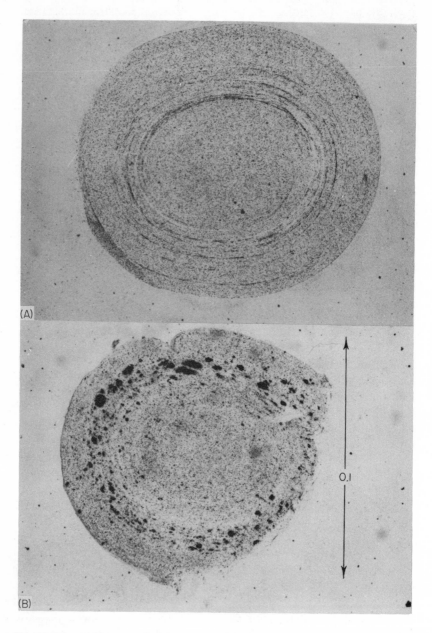

FIG. 12. Observed pattern of mixing over cross section of extruded rod. (A) $\mu_b/\mu_a = 1.3$ and (B) $\mu_b/\mu_a = 12$ [reprinted from *Ind. Eng. Chem.* (M5)].

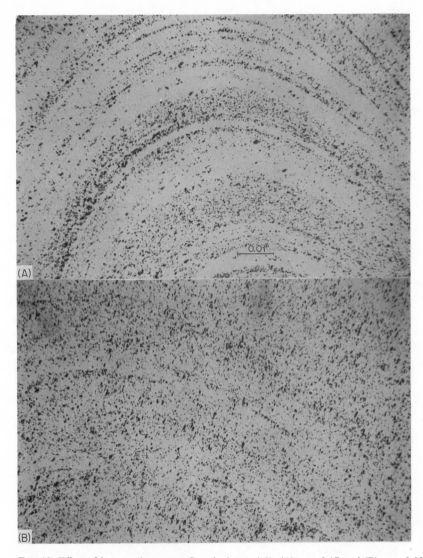

FIG. 13. Effect of increased pressure flow ($\mu_b/\mu_a = 1.3$). (A) $a = 0.17$ and (B) $a = 0.30$ [reprinted from *Ind. Eng. Chem.* (M5)].

E. COMPOUNDING PIGMENT CONCENTRATES

These concepts of mixing also have been applied to the compounding of carbon black with polyethylene. The compounding route is shown by the following.

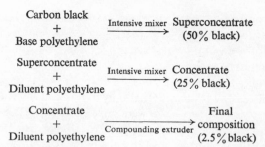

The *dispersion*, or *deagglomeration*, of the pigment is carried out at high pigment loading to form a superconcentrate. The high pigment content raises the viscosity so that high shear stresses are developed in the superconcentrate (Fig. 14). The first, and more difficult *mixing* problem is to dilute the super-concentrate to about half the initial pigment concentration. Refer again to Eq. (11) which relates striation thickness to process variables. In this case the viscosity ratio is superconcentrate viscosity divided by diluent resin viscosity.

Remember that striation thickness is a measure of the average size of regions occupied by one component. Equation (11) states that if other variables are held constant, the number of regions of undiluted superconcentrate above a given size is proportional to the viscosity ratio. That is, if the viscosity ratio could be reduced, we would expect to make a more homogeneous composition for the same imposed shear strain. The diluent resin of the second and third compounding stages is ordinarily selected on the basis of the physical properties required in the product. However, the base resin of the superconcentrate only makes up about 2.5 % of the final product, so that in some cases its contribution to final properties is not important. The viscosity of the superconcentrate can

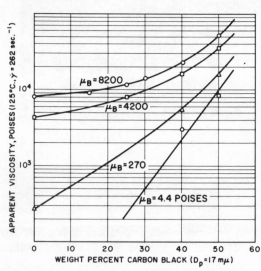

FIG. 14. Comparison of viscosities of concentrates and base resins.

be reduced 5- or 6-fold by choosing base resins of lower viscosity (Fig. 14). This reduction in viscosity does not affect the breaking up of agglomerates of pigment, but for a given diluent resin it does affect the goodness of mixing achieved when the superconcentrate is diluted.

In Fig. 15 are shown observations for a series of experiments in which $l_b/\gamma Y_b$ was held nearly constant, so that only the viscosity ratio was a variable. Superconcentrate and diluent resin were mixed in a size BR Banbury[2] mixer operated at a constant rotor speed (76 r.p.m.) for a fixed time interval (15 min.) after melting of the stock (assumed to coincide with a stock temperature of 110°C.). In Fig. 15 the population of particles of unmixed 50% super-concentrate is plotted against viscosity ratio. The particles were isolated from the mixtures of superconcentrate and diluent resin by passing the molten mixtures through fine screens. The number of particles increased nearly 100-fold as the viscosity ratio rose 10-fold.

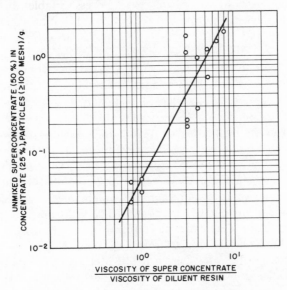

FIG. 15. Correlation of goodness of mixing with ratio of viscosities of superconcentrate and base resin.

F. USE OF RHEOLOGICAL DATA

To predict mixing behavior from rheological data the correlation shown in Fig. 15 is a useful empirical correlation. However, it is not the best type of correlation because the viscosities used in the viscosity ratios were measured at one shear rate and one temperature. Consequently these viscosity ratios, while consistent within this group of experiments, were not the ratios that

[2] Registered trademark of Farrel Corp., Ansonia, Conn.

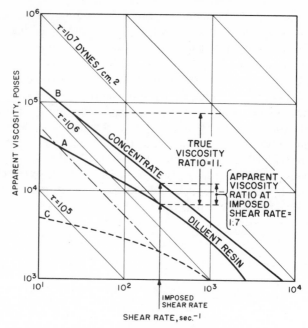

FIG. 16. Method of estimation of true viscosity ratio.

really existed in the mixing device. Figure 16 shows how to estimate more realistic viscosity ratios from capillary rheometer data. To do this, a full rheological description is needed, so that the effects of shear rate (or shear stress) and of temperature on the viscosities of the components are known. A convenient rheological description consists of a plot of apparent viscosities, at the temperature of mixing, vs. shear rate. To be really precise, of course, we should go further, and derive the true (rather than apparent) shear stress-shear rate relationship from the capillary rheometer data. This can be done by the method of Rabinowitsch (R1) [also given by Metzner (M3) and by Bird *et al.* (B2)]. However, in view of the difficulty of describing the distributions of shear rate and temperature that obtain in practical mixing devices the usefulness of refining the viscosity relationships seems doubtful.

Suppose it is desired to mix a minor component described by the upper line (*B*) with a major component described by the lower line (*A*) of Fig. 16. Enter the diagram on the abcissa at the (average) shear rate prevailing in the mixing device (shown as 242 sec.$^{-1}$) and read up to the line (*A*) that describes the major component (8 × 10^3 poises). Because the shear stress is the same in both components, proceed along the diagonal isobar to intersect the line describing the minor component (*B*). The viscosity ratio for the mixing of these plastics at this temperature and imposed shear rate can then be computed as the quotient of the two ordinates [(9 × 10^4)/(8 × 10^3)]. To be really precise, we would have to know the distribution of shear rate and temperature in the mixing device, and determine true viscosity ratios along a number of streamlines. This is impractical for most mixing problems.

Figure 17 illustrates a situation in which the two components are predicted to be, literally, *immiscible* under the given conditions of shear rate and temperature. This is more apt to occur if the minor component is *plastic* (M3, P1) as shown here by graph *C*. At an imposed shear rate ($\dot{\gamma}_a$) of 242 sec.$^{-1}$ the

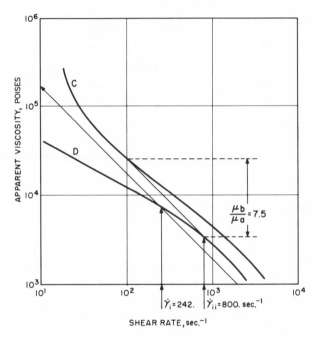

FIG. 17. Prediction of immiscibility from a rheological diagram.

isobar drawn from the curve for the major component (*D*) does not intersect the curve for the minor component (*C*). Hence the particles of *C* will not be sheared, and no mixing will occur.

By raising the imposed shear rate in the matrix fluid (*D*) to $\dot{\gamma}_a = 800$ sec.$^{-1}$, the isobar can be caused to intersect the graph of material *C*. The true viscosity ratio under these conditions is then $\mu_b/\mu_a = 7.5$.

In principle the dependence of viscosities of any two materials on shear rate and temperature could be expressed mathematically, and conditions of shear rate and temperature that would minimize the energy required for their mixing established by computation. Practical mixing conditions would, however, represent a compromise among the minimization of the viscosity ratio, the energy expended, and the residence time (mixer volume).

G. Energy and Time

It is important to note that *shear stress* is irrelevant to the mixing of viscous materials that do not have a yield point. Mixing devices operate by setting up velocity gradients within the material, i.e., by imposing a shear rate on the major component. The shearing strain necessary to effect a desired goodness of mixing can be accumulated at any shear rate. The penalty for high shear rate is high energy consumption, for low shear rate, very long residence times. Mixer design, in the simplest terms, consists in compromising between energy consumption on the one hand, and equipment size and material flow rate on the other. It should be remembered, however, that because

$$\gamma = \dot{\gamma}\theta$$

shear strain = (shear rate) (time)

cm./cm. = (cm./cm.-sec.) (sec.)

and

$$Z = \dot{\gamma}^2\mu = \dot{\gamma}\tau$$

power = (shear rate)^2 (viscosity)

= (shear rate) (shear stress)

ergs/cm.^{3}-sec. = (cm./cm.-sec.)^2 (dyne-sec./cm.^{2})

= (cm./cm.-sec.) (dynes/cm.^{2})

and

$$E = Z\theta = \gamma^2\mu/\theta$$

energy = (power) (time) = (shear strain)^2 (viscosity)/time

ergs/cm.^{3} = (cm./cm.)^2 (dyne-sec./cm.^{2})/(sec.)

the energy required to produce a given shear strain in a system of given viscosity is inversely proportional to the time over which the strain is imposed. Another advantage of operation at low shear rate is the increased opportunity for heat removal, relative to the rate of heat generation.

III. Equipment

A. Effect of Viscosity on Design

Equipment for mixing viscous materials is characterized by high power supplied per unit volume and by relatively small volume, i.e., small separation of "impeller" and vessel walls. In low viscosity fluids momentum may be transferred from a rotating member, e.g., a propeller, throughout a space measuring several cubic impeller diameters. As viscosity rises, however, the volume of fluid the impeller is able to set in motion diminishes, until, in the range of viscosity considered here, the thickness of the layer of fluid affected by the rotor is a fraction of the rotor radius.

B. MIXER TYPES

No attempt will be made to cover all of the devices employed to mix viscous materials. A few major types have been selected from current commercial equipment. The various types of equipment are arranged below, roughly in order of descending energy input per unit volume.

Banbury mixer, batch
Continuous mixer (Banbury type)
Roll mill
Ko-Kneader[3]
Extruders
Frenkel C-D mixer
Double-blade batch mixers

This order in general corresponds with a descending level of viscosity of materials that may be processed, although adjacent types are not very different in this respect. These types of equipment will be described below.

1. Banbury Mixers

The Banbury mixer is a heavy duty machine, capable of masticating rubber and compounding pigment concentrates in plastics or rubber. The mixing chamber is figure-eight shaped (Fig. 18). A spiral-lobed rotor occupies each half of the chamber. The rotors turn with a slight speed differential so that material is sheared between them as well as between rotors and the wall. Stock is charged to the mixing chamber through a vertical chute in which travels an air- or hydraulic-powered ram. The lower face of the ram forms the upper boundary of the mixing chamber. The ram pressure holds the stock down in the "bite" of the rotors. The finished batch is discharged through a slide or drop door. A sheeting mill or a pelletizing extruder are normal auxiliaries for a Banbury mixer.

The batch size must be held within limits, since for best performance the chamber should be 60–80% filled, while the ram is "bottomed." Both rotors and the chamber walls may be heated or cooled. Steam heat and water-spray cooling are frequently used.

Figure 18 shows a cross section of a Banbury mixer, while a photograph of a large commercial unit appears in Fig. 19.

In Table V are tabulated typical data for Banbury mixers (F1).

2. Continuous Mixer

A relatively new development is the continuous mixer. The mixing elements and action are much the same as in the older, batch, Banbury mixer, but the stock flows continuously along the length of the rotors, as in a twin-screw

---

[3] Registered trademark of Baker Perkins Co., Saginaw, Mich.

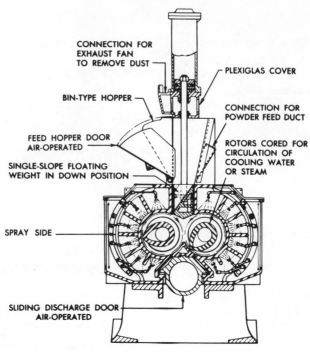

CONNECTION FOR
EXHAUST FAN
TO REMOVE DUST

PLEXIGLAS COVER

BIN-TYPE HOPPER

CONNECTION FOR
POWDER FEED DUCT

FEED HOPPER DOOR
AIR-OPERATED

ROTORS CORED FOR
CIRCULATION OF
COOLING WATER
OR STEAM

SINGLE-SLOPE FLOATING
WEIGHT IN DOWN POSITION

SPRAY SIDE

SLIDING DISCHARGE DOOR
AIR-OPERATED

FIG. 18. Cross-sectional view of size 11 Banbury mixer (courtesy of Farrel Corp.).

Table V

Specifications of Banbury Mixers[a]

Size of machine	Approx. capacity (lb.) (specific gravity = 1)	Standard mixing		Approx. over-all dimensions (ft.)		
		Rotor (r.p.m.)	Motor (hp.)	Length	Width	Height
BR	2.5	230/155	25/17	8	3	6.6
OOC	4	125/16.5	30/15	9	4.5	7.9
I	25	120/60	100/50	12	5.8	11.5
3A and 3D Unidrive	100	70/35	400/200	17	10.5	15
9D Unidrive	270	43/21.5	500/250	20	11.5	13
11 and 11D Unidrive	340	40/20	800/400	24	14	17.3
27 and 27D Unidrive	750	32	1500	36	15	22

[a] Data provided by Farrel Corp. (F1).

FIG. 19. Size 11 Banbury mixer (courtesy of Farrel Corp.).

extruder. The rotors are extended on one end by short screws which feed the material into the mixing zone. At the opposite end of the mixer the stock is discharged through an adjustable slit die. Adjustment of the die opening controls the pressure in the mixing chamber. Energy input is varied by controlling output per turn (feed rate/rotor speed), and discharge pressure. The extruded ribbon may be cooled and cut with a guillotine shear or sent to a pelletizing extruder.

As is true for the older Banbury mixer, the rotors are fully supported by bearings at each end, and both rotors and chamber may be heated or cooled.

Figure 20 shows a continuous mixer with the barrel retracted to expose the screws for cleaning. The feed inlet is at the left end of the rotors. Capacity data are given in Table VI (F2).

3. Roll Mill

The two-roll mill is also designed for materials of the highest viscosity. The two contrarotating, differential-speed rolls subject material in the narrow clearance to intense shear. With a batch mill the mixing achieved depends on the skill of the operator, who repeatedly has to slit the layer of stock, peel away strips, and lay them across the face of the roll. This redistribution is required

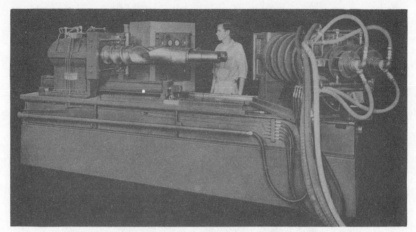

FIG. 20. Farrel continuous mixer, size 9FCM (courtesy of Farrel Corp.).

Table VI

Maximum Production Rates of Continuous Mixers[a]

Material Rotor diam. (in.)	Production rates (lb./hr.)			
	4	6	9	14
Polyvinyl chloride	1200	5000	8000	20,000
Polyethylene	600	3000	5000	10,000
(color master batch)				
Tire tread master batch	1000	3500	6000	10,000
(rubber and carbon black)				

[a] Data provided by Farrel Corp. (F2).

because there is virtually no lateral transport on a mill. Modern rubber and plastics mills, however, have a number of auxiliaries which make them close to automatic, and allow continuous operation. For rubber mills these devices include continuous strip feeders, mill blenders (Fig. 21), which take off a strip of stock and feed it back to the mill in crisscross fashion, and continuous sheeting and stacking machinery (A1). In place of the operator's knife and hands, "curling wheels" continuously roll up strips of stock, and enforce reorientation and kneading of the material. Figure 22 shows a pair of curling wheels in action, while Fig. 23 illustrates their operation.

In the case of plastics, granular stock can be fed continuously at one end of the mill, while "ploughs" turn and redistribute the molten layer as it moves along the mill. Mixed material is continuously stripped off at the far end. Figure 24 shows a modern automatic mill with ploughs lifted away from the roll surface.

FIG. 21. Mill blender (courtesy of The Akron Standard Mold Co.).

FIG. 22. Curling wheels on 26 by 84 in. mill (courtesy of The Akron Standard Mold Co.).

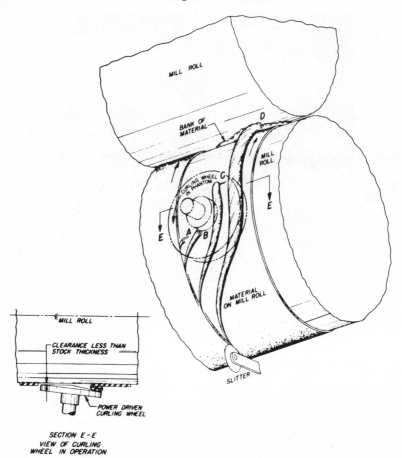

FIG. 23. Action of power driven curling wheel (courtesy of The Akron Standard Mold Co.). *A*, start of pick up and curl; *B*, approximately 1/10 sec. later, curl starting to form; *C*, another 1/10 sec. later, curl nearly formed; *D*, curl completed and advancing into mill bank.

Unlike the Banbury mixer, in which the stock is completely contained, the mill depends on adhesion to hold the stock on the rolls as a clinging "blanket." On the other hand, the finished stock should strip reasonably cleanly from the roll surfaces. These two considerations set some limits on the type of material and conditions suitable for processing.

The roll mill has a good feature for some materials in that the work is all done at the bite of the rolls, and the material can cool as it travels around with the roll to the bite again. As a result very viscous materials can be mixed at lower sustained temperatures than with almost any other mixer.

The loading of a mill can be quite slow with granular materials, such as plastic powder or pellets, because several passes may be required before the

FIG. 24. High-speed automatic mill with ploughs (courtesy of The Aetna Standard Engineering Co.).

stock is softened enough to stick to the surface and form an adhering layer. Dusty pigments or fillers may be lost to the exhaust hood over the machine or to the floor. This can create a dirty plant. Independent heating and cooling of the two rolls is provided. Often a temperature difference is maintained between the rolls so as to hold all the material banded around one roll. A 22 by 60 in. batch mill holds a charge of about 100 lb. of rubber. A cycle of 30–40 min. may be used.

Multiroll mills (usually 3 or 5 rolls) are used for chocolate and paint materials, but these are comparatively soft pastes, and the function of the mill is to disperse particles in a fluid vehicle with relatively little mixing.

4. Ko-Kneader

The Ko-Kneader physically resembles a single-screw extruder. The screw rotates continuously, but, in addition, oscillates along the axis. In the feed zone the screw has one flight, but in the mixing zone there are several flights, interrupted at intervals. The interruptions coincide with rows of teeth, or lugs, set in the barrel. As the screw oscillates the flights pass close by the lugs, forming numerous zones of high shear rate. Both screw and barrel can be heated or cooled. For cleaning, the barrel is opened "clam-shell" fashion (Fig. 25). Because the oscillating screw causes the output to pulsate, the extrudate often is fed to a pelletizing extruder. Screws from 2 to 16 in. diam. are available. In Table VII some performance data for a 4-in. machine are listed.

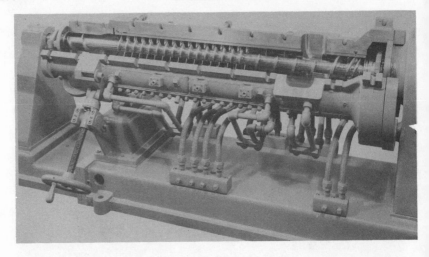

FIG. 25. Mixing chamber of the Ko-Kneader (courtesy of Baker Perkins Co.).

Table VII

Ko-Kneader Performance Data[a]

Compound mixed	Output (lb./hr.)	Residence time (min.)	Product discharge time (min.)	Net energy input (hp.-hr./lb.)
Carbon electrode paste	214	1.6–2.37	330	0.0062
Rocket fuel	420	—	85	0.011
Cereal	200	1.2	250	0.019
Battery paste	900	4.0	160	0.0037
Polyethylene				
Low density	167	1.88	320	0.191
Medium density	134	2.33	355	0.207
High density	143	2.18	360	0.185
Highest density	48	6.50	580	0.660
Polypropylene	134	2.32	360	0.173
Rubber				
Crude	60	5.20	320	0.338
Tread stock	180	1.73	240	0.139

[a] Provided by Baker Perkins Co.; 4 in. diam. machine.

5. Extruder

There are two major classes of extruders: single and twin screw. Of these, the single-screw type (Fig. 26) is the better suited for compounding of high viscosity materials insofar as energy input is concerned. This advantage is simply explained by geometry: It is possible to attach larger gears and thrust

Fig. 26. Steam-jacketed plasticating extruder (8.5 in. diam.) (courtesy of Black Clawson Co.).

bearings to a single screw than to two screws side by side. As a consequence single-screw machines generally operate with higher torque on the screw and greater shaft power input to the stock. The basic geometric disadvantage of twin-screw machines can be overcome by the use of stacked thrust bearings and specially designed gear reducers (W1, W2, W3). Such features add to the cost, however.

The conventional, helically flighted screw, whether single or paired is, as explained earlier, not an efficient mixing device. For a mixing task of any great difficulty, specially designed *mixing sections* must be provided in the melt zone of the extruder. For the single-screw machine the mixing zone is usually the last section before the stock is discharged. The multiflighted *mixing torpedo* (Fig. 27) (D3, S1), the *disk mixer*, or other types with rotor-stator elements

Fig. 27. Multiflighted mixing torpedo (courtesy of E. I. du Pont de Nemours & Co.).

Table VIII

Single-Screw Extruder Data[a]

Screw diam. (in.)	Output rates (lb./hr.)			Standard motor hp.	Thrust bearing basic dynamic load rating (lb.)	Max. standard screw speed (r.p.m.)	Heater capacity (kw.)
	Plasticized polyvinyl chloride	Polyethylene[b]	High impact polystyrene				
1.75	60–90	40–65	55–85	5–7.5	65,000	35–145	7–10
2.0	120–180	80–130	110–170	10–15	65,000	35–145	10
2.5	200–300	170–280	200–300	20–40	143,000	120–220	14–18
3.5	370–500	350–450	370–500	40–75	196,000	75–140	25–30
4.5	575–750	540–675	575–750	60–150	270,000	65–120	30–35
6	1100–1500	1000–1400	1100–1500	150–250	465,000	65–100	68–77
8	1900–3000	1700–2800	1900–3000	250–400	600,000	55–100	90–102
10	3000–5000	2800–4800	3000–5000	400–700	2,000,000	50–90	160–175

[a] Data from Prodex Corp. (P2); L/D = 20 or 24 (vented barrel).

[b] Melt index = 4.

(B1, D2, R2, S3), and the *planetary gear mixer* (H1, W4) are some devices that increase the mixing ability of a single-screw extruder. A rearward pumping section, installed in the middle of the screw, has also been employed (G1). Reducing the output per turn by means of increased discharge pressure (M2), by cooling the core of the screw, or by starve-feeding, is a way to improve the mixing performance of an existing extruder.

Output and other data for a series of single-screw extruders are shown in Table VIII (P2).

Twin-screw machines, in turn, are of two major types: those with *tangential, contrarotating screws* and those with *intermeshing* screws that, perforce, rotate in the same direction because the flights of each mesh with the channels of the other.

In the case of tangentially arranged screws, there is some exchange of material between the adjoining channels, and the screws do act somewhat after the fashion of the rolls of a roll mill. Also rearward pumping sections, and close-clearance "milling" sections are sometimes placed in the screw after the plasticating section (S4). Table IX (S4) presents some figures for contrarotating, twin-screw extruders.

Table IX

Capacities of Contrarotating Twin-Screw Extruders[a]

Screw diam. (in.)	Motor power (hp.)	Output (lb./hr.)	Product
2	10	120	Polystyrene
2	10	90	Cellulosic film
3.5	30	400	Polyvinyl chloride (plasticized)
3.5	30	300	Cellulosic film
14	500	2000–4000	Clay

[a] Data from Street (S4).

The inherent mixing ability of machines having intermeshing screws is greater since the melt is periodically displaced from the channel by the mating flight and forced into a figure-eight-shaped flow path. Stacked, triangular *kneading disks* (W1) or mixing blades shaped somewhat like the blades of a Banbury mixer are provided in this type of equipment to improve mixing performance. A side view of a devolatilizing, twin-screw machine (intermeshing screws) with kneading disks is shown in Fig. 28.

In Table X performance data are shown for a series of intermeshing, twin-screw machines (W1, W2, W3).

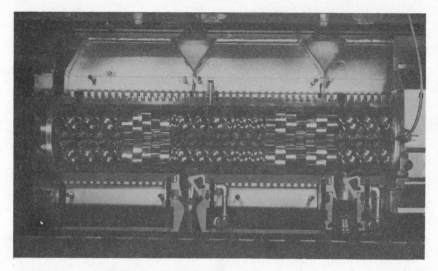

FIG. 28. Devolatilizing, intermeshing twin-screw extruder (courtesy of Werner and Pfleiderer Corp.).

Table X

Characteristics of Intermeshing, Twin-Screw Extruders[a]

Screw diam. (in.)	2.08	3.27	4.72
Screw length (in.) (flighted)	(variable)	43.3	59
Maximum r.p.m. (standard gearing)	300	150	150
Drive power (hp.)	48	80	200
Max. heating capacity (kw.)	18	24	44
Output (lb./hr.)			
Polyethylene			
Low density	380	330–550	1100–1500
High density	220–380	330–550	1000–1300
Polystyrene	290	270–500	1300–1500
PVC			
Plasticized	—	440–880	2200–2700
Rigid	—	220–400	—

[a] Data from Werner and Pfleiderer Corp. (W1, W2, W3).

6. Frenkel C-D Mixer

The Frenkel C-D mixer resembles the single-screw extruder in general arrangement, having a rotating screw inside a stationary barrel. In the C-D unit, however, both barrel and screw are divided into frustroconical sections, and both have machined, helical channels. The channel on the screw is of opposite hand to that on the barrel so that rotation of the screw moves material forward in both (Fig. 29).

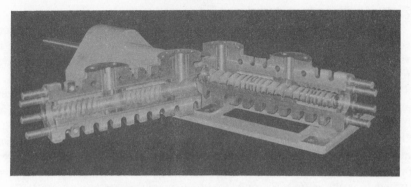

FIG. 29. Frenkel C-D mixer (courtesy of Frenkel C-D Corp.).

In a section the depth of the channel in the screw decreases from a maximum at the inlet to zero at the outlet, while the depth of channel in the barrel changes in exactly the opposite way (Fig. 30). The reverse arrangement, having increasing depth on the rotor, may also be used. Thus material is gradually transferred from one channel to the other while passing through the section. In both channels, the material circulates in the helical pattern described earlier for extruder screws. But as the stock is transferred, layer by layer, from the contracting to the expanding channel, material initially on the inner streamlines of the contracting channel forms the outer layers of the expanding channel. Thus all of the stock spends some time in a region of higher shear rate (outermost streamlines) and of greater proximity to a channel wall. This ordered rearrangement of the material fosters both mixing and heat transfer.

A wide variety of substances can be processed, including bakery mixes, pasty filter cakes, plaster, asphalt, clay mixes, soap, synthetic fibers, spinning solutions, and thermoplastic resins. Devolatilizing sections can be provided. Both rotor and barrel can be heated or cooled.

Diameter ranges from 2 to 10 in. Output of the latter unit is in the range of 5000–15,000 lb./hr., depending on the task.

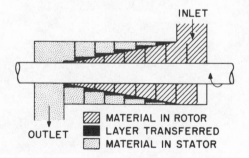

FIG. 30. Layer transferral action in C-D mixer (courtesy of Frenkel C-D Corp.).

7. Double-Blade Batch Mixers

For the majority of applications, the type with horizontal blades is preferred, since this design is suitable for processing materials spanning a wide range of viscosities (Fig. 31). The double-blade machine consists essentially of a rectangular trough, curved at the bottom to form two half cylinders or saddles. The two blades are driven by suitable gearing on one or both ends of the trough, and sweep the entire area of each half cylinder during each revolution. The blades revolve toward each other at unequal speeds, usually in the ratio of about 1.9:1, and produce mixing by imparting both transverse and lateral motion to the charge. Clearance between the blades and trough is close (about 0.03 in.) to eliminate stagnant regions. Intensive kneading action is provided as the material is pulled and squeezed against blades, saddle, and side walls.

Double-blade mixers fall in the class of medium intensity mixers best suited to mixes having viscosities of approximately 5.0 to 500 poises. Of the different styles of blades available the *sigma* (Fig. 31) is the most widely used. It has a good mixing action, readily discharges materials which do not stick to the blades, and is relatively easy to clean when sticky materials are processed. In some cases the overlapping sigma blade may prevent a tendency of the material to form a cylinder about the axis of each blade, which rotates without any mixing taking place. This is a problem when the batch must be started with a relatively small amount of material.

FIG. 31. Universal mixer with sigma blades (courtesy of Baker Perkins Co.).

The *dispersion* blade (Fig. 32) is used for dispersing fine particles in a viscous mass. It is the easiest blade of all to clean but does not have quite the over-all mixing ability of the sigma. Rubbery materials have a tendency to ride the blades, and a ram is frequently used to keep the material down into the mixing zone.

The *double naben* or *fishtail* type of blade (Fig. 33) is a good mixing blade,

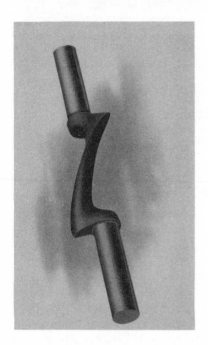

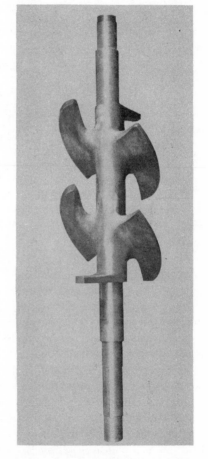

FIG. 32. Dispersion blade (courtesy of Baker Perkins Co.).

FIG. 33. Double naben blade (courtesy of Baker Perkins Co.).

but harder to clean than the sigma. Certain mixes which tend to "ride," that is, to form a lump which bridges across the sigma blade, are held in the mixing zone by the double naben blade.

The multiple-wing, overlapping blade of which the *beken* is the best known (Fig. 34) affords better performance for mixtures which start as tough, rubber-like material, because the blade cuts it into small pieces before plasticating it. This results in a smaller power peak and a lower maximum torque on the

FIG. 34. Beken blade (courtesy of Baker Perkins Co.).

machine. For very sticky masses discharge is difficult, and the beken is the hardest blade of all to clean.

The double-blade type of mixer, with either tangential or overlapping blades, offers a relatively large disengaging area, and consequently can be used for removing volatiles either at moderate pressure or under vacuum.

It is quite common with this type of mixer to have a power peak during the mixing cycle, during which the power requirement will be much higher than the average. This peak is the factor that determines the necessary strength of the machine and the size of the motor.

The trough provides the main area for heat transfer to or from the batch, and is usually jacketed only on the shell. However, on larger sizes the trough ends, and even the blades may be heated or cooled, Jackets are designed for water, steam, refrigerant, or high-temperature, heat transfer media. Electrically heated troughs are commonly used for mixing of carbon electrode pastes. Over-all heat transfer coefficients will in almost every case be controlled by the film coefficient of the material being mixed, and usually lie in the range of 40–80 B.t.u./hr.-ft.^{2}-°F.

The discharging of the finished mixture from the machine is frequently tricky, and various techniques have been devised to suit the types of mix involved. Free flowing or liquid mixes are usually discharged from valves in each portion of the trough. Stiff mixes and pastes may be discharged through a hinged door, and the batch assisted through the opening by the movement of the blades. Another common method used to remove the batch is by tilting the trough and working or pouring the mixture over a discharge lip formed in the trough edge.

The trough design affects the maximum power that can be transmitted to the charge and the ease of filling or discharge.

Kneaders are available in size ranges from laboratory bench scale models to production machines of over 2000 gal. capacity. Heavy duty units are available for batches of up to 1000 lb., and may require motors delivering more than 2000 hp. Table XI presents performance data, while Table XII shows typical size and power ranges for sigma-blade mixers.

Table XI

Typical Performance of Double-Blade Batch Mixers[a]

Charge	Type blade	Batch size (lb.)	Cycle time (min.)	Product discharge temp. (°F.)	Net work on product (hp-hr./lb.)
Refrigerator enamel	Dispersion	1500	130	150	0.12
Fiber glass in polyester resin	Single curve	12	15	90	0.012
Polyvinyl chloride plastisols	Sigma	24	40	90	0.016
Carbon electrode (anode)	Beken	60	30	305	0.025
Brake lining	Sigma	9	40	104	0.014
Vinyl record mix	Dispersion	9	6	300	0.05
Polyisobutylene— wax coatings	Dispersion	8.5	120	200	0.47
Dog food	Sigma	10	7	85	0.003

[a] Data from Baker Perkins Co.

Table XII

Specifications of Double-Blade Batch Mixers[a]

Capacity in gal. Working	Total	Over-all Dimensions (ft.) Length	Width	Height	Drive power (hp.)	Weight (lb.) (less motor)
$\frac{1}{4}$	$\frac{1}{2}$	—	—	—	$\frac{1}{2}$	375
$4\frac{1}{2}$	7	—	—	—	5	1,350
10	17	—	—	—	10	3,500
20	35	8	6	5.5	20	6,000
50	85	10.5	6.8	7	40	9,000
100	170	11.5	8.8	8.5	50	17,000
150	260	13.5	9	9	60	18,500
200	350	14.8	9.8	9.8	70	24,000
300	550	16.3	10.8	10.8	100	28,000

[a] Data from Baker Perkins Co.

Vertical-shaft mixers comprise the other major type of double-blade mixer, and include planetary mixers and Pony mixers, as well as heavier duty, twin-shaft machines. Most designs include provision to remove the mixing blades from the mixing can either by tilting or by vertically lifting the assembly from the cans. Some designs have removable mixing cans while others have a tilting arrangement for pouring the product out of the cans. With removable cans the time for cleaning between batches is minimized, and production is greater than for comparable horizontal mixers. One other advantage is the elimination of contamination problems, because there is no contact between the packing glands and the material being mixed. Figure 35 shows a 150-gal. machine. Table XIII presents capacity and power data for this category.

IV. Scale-Up Techniques

It is common practice to arrive at the type and size of machine required for a given production task by performing that task in a small scale mixer, and

FIG. 35. Vertical mixer (150 gal.) (courtesy of Baker Perkins Co.).

scaling up the observed values of production rate and power consumption along with the physical dimensions. The maintenance of some sort of geometric similarity, of approximately constant shear rates, residence times, and the like, greatly simplifies the art of mixer design. Most machine vendors offer laboratory models on which small scale tests may be run. Semiempirical rules for scaling dimensions, and for predicting the performance of the full scale device, have been developed for many types of mixers. In some cases performance data are known for a range of sizes, and for a variety of materials. This type of experience is most helpful to reinforce predictions.

Scale-up practices for some common types of mixers are described in the following paragraphs.

Table XIII

Specifications of Batch Mixers with Vertical Blades[a]

Working capacity (gal.)	Total capacity (gal.)	Drive power (hp.)
$\frac{1}{8}$	$\frac{3}{16}$	$\frac{1}{4}$
1	1.5	1
5	7.5	3
20	30	7.5
150	225	40
300	450	60

[a] Data from Baker Perkins Co.

A. SINGLE-SCREW EXTRUDERS

Hydraulic theory leads logically to scale-up of extruder dimensions according to *geometric similarity* (C3). That is, each dimension (flighted screw length, channel width and depth, etc.) should be scaled by the ratio of diameters, while the helix angle (or ratio of flight lead to diameter) should remain constant. If the screw speed is held constant this procedure leads to the increase of open discharge flow with the cube of the ratio of diameters. At the same time the shear rate and residence time are not changed, so that the imposed shear strain should be constant. This approach might be suitable for some melt pump designs, although it leads to rapidly increasing sensitivity of output to pressure. The pressure gradient that a screw can develop is inversely proportional to h^3.

In the case of plasticating extruders another serious objection to maintaining geometric similarity is the inability of a screw of large absolute channel depth to completely melt and homogenize the stock. At large channel depths thermal equilibration within the material can become so incomplete that unmelted particles pass completely through the machine. The distribution of heat generated by friction in the outer layers of solid or partially plasticated stock, or conducted in through the barrel wall, requires more time the deeper the bed of solids that must be melted. This process, combining *mixing* of portions of unequal temperature and *thermal conduction*, governs the uniformity of temperature of the extrudate, and profoundly influences the mixing performance of the extruder.

Mixing performance suffers as uniformity of stock temperature deteriorates, because, very simply, differences in temperature create differences in viscosity. The goodness of mixing achieved may be determined by the shear strain imposed after the point at which the last particle of minor component is melted. For any element of volume, mixing cannot begin until melting is over. Consequently, calculations of shear strain from hydraulic theory do not suffice to describe behavior of plasticating extruders, unless melting is fairly complete in the first half of the screw length.

A common practice among extruder vendors is to scale up channel depth, and simultaneously reduce operating speed of plasticating extruders, by the square root of the screw diameter (M1). This procedure leads to increases in output with D^2, and in residence time and shear strain with $D^{0.5}$. Shear rate remains constant, while the response of output to pressure, although increased, is held to a tolerable level. Figure 36 shows the thread depths (metering zone) recommended by one plastics manufacturing concern (D4). Figure 37 presents data published by a number of manufacturers on output (of polyethylene) and installed drive power for single-screw extruders.

B. TWIN-SCREW EXTRUDERS

Scale-up of output and power for one series of intermeshing, twin-screw extruders is illustrated in Fig. 38 (W1, W2, W3). In this case both quantities increase faster with diameter than is ordinarily the case with single-screw machines. The relatively long melt zone together with the superior mixing action of this type of extruder permit scale-up of screw dimensions closer to geometric similarity.

C. CONTINUOUS MIXER

Scale-up of the Farrel continuous mixer is carried out with an exponent of diameter of *ca.* 2.6 (Fig. 39). The value of installed drive power per unit of throughput ranges from 0.1 to 0.2 hp.-hr./lb. (for the products shown here).

D. DOUBLE-BLADE MIXERS

1. Prediction of Power

Experience with double-blade mixers has led to the relation:

$$Z \propto R_t \cdot L_t \cdot B \cdot \bar{N}$$

where $\bar{N}$ is the average speed of the two blades. This relation holds true when scaling up or down so long as the machines have approximately the same configuration of blades, whether they are tangential or overlapping.

It also has been found that extrapolations from laboratory size mixers smaller than 2.25 gal. working capacity are not reliable, because of the difficulty of measuring power for the small machine with sufficient accuracy.

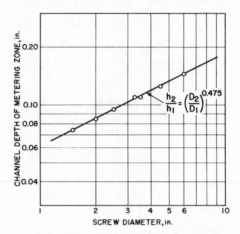

FIG. 36. Relationship of channel depth in metering zone to diameter for single-screw extruders [data provided by E. I. du Pont de Nemours & Co. (D4)].

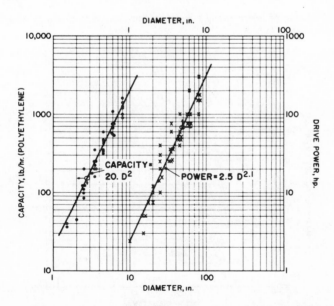

FIG. 37. Output and power of single-screw extruders as functions of diameter.

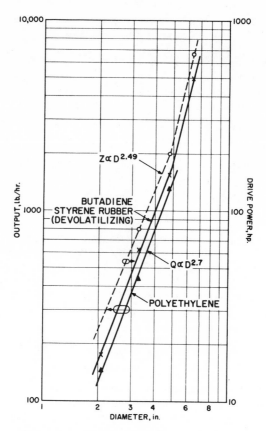

FIG. 38. Scale-up of intermeshing twin-screw extruders [data provided by Werner and Pfleiderer Corp. (W1, W2, W3)].

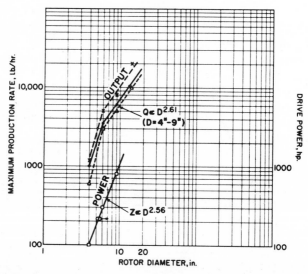

FIG. 39. Scale-up of output and power of Farrel continuous mixer [data provided by Farrel Corp. (F2)]. KEY: (X) polyvinyl chloride, (△) polyethylene-color master batch, and (●) rubber-carbon black master batch.

To obtain the net horsepower of a test mixer we must have the "no-load horsepower" of the machine. This is usually recorded prior to the test, but with the machine thoroughly warmed up by running empty for at least one half hour and with cooling or heating medium circulating in the jackets as the test may require.

EXAMPLE

Predict net horsepower required for a 100-gal. production mixer from test data on a 2.25-gal. laboratory mixer:

Test Data		Known Data
2.25 gal. mixer		100 gal. mixer
$R_{t1} = 2.75$	Trough radius (in.)	$R_{t2} = 9.38$
$L_{t1} = 10$	Trough length (in.)	$L_{t2} = 33$
$B_1 = 1.25$	Blade wing depth (in.)	$B_2 = 5.5$
$\bar{N}_1 = 43$	Average r.p.m.	$\bar{N}_2 = 25.$
Gross hp. = 1.9		
No-load hp. = 0.4		
$Z_1 = 1.5$	Net horsepower	$Z_2 =$ unknown

where r.p.m. − fast blade = 56   r.p.m. − fast blade = 30
      r.p.m. − slow blade = 30   r.p.m. − slow blade = 20

$$\bar{N}_1 = \frac{56 + 30}{2} = 43 \qquad \bar{N}_2 = \frac{30 + 20}{2} = 25$$

$$Z_2 = \frac{Z_1 \times R_{t2} \times L_{t2} \times B_2 \times \bar{N}_2}{R_{t1} \times L_{t1} \times B_1 \times N_1}$$

$$Z_2 = \frac{1.5 \times 9.38 \times 33 \times 5.5 \times 25}{2.75 \times 10 \times 1.25 \times 43} = 43 \text{ hp. (net)}$$

From this calculation we may safely conclude that the 100-gal. production mixer, designed for and driven by a 50-hp. motor, will fill the requirements. The 7-hp. allowance for no-load horsepower is more than ample for this size of machine. Indeed the no-load power seldom exceeds 10 hp. even for larger mixers up to 1000-gal. working capacity.

The same method is used for scale-up of single- or double-blade continuous mixers. Of course, for single-blade mixers, the actual speed of the blade is used, rather than an average r.p.m.

2. Prediction of Cycle Time

A complete cycle of mixing includes the times for charging, mixing, and discharging. The last includes scraping of trough and blades if the batch is of a sticky nature.

Experience with the product being mixed will dictate the time allotted for charging and discharging. Goodness of mixing is dependent on the number of revolutions of the blades, regardless of the size of mixer. In other words, if the average speed of a laboratory mixer is 43 r.p.m., and the average speed of a production mixer is 25 r.p.m., then the mixing time of the production machine will be 43/25 times greater than that of the laboratory size machine.

EXAMPLE

Laboratory mixer, 2.25 gal.		Production mixer 100 gal.
Charging	2 min.	10 min.
Mixing	20 min.	20 × 43/25 = 35 min.
Discharging		
and cleaning	3 min.	15 min.

When heating or cooling is needed, the problem becomes more involved, because the jacketed areas of larger machines are smaller in relation to the volume of material being mixed. Therefore, the cycle time is not only dependent on the speed of blades but also on heat transfer rate as well.

Since in most cases the required heat transfer data such as specific heats, film coefficients, etc., are not available, the most practical method of predicting the cycle time in a production mixer is from a test in a laboratory mixer. Since the initial and final batch temperatures, coolant types, and other variables affecting heat transfer should be the same at both scales, the cycle time will be, approximately at least, proportional to batch size, and inversely proportional to power input and heat transfer area:

$$\theta_2 = \theta_1 \left(\frac{W_2}{W_1}\right) \left(\frac{A_1}{A_2}\right) \left(\frac{\bar{Z}_1}{\bar{Z}_2}\right)$$

where

$$\theta = \text{mixing plus heating time, and}$$

$$\bar{Z} = \text{time-average power input.}$$

To this mixing time must be added the charging, discharging, and cleaning times.

If the mixer is used to devolatilize the charge, the disengaging velocity of the vapor may be the limiting factor. Consequently, any laboratory test should be run as near the maximum disengaging velocity as possible. When this is done

$$\theta_2 = \theta_1 \left(\frac{W_2}{W_1}\right) \left(\frac{\sum_1}{\sum_2}\right)$$

where θ = mixing time.

V. Evaluation of Goodness of Mixing

Tests for "goodness of mixing" are usually quite specifically related to some criterion of performance. The electrical properties, e.g., dielectric strength, dissipation factor, are used to characterize insulating compounds. Environmental stress crack resistance, weld-line strength, or low temperature brittleness are used to measure uniformity of polymer mixtures.

Somewhat more fundamental assessments are made by microscopic examination of microtomed sections (see Figs. 12 and 13) or of molded or extruded films. Lumps and streaks may be counted or measured, or the specimen may be ranked against a series of standards. "Tracer" particles may be introduced in one of the components and their distribution followed (Figs. 12 and 13). Films may be scanned with a microdensitometer, and a trace of optical density recorded.

The American Institute of Chemical Engineers has set up a "test procedure for paste and dough mixing equipment" (A2) which should help to standardize methods.

VI. Summation

The technology of mixing of viscous material is highly advanced in many areas of commercial practice, but only rudimentary as a field of engineering. The engineering treatment will require a synthesis of rheology, hydraulics, heat transfer, and equipment geometry. The first three subjects, at least, are susceptible to mathematical analysis. The ability to predict mixing behavior from rheological data and heat transfer performance from thermal properties and machine geometry promises rewards of increased mixing efficiency and surer scale-up. The effects of differences in melting points and activation energies of the components need elucidation. The first effect shifts the time at which flow of one component begins. Differences in activation energies change the temperature at which a favorable viscosity ratio obtains. The utility of simplified mathematical models such as an "isothermal-isoclinal mixer," operating at constant temperature and shear rate, should be explored. What are the consequences if the components differ in elasticity? Should shear rate and temperature be "programmed" during the mixing cycle? Questions abound. Mixing remains in a condition similar to distillation before the work of McCabe and Thiele.

Acknowledgments

The authors wish to acknowledge the assistance of Mr. D. A. Wheeler and Mr. L. Yablonski of the Baker Perkins Company in the preparation of the section of this chapter on double-blade mixers.

List of Symbols

a ratio of pressure flow to drag flow in plane parallel to channel

$$\frac{Q_{\mathscr{P}}}{Q_{\mathscr{D}}} = \frac{h^2}{6V\mu}\left(\frac{dP}{dz}\right)$$

A area for heat transfer
B depth of blade wing of double-blade mixer
c ratio of pressure flow to drag flow in transverse plane

$$\frac{h^2}{6T\mu}\left(\frac{dP}{dx}\right)$$

$\cos\alpha$ a direction cosine of normal to original interface at point (x, y, z)
D outside diameter of screw; diameter with subscript
f local velocity in axial direction
F dimensionless function of H
h channel or thread depth
H reduced height in channel, y/h
k proportionality constant between forward velocity and position in channel
l cube side length
L length of screw; length in axial direction with subscript
N screw speed, revolutions per unit time
P pressure
q local volumetric rate of flow
Q volumetric rate of flow
r striation thickness, average separation of two like interfaces between two components; radius with subscript
R outside radius
s local velocity in plane transverse to channel
s' cube side area
S total interfacial area between components of mixture
T component of barrel surface velocity perpendicular to flight
u displacement vector of a point on original interface
U circumferential velocity of extruder screw $= \pi DN$
v local velocity in plane parallel to channel
V component of barrel surface velocity parallel to channel; total volume of system
w width of channel
W weight of mixer charge
x, y, z coordinate axes
x coordinate axis perpendicular to flight
y height in channel measured perpendicular to screw root; coordinate axis perpendicular to screw root
Y volume fraction
z coordinate axis parallel to channel
Z power

GREEK LETTERS

γ net amount of shear supplied; shearing strain
$\dot{\gamma}$ shear rate
θ residence time

ρ reduced radius
μ viscosity
Σ area for disengagement of volatiles
τ shear stress
ϕ screw helix angle

SUBSCRIPTS

a major component
b minor component
B base resin, as for a pigment concentrate
C channel
D die
E unflighted screw extension
P referring to a particle
R extruded rod
t trough (of double-blade mixer)
T transverse plane
V parallel plane
0 referring to original state of system
1 initial position in channel, referring to streamline beginning at this position; laboratory scale
2 referring to streamline complementary to 1; plant scale
λ axial direction

SUPERSCRIPT

Overbar average of the quantity.

References

(A1) Akron Standard Mold Company, Akron, Ohio.
(A2) A.I.Ch.E. "Test Procedure for Paste and Dough Mixing Equipment." American Institute of Chemical Engineers, New York, 1964.
(B1) Beck, H. G. (to the General Tire and Rubber Company), "Method of and Apparatus for Homogenizing Plastic or Plasticizable Materials." U.S. Patent 2,813,302 (November, 1957).
(B2) Bird, R. B., Stewart, W. E., and Lightfoot, E. N., "Transport Phenomena," p. 67. Wiley, New York, 1960.
(C1) Carley, J. F., and Strub, R. A., *Ind. Eng. Chem.* **45**, 970–973 (1953).
(C2) Carley, J. F., Mallouk, R. S., McKelvey, J. M., *Ind. Eng. Chem.* **45**, 974–979 (1953).
(C3) Carley, J. F., and McKelvey, J. M., *Ind. Eng. Chem.* **45**, 985–988 (1953).
(D1) Danckwerts, P. V., *Appl. Sci. Research* **A3**, 279–296 (1951–1953).
(D2) de Laubarede, L. M. H., "Homogenizing Device for Extruding Machine." British Patent 738,784 (October, 1955).
(D3) Dulmage, F. E. (to The Dow Chemical Co.), "Plastics Mixing and Extrusion Machines." U.S. Patent 2,753,595 (July, 1956).
(D4) Du Pont Company, Wilmington, Delaware, "Suggested Screw Design for the Extrusion of Alathon[4] Resins." Information Bull. X-100.

[4] Registered trademark of E. I. du Pont de Nemours & Co., Wilmington, Del.

224 H. F. Irving and R. L. Saxton

(F1) Farrel Corp., Ansonia, Connecticut, "Banbury Mixers." Bull. 207B.
(F2) Farrel Corp., Ansonia, Connecticut, "Farrel Continuous Mixers." Bull. 227.
(G1) Gregory, R. B., and Heston, E. E., "QLT, A New Concept in Extrusion." *SPE Tech. Papers* X, XXIV-1 (1964).
(H1) Henning, G. E., and Weitzel, E. W. (to Western Electric Company), "Apparatus for Simultaneously Advancing and Milling Plastic Compounds." U.S. Patent 2,754,542 (July, 1956).
(M1) Maddock, B. H., Extruder scale-up theory. *SPE Journal* pp. 983–984 (November, 1959).
(M2) Maddock, B. H., Nalepe, H. J., and Zurkoff, B., Controlled pressure (valve) extrusion. *Rubber and Plastics* (December, 1958).
(M3) Metzner, A. B., *In* "Processing of Thermoplastic Materials" p. 41 (E. C. Bernhardt, ed.). Reinhold, New York, 1959.
(M4) Mohr, W. D., Saxton, R. L., and Jepson, C. H., *Ind. Eng. Chem.* **49**, 1855 (1957).
(M5) Mohr, W. D., Saxton, R. L., and Jepson, C. H., *Ind. Eng. Chem.* **49**, 1857 (1957).
(P1) Perry, J. H. (ed.), "Chemical Engineers Handbook," 3rd ed., p. 1197. McGraw-Hill, New York, 1950.
(P2) Prodex Corp., Fords, New Jersey, Bull. E-8.
(R1) Rabinowitsch, B., *Z. physik. Chem.* (*Leipzig*) **A145**, 1 (1929).
(R2) Ross, O. E., and Williams, J. L., "Mixing Head for Plastics Extruders." U.S. Patent 2,705,131 (March, 1955).
(S1) Schnuck, C. F., and Whittum, W. C., "Rotor for Blendors." U.S. Patent 2,680,879 (June, 1964).
(S2) Spencer, R. S., and Wiley, R. M., *J. Colloid Sci.* **6**, 133–45 (1951).
(S3) Stanley, T. A. (to Imperial Chemical Industries, Ltd.), "Improvements on or Relating to Mixing Apparatus." British Patent 841,743 (July, 1960).
(S4) Street, L. F., *India Rubber World* **123**, 58 (1950).
(W1) Werner and Pfleiderer Corporation, New York, "Twin-Screw Continuous Kneader and Compounder for the Plastics Industry." Brochure F.2186.1. IX. 63A.
(W2) Werner and Pfleiderer Corporation, New York, "Twin-Screw Extruder Type ZSK 53/L for Laboratory Use." Brochure F.2185 e 1. X. 63. Oe.
(W3) Werner and Pfleiderer Corporation, New York, "Continuous Drying of Rubber." Brochure F. 2204 e. 1,5. 1.64 B 10503.
(W4) Willert, W. H. (to Frank W. Egan and Company), "Plastics Extruder with Mixing Head." U.S. Patent 2,785,438 (March, 1957).

CHAPTER 9

Suspension of Solids

Emerson J. Lyons

General American Transportation Corporation
Chicago, Illinois

I. Introduction

Previous chapters have defined the broad scope of mixing, the elements by which it is achieved, and the flow patterns, velocities, and power obtained under various conditions. This information can now be related to the more circumscribed area of the suspension of solids.

A slurry is a dispersion of particulate solids in a continuous liquid phase which is sufficiently fluid to lend itself to circulation by a mixing device.

225

Slurries range in character from low-viscosity Newtonian systems to high-consistency psuedoplastic or plastic mixtures.

This chapter will define and discuss the types of suspension problems encountered in industry, the properties of slurries in the suspended state, and the types of equipment which are applicable. Selection of appropriate mixers and vessels based on slurry characteristics and bench scale tests will also be discussed. Reference will be made to specific industrial systems.

The first known illustration of a mechanical device for slurry handling appeared in Agricola's *De Re Metallica* (A1) printed in 1553. This woodcut pictured a four-blade version of the straight-blade turbine in common use today. It was mounted centrally in a round, wooden tank, was operated by water power through a set of wooden gears, and suspended crushed ore for the leaching and washing of gold. Its inclusion in Agricola's text would indicate that turbine and paddle mixers had been in use in the northern European mining industry for several prior years.

Essentially, the art of large-scale industrial mixing was nonexistent prior to the sixteenth century, the small batches of liquid or slurry being processed would be manually stirred with a rod or paddle. When mixing operations increased in importance as an industrial tool for new processes and larger scale operations, mixer design progressed from paddles through multiblade constructions to the turbines, propellers, and more exotic shapes in current use. The majority of applications in this early period were solids suspensions such as the leaching and treating of metallurgical slurries, salt dissolving and crystallizing, and the extraction of organic infusions in the growing dye industry.

The first known attempt to evaluate the variables affecting solids suspension was reported by White and co-workers (W2, W3) using a flat paddle in an unbaffled tank. They found that sand concentration (at any given location) decreased with an increase in sand particle size and increased with impeller speed up to a "critical point" which they suggested as a criterion for effectiveness of agitation. At this point the sand concentration reached a maximum. At higher agitator speeds, lesser suspended sand concentrations were obtained because of centrifugal separation of the sand. The use of baffles could have prevented this separation.

Another criterion for effectiveness of agitation was developed by Hixson and Tenney (H2) who used a four-blade turbine in an unbaffled tank. Using a grab sampling tube to measure sand concentrations, they calculated a "mixing index" as the arithmetic average of the "percentage mixed" values for each sample location. The percentage mixed values varied between 0 and 100 and were expressed at any sample location as the ratio of the concentration of the component present in smaller concentration than the average in the vessel as a whole, to this average concentration. The mixing index was found to increase with impeller speed to approximately 90 at which point it leveled off. They observed that the mixing index increased with the viscosity of the continuous phase.

Further study of solids suspension in viscous liquids by Hirsekorn and Miller (H1) showed "complete" suspension[1] was affected by vessel dimensions, liquid viscosity, particle size and settling rate. These authors observed that the portion of liquid nearest the surface was free of solids when complete suspension was achieved. The mixture below this slurry-liquid interface was essentially uniform. Zwietering (Z1) reported a similar solids-free surface layer at complete suspension, which was further verified by Weisman and Efferding (W1). By increasing the speed above that required for complete suspension, solids could be dispersed throughout the entire liquid volume. The ultimate, maximum suspension would be limited, however, by swirl in an unbaffled tank as mentioned above.

Oyama and Endoh (O2) and Hixson and Tenney (H2) also observed a critical speed above which particle concentration became essentially constant. They used a light transmission method to determine the degree of suspension of sand and resin particles. Calderbank and Jones (C1) used a similar light transmission method.

II. Variables Which Affect Uniformity of Solids Suspension

Vessel geometry, impeller construction, and operating speed have already been mentioned as significant factors in the design of an effective mixer for suspending a slurry homogeneously. These properties of solid-liquid systems also affect uniformity of solids suspension:

1. particle density;
2. solids concentration;
3. density of liquid phase;
4. size, size range, and shape of solid particles;
5. viscosity of liquid phase; and
6. hindered settling as a corollary to items 2 and 3 above.

As a practical example of hindered settling, a high-density slurry of powdered iron in water can be more readily maintained in a homogeneous state than a low concentration of coarsely ground chalk. In spite of the greater gravity difference between powdered iron and water, particle size and concentration create a "hindered settling" environment evidenced as an apparent viscosity of the iron slurry, which condition is not present with the free-settling chalk-water mixture.

---

[1] It should be noted that "complete" suspension refers to the instant at which all solids are in circulation and none is resting on the vessel bottom.

III. Scope of Application

Solid suspensions encompass the complete range of systems from simple, particulate solids in a liquid, to flocs, semidissolved crystalline structures, and slurries of fibrous and mat forming materials. Suspensions include both soluble and insoluble solids in either low- or high-viscosity liquid media. Typical industrial applications are listed below.

A. Simple Suspension

This category covers the uniform distribution of an essentially insoluble solid in a liquid medium for the purpose of uniform withdrawal of slurry from the vessel (sand in water), adsorption (coagulant floc or ion-exchange resin in solution to be clarified), or resuspension after decantation (washing of crystals or precipitates such as cellulose acetate).

B. Particle Reduction

Shear forces greater than those which are present in simple suspension may be imposed to reduce agglomerates to the original individual particles. Examples are repulping of filter cake and fiberizing wood-pulp stock. Actual reduction of particle size such as wet grinding of pigments or shredding of newsprint may be the objectives of a process.

C. Dissolution or Leaching

Suspension may be used to obtain rapid dissolution of a wholly or partially soluble solid in a liquid. The mechanics of these functions will be explored later in this chapter.

D. Reaction

When chemical reactions are involved, the intensity of agitation may vary from simple suspension to a more vigorous, turbulent recirculation to reduce the resistance of the film surrounding the solid particle. Suspension of a solid catalyst in a reaction mixture and dissolving of lead in caustic to make sodium plumbite are examples.

E. Solids Formation

When a solid is produced from a liquid mixture, very mild agitation may be required for production of a floc or moderate shear forces may be needed to control particle size for crystallization from a magma. Vigorous agitation may be required in polymer formation to keep the polymer particle size small. An example is the production of rubber curd from latex.

IV. Impellers and Circulation Patterns

All of the common types of mixing devices are used for solid suspension. Selection of suitable devices is a function of vessel shape and the physical operation to be performed. A description of the various elements and the circulation pattern they generate was given in Chapters 3 and 4 of Volume I. Reference is also made to this subject by Calderbank in Chapter 6 of the present volume and in an earlier article by Lyons (L1). The fluid circulation pattern becomes most critical with solids present. Satisfactory design for the extremes of free fall and hindrance in settling necessitates a knowledge of the physical characteristics (see Section IV, C) of the mixture and the operating characteristics of the impeller types available.

A. IMPELLER CHARACTERISTICS

Impellers for suspension fall into two basic categories depending on the method by which they generate motion in the mass. The first and simplest is represented by the several types of paddles such as the single or multiple straight-blade designs, gates, and anchors. Each provides mass motion by physically pushing the material in its path and entraining the adjacent mass in a generally rotary motion. Little or no velocity head is imparted to the fluid. The second, more commonly used types include propellers, turbines, cones, and disks, each of which develops a velocity head causing the fluid mass to be discharged at a velocity greater than the surrounding mass. The impelled fluid, plus entrained adjacent material, flows in a definite pattern through the vessel and returns to the "eye" of the impeller. Most slurries are of a non-Newtonian nature and are dependent on an imparted velocity in excess of a "critical" value to insure flow of the combined mixture. In propellers the stream is axial; in all other types, it is a centrifugally developed, radial discharge. The pitched turbine uses a combination of both axial and radial forces to produce a diagonal flow with an angle dependent on the degree of pitch, number of blades, blade width, and rotational speed.

Selection of the preferred geometry will depend on the concentration of solids in the slurry, flow characteristics of the mixture, vessel shape, and physical operation to be achieved. A mass which cannot be pumped centrifugally obviously must be transported by a paddle-type element. Pumpable slurries will be transported more freely by one type of impeller than another as a function of the internal resistance to flow of the slurry and of the circulation pattern to be achieved (see Section B).

Liquid and air jets are also used for conveying slurries vertically in vessels and tanks. One popular design in the minerals industry is the Dorr agitator (Fig. 1) adaptable to large diameter tanks. A central air lift conveys the slurry to a radially rotating launder or trough above the liquid surface which distributes the slurry uniformly and allows it to gravitate down to the tank bottom for recycling. An air lift is less frequently used today since large single or multiple impellers have been successfully applied to large tanks. The Pachuca tank used in ore leaching functions in the same manner as the Dorr agitator with a central air lift but without a launder or a bottom rake. It is normally a tall, cone-bottomed tank with a height-to-diameter ratio of two or higher. Air lifts are also used to convey slurry from one vessel to another in the countercurrent leaching of uranium ores.

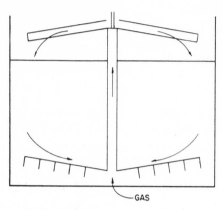

FIG. 1. Dorr agitator—curved rakes at bottom of vessel move solids to hollow shaft where gas lifts slurry to rotating distributor arms.

B. CIRCULATION PATTERNS

Circulation within a vessel for the purpose of solids suspension is one of two kinds: "upwardly directed," which physically lift the dispersed phase, or "universally directed," which distributes the solids into the fluid.

The patterns developed by various impellers are illustrated in Figs. 1 to 12 (see also Chapter 4 of Volume I). The types shown in Figs. 2 to 6 produce an upward lifting stream. The turbines in Figs. 5 and 6 are the most effective types for the difficult suspensions of fast settling, granular solids in a low-viscosity fluid such as sand in water. The turbine in the position directly on the vessel bottom provides the maximum stream velocity at the vessel location where maximum lifting velocity is required. Further, the discharge is radial, sweeping the vessel bottom and making only one directional change at the vessel wall to flow vertically. Stator blades (Fig. 5) provide a straight, radial discharge in all directions from the center, hence an essentially vertical pattern without a rotational component. Loss of velocity in the stators is less than 5 % when properly designed. With wall baffles (Fig. 6) some rotary motion is observed between impeller and vessel wall. The raised turbine (Fig. 7) produces the familiar figure-eight pattern and depends on the lower recycle stream to

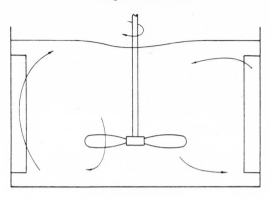

FIG. 2. Propeller.

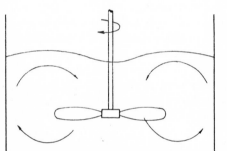

FIG. 3. Propeller in non-Newtonian mixture.

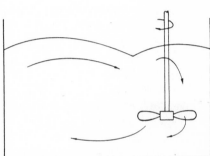

FIG. 4. Off-center propeller.

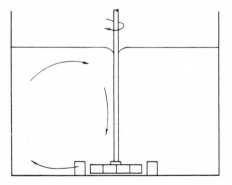

FIG. 5. Turbine with stator blades.

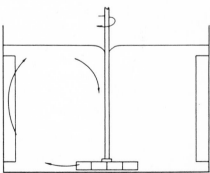

FIG. 6. Bottom-mounted turbine in baffled vessel.

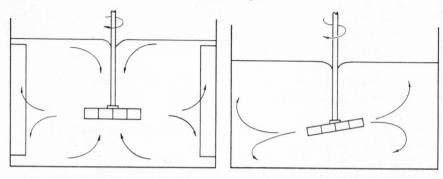

FIG. 7. Raised turbine in baffled vessel.   FIG. 8. Tilted turbine in non-Newtonian slurry.

lift slurry up to the bottom eye of the impeller for redistribution. This pattern is adequate for the average industrial application.

When located on the vessel bottom, a turbine with back-sloped blades is an effective, fibrous slurry circulator. When located higher in the vessel, a turbine may be tilted (see Fig. 8) to extend the circulation pattern of a viscous slurry. The depth of the figure-eight pattern is thereby increased and the vertical rate of interchange of the mixture through the plane of the impeller is multiplied, the factor being dependent on the character of the slurry. The horizontal discharge of the turbine makes it more effective on slurries, particularly fibrous, in shallow, large diameter tanks, than other impeller types. Two typical examples are the suspension and chemical treatment of paper pulp in stock and "broke" chests, and of cotton linters prior to acetylation.

Figure 9 shows a specialized turbine used primarily in the dispersion of sludge in waste treatment service in which sludge concentration and density are low and the volume treated is large.

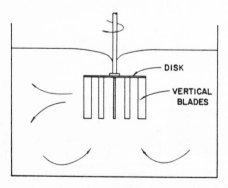

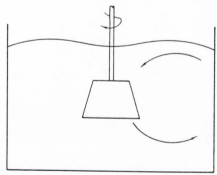

FIG. 9. Vorti-mix turbine (see footnote 2, p. 251).   FIG. 10. Cone mixer.

The high-speed disk is a circular, horizontal plate with shallow, radial vanes at the periphery, or circumferential convolutions simulating blades, which is operated in the 1800 to 3500 r.p.m. range. It is dependent on a high-velocity radial discharge to effect suspension. It is limited in size and volumetric capacity. The disk is particularly applicable to non-Newtonian mixtures of fibrous or mat-forming materials such as cellulose acetate or nitro-cotton, wood or paper pulp in small-scale equipment. Imposed stream velocities are sufficiently above the critical velocity of such slurries to momentarily convert them into free-flowing fluids.

The cone impeller in Fig. 10 is especially well suited to fibers because of its smooth contour and the vertical separation of intake and discharge points. It does not have sharp corners opposed to the line of stream flow which cause fiber hang-up. The cone impeller also has a greater height of circulation pattern than a turbine at the same slurry circulation rate.

The propeller, Figs. 2 to 4, projects a stream downwardly which scours the vessel bottom and rises in the outer, annular space to the surface. To be effective, it must impart sufficient velocity to the entire volume discharged from the impeller to pass through a 180° directional change with a residual velocity exceeding the settling rate of the solids. By using a large-diameter three- or four-blade propeller with a wide blade face, velocity may be imparted to a large mass of slurry on each rotation. For $T/D = 4$ the moving slurry column with its entrained annulus usually represents 30 to 40% of the cross-sectional area of the container and readily motivates a return flow pattern in the remaining space to the vessel wall. The propeller is more effective in slurries in which hindrance or apparent viscosity of the mixture assists in the support of the solid particles. It is frequently used in the suspension of fine particle metallurgical slurries. When slurry viscosity is increased, propellers lose circulation capacity sooner than curved-blade turbines.

A propeller can also be used effectively for fibrous-slurry pumping. It, too, has an essentially corner-free contour in the line of stream flow.

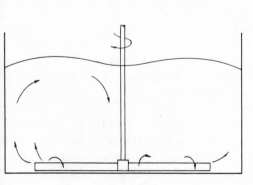

FIG. 11. Paddle mixer.

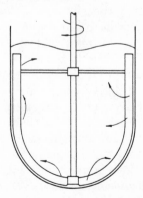

FIG. 12. Anchor mixer.

Each of the above types is readily adaptable to the pumpable class of dispersions which are partially self-supporting due to viscosity or hindered-settling effects. The paddle types in Figs. 11 and 12, though they do not produce stream velocities sufficient for suspension, can be applied to this category of self-supported solids. The anchor (Fig. 12) and multiblade paddle are used in high-viscosity mixtures, usually non-Newtonian, which cannot be caused to flow, such as polymers of the Bingham or pseudoplastic type. The paddle is frequently used for simple suspensions in large tanks in which homogeneity is not required but the solids are sufficiently light and flocculent to be easily dispersed. The flocculater (Fig. 13) is typical of this class. Homogeneity is required, but low liquid turbulence is needed to preserve the floc. The horizontally mounted paddles physically move from the bottom to the top of the channel providing gentle lifting and circulation throughout the vertical cross section.

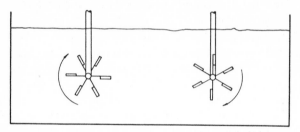

FIG. 13. Paddle-wheel flocculator.

A free-liquid layer between the suspended-pulp surface and the liquid surface as the critical suspension state is achieved, illustrates the limited circulation pattern of impellers. The suspension height is determined by the impeller speed (discharge volume and velocity) and by the viscosity or con-sistency of the slurry. Figures 14 to 16 illustrate the change in pattern with increasing viscosity for various impeller types. In each instance, the vertical pattern tends to flatten with increasing viscosity.

In Fig. 15, the height of the suspension, h, represents the solids at the critical suspension speed with the vessel bottom swept clean as several in-vestigators have observed. The increment h' represents the clear layer of liquid between slurry interface and liquid surface. The slurry interface (hence, the height of slurry suspension) can be raised to equal $h + h'$ by increasing the impeller speed and the recirculating stream velocity. An alternative action would be to reduce the depth of slurry to h, all other dimensions remaining constant. A third alternative would be to change the tank and impeller dimensions at the same slurry volume so that the suspension-height to impeller-diameter ratio, h/D, and the impeller-diameter to tank-diameter ratio, D/T, remain the same.

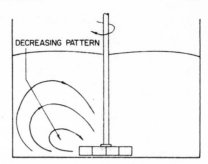

Fig. 14. Effect of slurry viscosity on solids suspension height for a bottom-mounted turbine.

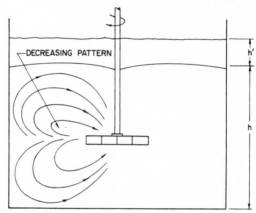

Fig. 15. Effect of slurry viscosity on solids suspension height for a raised turbine.

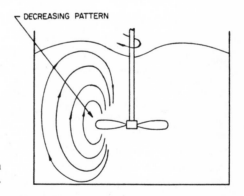

Fig. 16. Effect of slurry viscosity on solids suspension height for a propeller.

The latter procedure is normally followed in commercial design for an obvious advantage. Since power increases roughly as the cube of the speed (turbulent regime), maintaining the speed at the minimum (critical) for suspension and altering the vessel and impeller geometry conserves the power required for complete suspension. In this example critical speed would be selected to produce a uniform dispersion (distribution) as well as a complete suspension within the slurry layer.

C. SLURRY CHARACTERISTICS

Suspension of solids in a liquid medium will be obtained when the rising velocity of the liquid phase equals or exceeds the settling velocity of the particles. Fluid velocity and direction are a function of impeller selection, as previously described, and of vessel contour. Particle settling velocity is a function of gravitational force, fluid drag, and several hindering factors referred to below.

Settling velocities of particles may be calculated from empirical correlations of drag coefficient and particle Reynolds number. These dimensionless groups are defined as follows:

Drag coefficient:

$$C = \frac{gm_p(\rho_p - \rho)}{\rho\rho_p A_p u_t^2} \tag{1}$$

Particle Reynolds number:

$$(N_{Re})_p = \frac{D_p u_t \rho}{\mu} \tag{2}$$

The development of this correlating method appears in the *Chemical Engineers Handbook* (P2). Examples of drag coefficient and Reynolds number correlations are found in references (P1, E2, T1).

The following relationships between C and N_{Re} summarize the results of a correlation (P2) for settling velocities for spheres.

	Reynolds number range	*Correlating equation*
Stokes' law (laminar settling)	$0.0001 < (N_{Re})_p < 2.0$	$C = 24/(N_{Re})_p$
Intermediate law	$2.0 < (N_{Re})_p < 500$	$C = 18.5/(N_{Re})_p^{0.6}$
Newton's law (turbulent settling)	$500 < (N_{Re})_p < 200{,}000$	$C = 0.44$

An obvious condition affecting slurry suspension is the density difference between liquid and solid. When all other factors are constant, the settling rate bears a directly proportional relationship to this differential.

Another major influence is the size and shape of the solid particles. Most of the work reported on the effect of particle size has been qualitative (H1, H2, R1, W1, W3) since the slurries studied were of mixed diameters. One paper (L1) covers the practical aspects and limits of commercial-scale equipment based on industrial experience. Porcelli and Marr (P3) studied the suspension of discrete particle sizes in a propeller-stirred baffled vessel but they did not obtain a satisfactory correlation of results. They suggest gross flow rate (propeller discharge plus entrained flow) as the major function in single-particle suspension, represented by the factor q_T/u_t, and ascribe variations in their data to turbulence effects of different propeller sizes.

Industrial slurries normally involve a broad band of particle size, often extending from fractional-inch granules to slimes in the micron range. The following generalizations may be made concerning the effect of particle size on the terminal velocity of settling solids, and hence on the fluid velocity required for particle suspension:

1. Terminal velocity increases with an increase in
   a. density of the discrete particle, and
   b. mass of the discrete particle.
2. Terminal velocity decreases with an increase in
   a. cross-sectional area of the particle,
   b. irregularity or roughness of the particle surface,
   c. spread in the size spectrum of the particles,
   d. concentration of solids in the slurry,
   e. free space between particles, and
   f. viscosity of the fluid medium.

Conditions 2c, d, e lead to a state commonly termed "hindered settling," which is used advantageously in minerals practice for size separation and classification (G2). In true free settling, the minimum allowable distance between particles is 10 to 20 particle diameters, equivalent to a volume dilution of water to solids of 1 to 10. Interparticle distance falls off rapidly with decreasing dilution (Fig. 17). It therefore follows that an increase in the per cent of solids, represented by a larger number of particles, rapidly reduces the interparticle distance and rapidly increases the frequency of particle collisions.

The lowest concentration at which hindered settling occurs is not readily

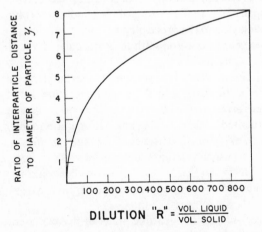

FIG. 17. Effect of dilution on interparticle distance.

defined. As a rule of thumb, it can be assumed to exist with solids concentrations starting at about 40% by volume and higher. Hindered settling will occur with a lower per cent of solids under laminar (Stokes) flow than in turbulent (Newtonian) flow conditions, and it will vary with particle size and size distribution, and with the attractive forces between particles. Thomas (T4) has indicated a direct relation between the hindered-settling characteristics of small particle suspensions and the concentration and degree of flocculation of the suspension.

Hindered settling is evidenced by a reduction in settling rate and by non-Newtonian behavior of the slurry. With suspensions of particles in the micron range, the above tendency is abetted by agglomeration. Under laminar flow (Stokes) conditions, the maximum particle diameter for flocculation (agglomeration) is given (G1) by the equation:

$$a = 5.06l^{1/5}\left[\frac{y + 1 - \sqrt{y(y + 2)}}{y^2(y + 1)}\right]^{1/5} \tag{3}$$

where a = the particle diameter in microns,

l = the settling depth in centimeters,

y = ratio of the distance between particles to the particle diameter.

Particles above the diameter, a, will settle at least the distance, l, before flocculating even if the probability that colliding particles will adhere is unity.

The hindered-settling characteristics of a slurry are particularly applicable to the flotation of minerals, which is a special aspect of solids suspensions not included in the scope of this chapter. However, the conditions elaborated herein apply equally well to flotation which is not limited to the laminar-flow range. Flotation involves a three-phase gas-liquid-solid state with a carefully controlled, physical environment.

Typical commercial applications incorporating flocculation as a factor in fine particle suspensions are the mixing of clay and water (blunging), washing and bleaching of clays, and the processing of starch. Michaels and Bolger (M3) in their study of shear rate in the settling of flocculated kaolin include many references on flocculation.

Crystal form influences hindered settling. For example, needlelike or plate-type crystals are particularly subject to agglomeration and to interparticle friction. Irregular forms, such as titanium sponge or crushed glass (platelettes with sharp, corrugated edges), exhibit an agglomeration which is purely due to structural interlocking of the particles. The resultant effect is an obstruction to free flow of the suspended mass and an increase in needed power. For example, a 50% slurry of −325 mesh, ground-silica glass in water required 6.2 hp. for circulation whereas the calculated power based on the known characteristics of the slurry was 3.3 hp.

Low bulk density fibrous and leaflike materials have a viscous resistance to flow which is largely due to interparticle friction and particle agglomeration.

Except in very thin suspensions, it is impossible to circulate a mass of this type as a uniform distribution in a liquid with a free-flowing separation of the fibers. A relatively high circulation velocity in excess of the critical velocity of the mass is necessary to create and maintain flow, which is usually laminar in character under the influence of the elongated particles. Fibrous slurries are typical of suspensions, mentioned previously, which behave as non-Newtonian fluids.

The viscosity of a slurry may be described in several ways (M2, T3, W4), such as:

1. viscosity of the suspending liquid;
2. suspension viscosity, a function only of solids concentration for Newtonian suspensions;
3. apparent viscosity, a function of both shear rate and solids size, shape, and concentration for non-Newtonian suspensions; and
4. limiting viscosity at high shear rates, a function only of solids characteristics of non-Newtonian suspensions.

Non-Newtonian viscosity, as used above, refers to Bingham-plastic and pseudoplastic flow characteristics. Most slurries exhibit the property of becoming less viscous as the shear rate is increased beyond a limiting value as a result of increased fluid velocity and turbulence. Dilatant mixtures are less frequently encountered and approach a condition of immobility as solids concentration increases, which removes them from the class of flowable materials that can be circulated by impeller-type mixers. Starch and sand slurries exhibit dilatancy above a minimum solids concentration.

D. SLURRY SUSPENSION DATA

A number of experimental studies of the variables affecting uniformity of particle suspension have been carried out. A variety of solids, liquids, and impellers were used in baffled and unbaffled vessels. White and co-workers (W2, W3) studied sand-water suspension in unbaffled tanks. They determined a critical point at which local particle concentration leveled off with increasing speed. They concluded that swirl in an unbaffled vessel inhibited production of a uniform concentration. Hixson and Tenney's mixing index similarly leveled off at 90 (values from 0 to 100) with increasing pitched-blade turbine speed when suspending sand and water in an unbaffled tank.

Zwietering (Z1), in an extensive investigation of impeller types, used sand and sodium chloride with a variety of liquids in baffled tanks. He defined a critical speed, N_c, at which no solids remained on the vessel bottom, and noted that this state did not necessarily provide homogeneity. Correlation of his data yields the equation:

$$N_c = \psi \left(\frac{T}{D}\right)^t \frac{g^{0.45}(\rho_p - \rho)^{0.45}\mu^{0.1}D_p{}^{0.2}(100R)^{0.13}}{D^{0.85}\rho^{0.55}} \qquad (4)$$

where R is the weight ratio of solid to liquid, and ψ and t have values dependent on impeller type and relative blade height; ψ ranged from 1.0 to 2.0 (averaging 1.5) and the exponent, t, is approximately 1.4.

Pavlushenko et al. (P1) arrived at a similar relationship using sand and iron ore with a variety of liquids at a 1 to 4 weight ratio. Tests were run in an unbaffled vessel with three-blade square-pitch propellers. Their correlating equation for critical speed, N_c, is as follows:

$$N_c = 0.105 \frac{g^{0.6} \rho_p^{0.8} D_p^{0.4} T^{1.9}}{\mu^{0.2} \rho^{0.6} D^{2.5}} \tag{5}$$

The values of ρ and μ are for the liquid medium. Dilute suspensions of insoluble solids in viscous liquids (to 50,000 c.p.s.) were investigated by Hirsekorn and Miller (H1) with paddle agitation.

Weisman and Efferding (W1), using single and multiple turbine impellers for thorium oxide and glass bead suspension, took special notice of the slurry and liquid surfaces. They studied the effect of solids concentration, particle size, impeller elevation, and system geometry on height of the slurry interface. Their "suspension criterion," s, describing a condition similar to the critical speed of Zwietering and of Pavlushenko, was correlated by

$$S = \frac{ND^{0.85}}{\nu^{0.1} D_p^{0.2} (g\Delta\rho/\rho)^{0.45} (100R)^{0.13}} \tag{6}$$

Oyama and Endoh (O2) developed an impeller power per mass of liquid relation for their so-called critical impeller speed at which "complete" suspension was attained:

$$\left(\frac{P_c g_c}{M}\right)^{2/3} = \frac{g D_p^{1.3} (\rho_p - \rho)}{16\rho} \tag{7}$$

where P_c = impeller power at critical speed,
M = weight of slurry in the vessel,
D_p = particle diameter,
ρ_p = solids density,
g = gravitational acceleration,
g_c = a gravitational conversion factor.

Their conclusions were limited by the very small percentage of sand and resin solids dispersed.

The above investigations extended from laminar to turbulent conditions. In each study, it was noted that additional speed above the critical was required for complete distribution of solids phase throughout the entire fluid mass. Each of the correlating relationships is valid for the conditions used in its derivation. However, all of the studies mentioned were conducted on a relatively small scale, and extrapolation to commercial scale will necessitate a re-examination of the applicable constants. Velocity loss due to entrainment of fluid by a high velocity stream and viscous drag do not scale up as a straight-line function of size in geometrically similar larger vessels. The velocity decreases with physical distance traveled by the slurry from the point of maximum velocity at the impeller tip, and the circulation pattern shrinks relative to impeller diameter as equipment size is increased. A higher discharge velocity or a larger impeller-to-tank-diameter ratio is required to maintain a similar flow pattern in the larger vessel.

E. POWER

Power theory and equations for mixers handling non-Newtonian fluids apply to solids suspended in liquids. These relationships were covered in Chapter 3 of Volume I where it was pointed out that equations and correlations for non-Newtonian fluids have not achieved the degree of accuracy or comprehensiveness as those for Newtonian fluids.

V. Effects of Vessel and Auxiliary Equipment on Suspension

Reference has been made to circulation patterns and tank geometry as important physical factors in the successful suspension of solids. Each of the major types of impellers has a flow pattern associated with it. The tank geometry should conform to this pattern to obtain a minimum impedance to the natural direction of stream flow. Optimum design results from a proper balance of impeller type and tank shape for the specific slurry to be treated. More than one combination is serviceable in any slurry system but, of these selections, one will usually be most appropriate for adaptation to process requirements or physical plant limitations or power economy.

A. TANK-TO-IMPELLER DIAMETER RATIOS

The size of the impeller bears a rather obvious relationship to the vessel diameter and the concentration, density, and particle characteristics of the solid. Referring to Figs. 14 to 16 each diagram shows typical flow contour lines from the impeller discharge and return. As slurry density and corresponding apparent viscosity increase, the illustrated circulation pattern or sphere of influence contracts toward the impeller leaving the outermost contours immobile. Impeller diameter can be increased to enlarge the circulating mass until it again encompasses the stagnant zone at the wall, both for a rising stream from a propeller (Fig. 16) and for a radial turbine stream (Figs. 14 and 15) which deflects vertically at the wall and folds back to the impeller intake. A maximum diameter is reached when the annular free space between impeller tip and wall becomes too restricted to permit free flow of the rising propeller stream or easy transition of the radial turbine discharge to a vertical direction with minimum velocity loss due to impingement on the wall surface.

Selection of the optimum tank-to-impeller-diameter ratio, T/D, is based on experience with slurries of like characteristics. This ratio may be predetermined by bench scale tests with or without assistance of the curves in Fig. 18. Formulation of a general equation to determine this ratio has not been attempted, except for a few very limited and specific slurry systems, and is impractical considering the number of variables involved.

The curves in Fig. 18 are based on experience with a wide variety of industrial applications, and may be used with reasonable accuracy for turbine size selection or as "order of magnitude" sizing for propellers. They will apply to slurries of particulate solids other than fibrous materials at an average solids concentration of 20 to 40% by weight. As an example, assuming a slurry of 120 mesh solids of 3.6 specific gravity in water, the curves indicate a 2.6 tank-to-impeller-diameter ratio based on specific gravity and a 4.65 ratio based on particle size. The arithmetic mean of these values, 3.63, is the ratio indicated for design. This value can be used directly for design, or for bench scale model tests, when solids concentration exceeds the indicated limits or when the solids

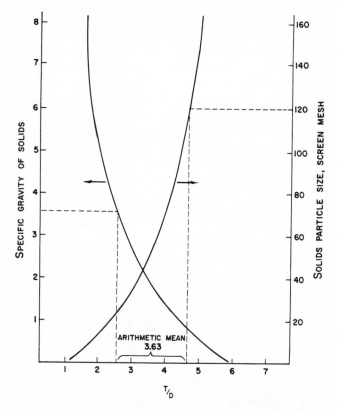

FIG. 18. Effect of specific gravity and particle size on optimum T/D ratio.

exhibit peculiarities of form affecting settling rate. With less than 10% solids concentration, the ratio would be used directly or increased over the value indicated by the chart and the impeller rotational speed and discharge velocity increased to insure lifting of the "unsupported" particles. Since the curves are based on slurry suspension with back-sloped turbines, the effect of increased discharge velocity can also be achieved at the same impeller speed by reducing the blade curvature to obtain a straight radial blade, but at a lesser power efficiency.

With greater than 40% solids concentration, the effect of hindered settling and of increasing apparent viscosity within the slurry will also tend to increase the optimum T/D ratio. This increased apparent viscosity provides support for the dispersed solids, reduces settling rate, and requires less energy from the recirculated fluid to maintain suspension. A suggested method of bench scale testing, using the T/D ratio as a guide, is mentioned in Section VI. Similar support for the suspended solids is provided by viscosity of the liquid phase, as previously mentioned. Apparent slurry viscosity, as used here, would be in the range of 1 to 15 poises, Newtonian or non-Newtonian (but not dilatant).

A further correction of the tank-to-impeller-diameter ratio can be made (Fig. 19) if the measured apparent viscosity increases above 15 poises. Dilatant slurries or suspensions demonstrating a "thixotropic" reaction can be evaluated only by bench scale testing. It is practically impossible to put a numerical figure on slurries in which the particles provide an impedance to flow as a result of fibrosity or friction due to surface irregularities. In any case,

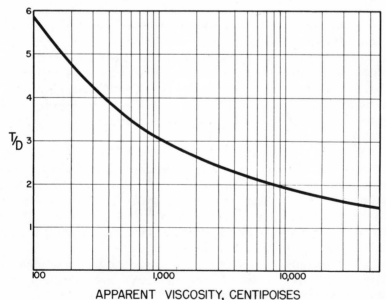

APPARENT VISCOSITY, CENTIPOISES

FIG. 19. Effect of viscosity on optimum T/D ratio.

as indicated by the viscosity curve, a tank-to-impeller-diameter ratio bel
would not be used because at this point the impeller diameter has reach
maximum beyond which free flow of the return streams at the wall of the ve
essentially disappears into noncontinuous eddies.

The following statements are a summary of the above comments: A mean
tank-to-impeller-diameter ratio for suspension of solids may be derived from
Figs. 18 and 19 and applied directly to equipment design. The engineer with
practical experience in slurry mixer operation will sense how slurry will move
in a vessel in response to a stirring device and will be able to apply the ratios
more broadly before resorting to laboratory tests. However, tests should be
conducted on fibrous, irregularly shaped or dense (over 40%) mixtures,
particularly if they exhibit a dilatant or thixotropic shear pattern.

B. BAFFLES

Baffling is essential in solid suspension to direct flow streams vertically and
prevent the particle size classification that results from uncontrolled swirl.
Tank baffling is discussed by Lyons (L1) and by Leedom and Parker (Chapter
11). Stator blades are illustrated in Fig. 5, and typical wall baffling is shown in
Figs. 6 and 7.

As the effect of viscous drag in the slurry increases, a viscosity level is
reached at which the wall baffles must be angled in the direction of the impeller
rotation in the tank to reduce swirl without directly impeding the recycled
stream returning from the wall to the impeller. At a still higher viscosity level,
the drag becomes sufficient to provide its own baffling effect and create a
spiraling stream pattern which returns to the impeller after a partial rotational
circuit in the tank. This pattern is inherent in the flow contours of Figs. 14 to 16
in which the effect of drag on vertical circulation pattern is depicted.

For Newtonian fluids, the approximate transition viscosity is 20,000 cp. for
changing from radially directed to angled vertical baffles, and 60,000 cp. for
changing from angled baffles to no baffles (L1). When plastic or thixotropic
properties are present in slurries, the transition would occur at lower values of
the apparent viscosity.

A square or rectangular tank is sometimes used on free-flowing slurries. At
viscosities above 5000 cp., the square corners provide sufficient baffling to
prevent swirl and force vertical flow. At low viscosity (under 5000 cp.),
supplemental vertical baffles centered on each of the four walls may be
necessary to control swirl. If the slurry is not free flowing, a square tank is not
satisfactory as the slurry will stagnate in the corners preventing complete
recirculation and permit the center of the mass to rotate.

Sometimes in round tanks, the vertical mixer shaft (or angle-mounted
propeller shaft) will be located off center (one-fourth to one-half of the tank
radius) for circulating thin slurries. The resulting distorted flow patterns
reduce swirl and produce vertical movement.

TANK SHAPE

The condition of complete suspension in which a layer of clear liquor exists above the suspension zone was discussed in Sections I and IV. It was suggested that the condition of complete homogeneity of a true suspension could be obtained by reducing the batch volume to that of the suspension volume, or increasing the tank and impeller diameters to reduce the charge depth to that equal to the normal height of complete suspension. Homogeneity could also be obtained by increasing the impeller speed but this would require additional power. Selection of the correct vessel shape, therefore, serves to provide a true suspension of the entire mass at minimum power.

The ratio of vessel height to vessel diameter should be selected on the basis of the height of the circulation pattern. As a rule of thumb, the height of the circulation pattern is two impeller diameters for an average slurry of 20 to 40% solids concentration with a low resistance to flow. For lesser solids concentration, the height of pattern would be increased except under the condition of fast settling and widely dispersed (free-falling) particles.

The previously suggested bench scale test to evaluate tank-to-impeller-diameter ratio for materials exceeding 40% solids or of a thixotropic nature will also establish the height of the impeller pattern. With this data, the optimum tank shape for the most economical mixer design can be obtained. The designer has a choice between a larger vessel diameter with single impeller construction or a smaller diameter and greater tank depth with multiple impellers. The ultimate selection may be influenced by process considerations such as the method and location of feed and drainage, batch versus continuous flow, the requirements of other concurrent processing operations, such as heat transfer, and chemical reaction.

Multiple impeller installations as illustrated by Fig. 20 will function to provide a homogeneous distribution of slurry throughout a tall vessel. Tank-to-impeller-diameter ratio considerations remain the same. The number and location of impellers on the shaft are a simple geometric evaluation based on an overlapping of the vertical heights of the circulation patterns of the impellers. To assure that slurry circulated to the top of the circulation pattern of the bottom impeller will be picked up by the bottom half pattern of the next impeller in vertical sequence, this overlap should be equal to approximately one-fourth of the circulation depth on the side of each impeller adjacent to the interchanging streams as indicated by Z_1, Z_2, and Z_3 in Fig. 20. Fibrous slurries tend to form a concrete circulating stream which has well-defined outer boundaries. In order to penetrate this surface and provide intermixing, the adjacent impeller streams must overlap about one-third of the circulation depth on the side of each impeller. This type of multiple construction only applies to radial discharge impellers. Axial flow impellers provide a uni-directional vertical pattern which is not as suitable for slurry mixing in tall vertical tanks because more power is required by axial flow than radial flow turbines to obtain satisfactory suspension.

Multiple impeller installations of the type indicated by Fig. 21 involve the same considerations as a single impeller installation, taking into account tank-to-impeller-diameter ratios capable of insuring free flow at the wall most distant from the impellers. This construction has a distinct advantage in the processing of large volumes of metallurgical slurries, permitting vessel diameters two or more times greater than could be circulated by a single, central impeller. Also, it has an advantage in power economy. Normally, any two units will maintain sufficient fluidity for the process function to continue while the third unit is being repaired. In this construction, the eccentric location of the mixers eliminates swirl from the circulating mass. Baffles, if used, will be located on the wall adjacent to each mixer to provide a vertical direction to the impeller discharge and assist in a more uniform suspension of the solids.

It is common practice in wet phosphoric acid manufacture to create continuous, staged-mixer compartments by partitioning a large, circular tank into pie sections (see Fig. 22). The mixer shaft is located approximately at the center of area of the horizontal cross section of the compartment. The section is usually sufficiently angular to obviate the need of baffling. The tank "diameter" in the tank-to-impeller-diameter ratio is the mean of the distance

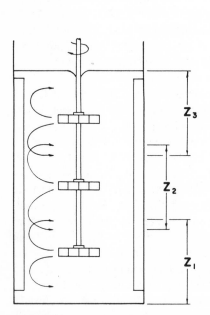

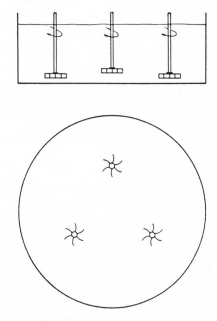

FIG. 20. Multiple turbine mixer.   FIG. 21. Multiple turbines for large diameter tank.

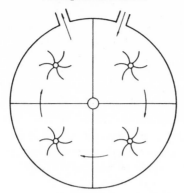

FIG. 22. Compartmented circular tank.

between the vertical shaft and the farthest corner, and the distance from the shaft to the nearest side wall.

D. SPECIAL CONSTRUCTIONS

The draft tube is frequently used as a conduit to control and assure vertical flow and distribution of solids. It is the basis of construction of a Pachuca tank for alumina digestion and similar light slurry treatment. An air lift in the central shaft tube provides vertical flow with return by gravity in the annulus between draft tube and vessel wall. The identical principle is applied in the Dorr agitator with the slurry being uniformly distributed over the annular space by a rotating launder (Fig. 1). However, both of these devices are less efficient in the average application than impeller designs which can now be adapted to larger vessels and can produce several times more rapid interchange throughout the mass with a more uniform distribution.

A propeller mounted in a draft tube (Fig. 23) duplicates the action of the air lift at a greater circulating velocity. The resultant forced circulation can be used to advantage for promoting heat transfer in vertical pipes or coils in the annular space. However, under average conditions heat transfer can be obtained at equal rates using a coil in a tank with an impeller and baffles. Where unusually high rates of heat transfer are required, the use of a series of radially mounted plate-shaped surfaces (Fig. 24) obviates the need for a draft tube and provides a large transfer surface. The plates serve as baffles forcing a linear flow over the transfer surface, and they provide a free path for slurry recycle without solids accumulation on obstructions or in low turbulence zones.

Processing of three-phase mixtures is typified by the flotation method of mineral beneficiation. Flotation is accomplished with a variety of impeller-agitated and air-circulated types of equipment (T2). Cell shapes are usually rectangular, 10 to 100 cu. ft. in volume, and well baffled.

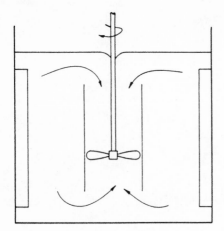

FIG. 23. Propeller in a draft tube.

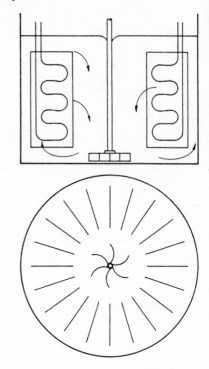

FIG. 24. Baffling by heat transfer plates.

The gas-liquid leaching of ores, such as ammoniacal treatment of nickel ore, has necessitated construction of large vessels with constant gas content in all parts of the slurry in the vessel. If gas is pressure fed by a sparger into the impeller discharge stream, uniformity of gas flow to all parts of the vessel occurs incidental to slurry distribution and the design parameters remain unchanged. The sparger is designed to exceed the impeller diameter by one to two inches. It is located flush with the top plane of the impeller blades and is perforated on the bottom side for downward discharge and drainage (see Fig. 25).

For larger gas volumes or reactions requiring intense shear forces between all three phases, a vaned disk or a gas-type impeller (see Fig. 26) is used with gas fed directly to the impeller. Impeller construction incorporates a horizontal plate or disk to prevent gas from rising through the low-velocity area of the impeller. Gas may be fed by a submerged bubble cap under the impeller (Fig. 26) or by induction (Fig. 27). The impellers are rotated at 20 to 40 % above normal turbine peripheral speeds (700 ft./min.) to achieve adequate gas distribution in the slurry and radial flow of the dispersed three-phase mixture. This increased speed also compensates for the reduction in density of the slurry due to gas displacement of part of the mass within the impeller and provides an impeller discharge volume at least equal to that of the impeller operating at normal speed without gas. The effect of adding gas is observable

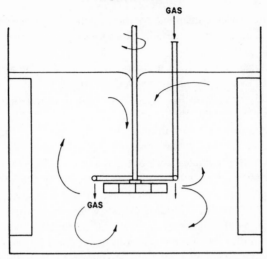

FIG. 25. Sparger gas feeder.

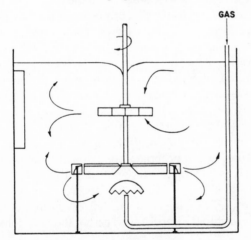

FIG. 26. Bubble-cap gas feeder and gas-type impeller (see footnote 3, p. 251).

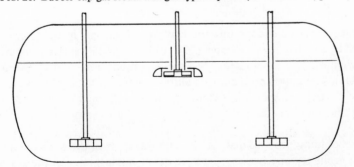

FIG. 27. Pressure leaching tank with self-inducing gas-type impeller (see footnote 3, p. 251).

as a reduction in power consumed by the impeller. The net reduction is a function of the quantity of gas circulated and will range approximately from 5 to 40%.

In the case of gas addition to a slurry, the necessity for a large volume circulation to attain complete mass movement in the vessel most frequently supersedes the design criteria for gas distribution. The impeller therefore is overdesigned with respect to the handling of the gas feed. Under such conditions, a 5% reduction in power upon addition of gas is reasonable.

Flooding of gas impellers is more pronounced in most slurries and viscous mixtures due primarily to coalescence of the gas bubbles before they reach the impeller. The larger bubbles will accumulate in the eye of the impeller sooner (i.e., at a lesser gas feed rate) than would be true with a nonviscous particle-free liquid. The condition of "flooding" exists when this accumulation of gas blows through or blocks the central zone of the impeller, thereby reducing liquid or slurry circulation to the small volume recycled by the impeller tips. This limited liquid movement is at low velocity and is usually restricted to whirls or eddy currents in the zone adjacent to the impeller periphery.

Initial coalescence of the gas bubbles can be minimized by introducing the gas in finely divided form over a broad area by a perforated sparger ring. Then viscosity and surface tension at the liquid-gas interface will help to keep the fine bubbles separated.

A similar condition of flooding frequently occurs in nongassed viscous solutions and slurries as a consequence of aspiration or cavitation. If the mass is permitted to vortex because of low-charge level or insufficient baffling, air may be aspirated into the eye of the impeller. Similarly, cavitation forces at the center of the impeller can form vapors in a slurry containing a volatile solvent. In each instance appropriate conditions of viscosity and surface tension will permit the air or vapor to accumulate into a large oblate, spherical bubble at the eye and effectively flood the impeller. Over-all circulation will disappear, the vortex will be eliminated, and power consumption will decrease. The only remedy is to stop the impeller and permit the gas bubble to rise to the surface.

E. CONTINUOUSLY FED EQUIPMENT

The theoretical considerations for continuous, staged mixing have been described by MacMullin and Weber (M1), Eldridge and Piret (E1), and others. The mechanics of continuous flow of slurries is briefly reviewed by Olsen and Lyons (O1). The major problem with slurries is maintenance of homogeneity such that the bypassing of solids from one stage to the next is prevented while insuring that the mass transported is a representative sample of the mixing stage.

Control of homogeneous distribution and transport of the mass is achieved by vessel geometry and baffling. Each vessel or mixing stage in the system is designed to produce true suspension and a total shear energy input compatible with the process requirements. Continuous mixing in a single vessel has the limitations inherent in the probability of short circuiting flow and minimum retention time. Basic to design of each vessel or stage is the chord baffle of Fig. 28. It prevents direct bypassing of liquid or solid from inlet to outlet and forces the feed to pass down through the impeller mixing zone at least once before flowing up and out behind the chord baffle. The cross-sectional area behind the baffle is established to make the rising velocity of the stream greater than the settling rate of the solids present.

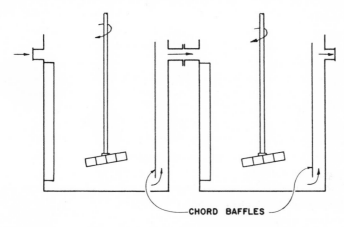

CHORD BAFFLES

FIG. 28. Two-stage continuous mixer.

Solid-liquid continuous systems are primarily concurrent flow using a single vessel or a vessel divided into compartments in series such as a circular tank with radial partitions dividing it into pie sections (Fig. 22), a vertical tank with horizontal baffles (Fig. 29), or a series of single vessels (Fig. 28). The continuous staged mixer equations of MacMullin and Weber (M1) and the refinements of subsequent authors all depend on complete uniformity of the mixture within each stage, hence, each mixer should be designed as previously suggested for complete suspension.

Countercurrent flow of solids and liquid is obtained by providing a separation step between each of the sequential mixer vessels or stages. Separation can be centrifugal or by sedimentation as in the Dorr CCD (continuous countercurrent decantation) system. A method developed by Infilco[2] uses an airlift to move solids up through a stepwise series of mixer tanks with extraction liquid moving by gravity overflow down through the tank sequence. The rotating disk contactor (RDC[3]), a disk and doughnut vertical extraction

[2] Infilco Division of the Fuller Company, Tucson, Arizona.
[3] Process Equipment Division, General American Transportation Company, Chicago, Illinois.

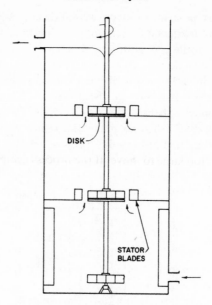

FIG. 29. Vertical-staged continuous mixer.

column, is also applicable to liquid-solid countercurrent treatment, the solids descending and liquid rising through the column.

VI. Operating Techniques

A greater degree of "art" still exists in mixing than in most other processing operations. A few observations are therefore appropriate on operating techniques developed through experience and not subject to theoretical prognosis.

A typical example is the impeller flooding, with or without gas feed, mentioned in Section V,D. When a gas feed is in excess of the flooding level, the corrective actions are:

1. Increase the impeller speed; this will provide a minor raising of the flooding level at an usually unjustified increase in power.
2. Increase the impeller diameter while maintaining the same peripheral speed; this will involve a moderate power increase.
3. Distribute the gas over more than one impeller.
4. Feed the gas by sparger in a circle closer to the impeller blade tips (0.7 to 0.9 times impeller diameter); this will prevent flooding but decrease the efficiency of gas dispersion.

When flooding results from aspiration of air from a vortexing liquid surface, the obvious corrections are to reduce or eliminate the vortex with tank baffling or by raising the surface level. When cavitation is the source of vapor flooding it is more difficult to control mechanically. A larger impeller at slower speed or vertical relocation of the impeller may be effective.

A condition frequently encountered in practice is the settling out and reslurrying of the solids due to power failure or as a planned decantation step. Reslurrying can present a serious problem if the solid particles are fine and dense and if the interfacial surface tension is low, permitting them to settle as a hard, dry cake. It also becomes difficult to reslurry particles possessing an irregular surface which allows them to become locked or matted into agglomerate masses too large to move. If the process geometry is such that the settled mass does not submerge the impeller, the solids will become gradually redispersed when the mixer is started. Lancing of the hard-packed sediments may be necessary to initiate suspension as movement of the mass can only be accomplished in the presence of enough fluid to permit a flow of slurry to start.

Frequently, by plan or by design necessity, at least the lowest impeller of a multiple impeller agitator will be submerged by the settled mass. Several procedures may be utilized to assist in suspending the mass, the simplest being injection of water or process liquid into the sediment over and around the impeller with a manual lance (Fig. 30). Or, the lance may be permanently

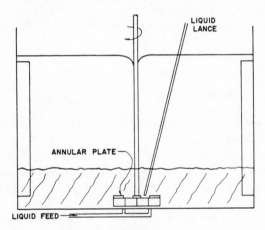

FIG. 30. Liquid lance for reslurrying solids.

installed with at least two injection ports in the tank bottom at the midpoint of the impeller blades as shown in Fig. 30. A bottom, center drain outlet through which liquid may be pumped back into the impeller will also serve. The purpose is to flood the zone of the impeller sufficiently so that a pumpable slurry is created and the impeller started. Once in motion, the recirculating slurry will collapse the cake above the impeller allowing free liquid entry progressively until the entire mass is reslurried.

A self-priming design applicable to most sediments and specifically built into decantation processes is also illustrated in Fig. 30. An annular plate ring is permanently attached to the top face of the impeller allowing a central orifice for liquid inlet. On settlement, the ring prevents solids from packing in the outer area of the impeller blades and creates a pyramidal zone between the ring and the vessel bottom which is semifluid. Enough liquid is retained in the impeller blades to allow them to be rotated until the entire mass is resuspended.

In the dissolving of solids a technique frequently used with turbine mixers takes advantage of particle attrition as an adjunct to mechanical shear to accelerate dissolution. This attrition occurs in the recirculation of dense slurries where the turbulent flow of the viscous mass creates shear planes between particle groups moving at different velocities. Interparticle friction at these shear planes has the effect of scrubbing the surface of the particles thus accelerating dissolution. An appropriate example is the dissolving of nitrocotton in lacquer manufacture or of cellulose acetate for spinning solutions. In the former case it would take a turbine dissolver $3\frac{1}{2}$ to 4 hours to produce a finished, fully dissolved lacquer of 20,000 c.p.s. viscosity starting with the complete charge of liquids and nitro-cotton. However, by withholding about one-third of solvent and diluents from the initial charge, the cotton will dissolve to form a heavy lacquer of approximately 150,000 c.p.s. in one hour due to the particle attrition effect. Dilution with the withheld solvent and diluents will produce the finished 20,000 c.p.s. lacquer in 30 minutes for a total cycle of $1\frac{1}{2}$ hours. This technique is known as the "holdback" method of dissolving.

VII. Extrapolation of Small-Scale Tests

Mixing is defined as the operation in which two or more materials (gas, liquid, and/or solid) are distributed throughout a mass in varying degrees of uniformity dependent upon the change and state to be accomplished. As applied in industry, mixing is less frequently used for the purpose of a simple blending operation, than for the promotion of some other operation such as heat transfer, reaction, leaching, and gas absorption. Although solid suspensions generally fall in the latter category, the scale-up parameters for the auxiliary operation are less critical than the scale-up parameters for producing a uniform suspension. Bench scale tests were previously suggested as the logical procedure for determining impeller-tank-circulation relationships for the specific slurry being processed. In the laboratory it is relatively simple to disperse all types of solid suspensions because adequate velocities are easy to achieve in small vessels, and the power used is relatively insignificant. The effects of velocity and power are more dramatic in the case of suspension than in other mixing operations and the differences in these characteristics in laboratory, pilot, and commercial scale operations are more exaggerated.

In a five-gallon or larger bench scale vessel, it is comparatively simple to maintain a clean swept bottom and complete suspension to the upper surface of the liquid with an impeller in the middle of the mixture. The lifting of solids by vertical suction of induced streams from a raised impeller, or by jet effect of an axial propeller stream, becomes increasingly difficult as the point of stream induction recedes from the vessel bottom. As described in Chapter 3 of Volume I on the functioning of various impeller types, the point of greatest effectiveness (highest velocity) is at the impeller tip. If this is the case, what is more logical than location of the impeller at the vessel bottom? An industrial installation should be the type exemplified by Figs. 5 and 6. However, the bench scale tests may be conducted with the impeller in a raised operation.

A. Bench Scale Tests

It is preferable to conduct bench scale tests with four-inch and larger impeller diameters. Tests with smaller impellers are questionable and require interpretation, as the suspension volume then is so small that the particle diameter to blade-width ratio changes significantly on scaling up to plant size vessels. Tests with smaller than a 4-in. diam. impeller should be interpreted by an experienced mixing engineer.

An ideal bench scale mixer is illustrated in Fig. 31. A 12 × 12 in. glass jar is equipped with four removable baffles, 1 in. wide and spaced 90° apart. Four, five, and six inch diameter impellers will suffice for the range of tests to be performed. Curved-, pitched-, or straight-blade turbines may be used. The

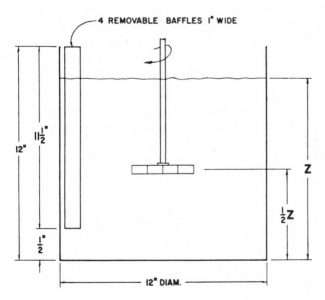

Fig. 31. Bench scale mixer.

impeller will be operated at 550 to 650 peripheral feet per minute and will be located at one-half the charge depth. The drive should be equipped with a dynamometer to determine the power consumed. For details on a laboratory scale dynamometer see Chapter 3 of Volume I. The power measured with such a dynamometer should be compared with the power calculated by the correlations in Chapter 3 for the suspending liquid. The ratio of the measured slurry power to this calculated power provides a multiplier which can be used to obtain the slurry power in large-scale equipment from a calculated power for the same impeller in the suspending liquid alone.

The basic parameters to be determined for extrapolation are the ratio of tank to impeller diameter, depth of charge, and impeller speed. Within these three variables, the entire equipment geometry is included. Their determination will establish the need for baffling, the optimum tank diameter to depth relation, and the number of mixer impellers to be used.

Bench scale tests should be set up as follows:

1. Using the known slurry characteristics, reference to Fig. 18 will give a close approximation of the tank diameter to impeller diameter ratio. Selection of the 4, 5, or 6-in. impeller can then be made. If the T/D is greater than three and an initial test with a 4-in. impeller shows velocities higher than necessary, an additional test should be performed in a larger diameter vessel or at slower rotational speed.

2. The charge level in the vessel, the vessel shape, and the ease of suspending solids determine whether the impeller should be mounted on the bottom of the vessel or at one impeller diameter above the bottom. The lower impeller position is preferred for low Z/T and difficult-to-suspend solids. With a bottom located impeller, the initial charge depth should be not more than $1.25D$ and the impeller to bottom clearance should be one-half the blade width. With a midmounted impeller, the charge level should be not more than $0.75D$ above the lower edge of the impeller.

The small physical dimensions of bench scale equipment provide a small recirculation path with little loss of stream velocity. In a large-scale commercial design, the recirculation path will be sufficiently long for the imparted stream velocity to deteriorate before returning to the eye of the impeller. It is obvious from the above that a lesser relative speed for the bench scale impeller will be equivalent to normal speed for a commercial scale unit. Turbine design is based on a "normal" speed of 700 peripheral feet per minute. Using the approximate impeller diameter for the commercial scale unit (determined from Fig. 18), the bench scale operating speed equivalent to "normal" can then be read from Fig. 32 for the large impeller diameter. The peripheral speed, v_p, in ft./min. can be converted to rotational speed, N, in r.p.m. for bench scale impeller (D ft. in diameter) by the following simple equation:

$$N = v_p/\pi D \qquad (8)$$

After establishing the rotational speed, the bench scale unit may be set up, charged, and the motor started. The first observation is to determine whether the chosen impeller provides good circulation to the vessel wall and a reasonable depth of recycle pattern. If circulation is excessive either a smaller impeller or a larger vessel diameter should be tested. If definitely insufficient, a larger impeller should be substituted. With what appears to be a reasonable circulation, the impeller location should be raised incrementally until circulation is just sufficient to recycle all material below the impeller level and maintain a complete suspension of solids off the vessel bottom.

When the impeller location is established, the volume of charge should then be increased to the point at which the surface flow reaches the acceptable minimum. At this point the impeller height above the bottom of the vessel, Z_i, has been established as a factor times the impeller diameter, and a similar factor has been determined for depth of slurry above the impeller.

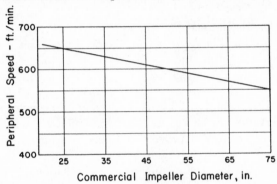

FIG. 32. Bench scale peripheral speeds equivalent to a plant-scale normal peripheral speed of 700 ft./min.

If a single impeller is used on the vessel bottom, only charge depth will be determined. If the first tests indicate a change in impeller or vessel diameter, the same factors must be determined for the new conditions.

It is preferable to retain the chosen impeller rotational speed if the factors just determined are geometrically reasonable. This will permit extrapolation to the operating speed for the commercial size unit at optimum power consumption inasmuch as a "normal" speed of 700 peripheral feet per minute is the rotational speed for optimum power consumption under average conditions, It is desirable, however, while the bench scale test is being run to make tests at 10% above and 10% below the selected peripheral speed, and to determine the impeller height and charge depth for these new speeds.

With an inverted plastic (dilatant) slurry, a satisfactory suspension may be obtained at a lower speed, hence, at a lower power consumption. With pseudoplastic mixtures, it is necessary to circulate at a rate above which the mixture becomes relatively fluid.

Power readings should be taken at the optimum geometric conditions in each instance and a multiplier determined which is the ratio of the slurry power consumption of the impeller to the calculated or observed power in water. As mentioned earlier, this multiplier can be used for calculating the approximate slurry power consumption of the commercial scale unit.

In designing equipment for the simple suspension of a low density or flocculent solid, the experienced engineer would base his design primarily on mechanical considerations and use experience to select paddle dimensions and rotational speed. A design for simple suspension may be checked in bench scale by using a similar vessel. A typical bench scale unit would be 18 in. diam., 12 in. deep with a 12-in.-diam. impeller having four 1-in.-wide blades operating at 30 to 50 r.p.m. The optimum conditions so determined could be scaled geometrically to plant requirements.

B. SCALE-UP OF TEST DATA

The optimum dimensional data obtained from the bench scale tests may be used to establish the same geometry for the plant scale unit. For the average slurry, the tank-to-impeller ratio, T/D, and depth factors, Z_i/D and Z/T, can be used at full value. With slurries having a high concentration of solids or fast settling solids, the factors would be reduced by 5 to 10% for extrapolations of 100 to 1000 times bench scale. The extent of this reduction becomes a personal observation through experience with similar slurries.

The commercially sized impeller would operate at its normal 700 ft./min. peripheral speed or at a higher or lower revised speed. Power can be calculated directly by the normal power equations or correlations and adjusted if a multiplier were observed to be necessary.

The geometry of a single impeller, either on the vessel bottom or raised in the charge, results directly from the bench scale data. If a tall vessel requiring multiple impellers is to be used, the bench scale data would determine location of the lowest impeller. It would also have established a height factor for free circulation above the plane of the impeller. One and one-half to 1.6 times this latter factor will establish the vertical separation between impellers in the multiple series. This insures approximately a 25% overlap of the circulating streams of the upper and lower impellers in a sequence and insures distribution of the slurry throughout the vessel.

Fibrous slurries are typical of those for which the bench scale geometric factors would be slowly decreased as the size of the commercial machine is increased. The pertinent observation in testing such materials is the point at which the bottom or surface circulation decreases sufficiently to permit a stagnant ring at the vessel wall. Impeller height and charge depth should then be readjusted to the point at which this stagnant ring disappears, for optimum operation. In extrapolating these optimums to plant scale, the height factors may be gradually decreased as mentioned above with increasing commercial unit size, or maintained constant with the operating peripheral velocity gradually increased to compensate for the effect of increased size.

An actual example of a commercial installation is as follows: A 70% by weight slurry of 200 mesh sintered ore solids with 6.0 specific gravity is to be suspended in leach liquor with 1.0 specific gravity. The slurry specific gravity is 2.4. The plant wishes to process this slurry in a series of vessels 6 ft. 6 in. diam. and 15 ft. charge depth.

Reference to Fig. 18 shows a T/D ratio of 1.6 for a solids specific gravity of 6.0 and 5.2 for 200 mesh, or a mean ratio of 3.4. The 4-in. impeller in the 12-in. bench scale jar gives a $T/D = 3.0$ ratio, which is suitable for the initial test. At $T/D = 3.0$, the plant impeller for a 6 ft. 6 in. diam. tank would be 26 in. Reference to Fig. 32 for a 26-in. impeller diameter gives a test speed of 640 ft./min. peripheral speed or 611 r.p.m.

From a test operating at this speed, it was determined that the height from vessel bottom to impeller center line was 1.7 times the impeller diameter, and from impeller to free-flowing charge level was 2.0 times the impeller diameter. In view of the good suspension obtained in the test and the desire to keep power consumption at a minimum, the bench test was repeated at 525 r.p.m. (86% of simulated normal speed).

Under the new conditions, satisfactory suspension was still obtained at $1.3D$ below and $1.5D$ above the impeller. These results were then translated into the plant units of 6 ft. 6 in. diam. by 15 ft. charge depth as follows:

T/D ratio of $3/1 = 78/3 = 26$ in. (use standard 27 in. impeller).

Using factors of $1.0D$ at bottom, $2.25D$ (1.5×1.5) between impellers, and $1.5D$ above the top impeller, a three impeller assembly gives a total height of 7.0×27 in. or 189 in. (vs. 180 in. required).

Standard gears provided a shaft speed of 87.5 r.p.m. (87.5% of normal for 27 in. impeller), a slight increase over the 86% bench scale rate.

Power was calculated for the plant size agitator at the suspending liquid density using correlations like those in Chapter 3 of Volume I. This power was multiplied by 2.4 to correct for the higher slurry specific gravity and by 3.0 for the three impellers.

List of Symbols

a particle diameter [Eq. (3)] of maximum flocculating size, microns
A_p cross-sectional area of particle, ft.²
C drag coefficient defined by Eq. (1)
D impeller diameter, ft.
D_p particle diameter, ft.
g acceleration due to gravitational force, 32.2 ft./sec.²
g_c gravitational conversion factor, 32.2 (lb./lb.$_f$)(ft./sec.²)
h suspension height for lowest rotational speed which sweeps the vessel bottom clean, ft.

h' depth of clear liquid layer above suspension height, h, ft.
l settling depth, cm.
m_p mass of particle, lb.
M slurry weight, lb.
N rotational speed, r.p.m.
N_c critical rotational speed at which no particles remain on the vessel bottom, r.p.m. [Eqs. (4) and (5)]
$(N_{Re})_p$ $D_p u_t \rho/\mu$, particle Reynolds number, dimensionless
P_c impeller power at the critical suspension speed of Eq. (7), hp.
q_T flow rate, ft.^{3}/sec.
R weight ratio of solid to liquid (volume ratio in Fig. 18)
S suspension criterion defined by Eq. (6)
t exponent in Eq. (4)
T vessel diameter, ft.
u_t terminal settling velocity, ft./sec.
v_p peripheral speed, ft./min.
y ratio of distance between particles to particle diameter
Z liquid depth, ft.
Z_i distance from impeller to bottom of vessel, ft.
$Z_{1,2,3}$ circulation height of impellers 1, 2, 3, . . ., ft.
μ viscosity, (lb.)/(ft.)(sec.)
ν μ/ρ, ft.^{2}/sec.
ψ constant in Eq. (4)
ρ liquid density, lb./ft.^{3}
ρ_p particle density, lb./ft.^{3}
$\Delta\rho$ $(\rho_p - \rho)$, lb./ft.^{3}

References

(A1) Agricola, "De Re Metallica," translated by H. C. Hoover and L. H. Hoover. Dover, New York, 1956.
(C1) Calderbank, P. H., and Jones, S. J. R., *Trans. Inst. Chem. Engrs. (London)* **39**, 363 (1961).
(E1) Eldridge, J. W., and Piret, E. L., *Chem. Eng. Prog.* **46**, 290 (1950).
(E2) Ellis, H. S., Redberger, P. J., and Bolt, L. H., *Ind. Eng. Chem.*, **55**, 18 (1963).
(G1) Gaudin, A. M., "Principles of Mineral Dressing." McGraw-Hill, New York, 1939.
(G2) Gaudin, A. M., "Flotation," 2nd ed., p. 48. McGraw-Hill, New York, 1957.
(H1) Hirsekorn, F. S., and Miller, S. A., *Chem. Eng. Progr.* **49**, 459 (1953).
(H2) Hixson, A. W., and Tenney, A. H., *Trans. A.I.Ch.E.* **31**, 113 (1934).
(L1) Lyons, E. J., *Chem. Eng. Progr.* **44**, 341 (1948).
(M1) MacMullin, R. B., and Weber, M., *Trans. A.I.Ch.E.* **31**, 409 (1934).
(M2) Metzner, A. B., *Adv. Chem. Eng.* **1**, (1956).
(M3) Michaels, A. S., and Bolger, J. C., *Ind. Eng. Chem. Fundamentals* **1**, 153 (1962).
(O1) Olsen, J. F., and Lyons, E. J., *Chem. & Met. Eng.* **52**, 118 (1945).
(O2) Oyama, Y., and Endoh, K., *Chem. Eng. (Tokyo)* **20**, 66 (1956).
(P1) Pavlushenko, I. S., Kostin, N. M., and Matveev, S. F., *Zh. Prikl. Khim.* **30**, 1160 (1957).
(P2) Perry, R. H., Chilton, C. H., Kirkpatrick, S. D., eds., "Chemical Engineers Handbook." McGraw-Hill, New York, 1963.
(P3) Porcelli, J. V., Jr., and Marr, G. R., Jr., *Ind. Eng. Chem. Fundamentals* **1**, 172 (1962).
(R1) Raghavendra, R. S., and Mukhenji, B. K., *Trans. Indian Inst. Chem. Engrs.* **7**, 63 (1954–1955).

(T1) Taggart, A. F., "Elements of Ore Dressing," p. 76. Wiley, New York, 1951.

(T2) Taggert, A. F., "Handbook of Mineral Dressing," Sect. 12. Wiley, New York, 1953.

(T3) Thomas, D. G., *A.I.Ch.E. J.* **8**, 267 (1962).

(T4) Thomas, D. G., *A.T.Ch.E. J.* **9**, 310 (1963).

(W1) Weisman, J., and Efferding, L. E., *A.I.Ch.E. J.* **6**, 419 (1960).

(W2) White, A. M., and Summerford, S. D., *Ind. Eng. Chem.*, **25** 1025 (1933).

(W3) White, A. M., Summerford, S. D., Bryant, E. D., and Lukens, B. E., *Ind. Eng. Chem.* **24**, 1160 (1932).

(W4) Wilkinson, W. L., "Non-Newtonian Fluids." Pergamon Press, Oxford, England, 1960.

(Z1) Zwietering, T. N., *Chem. Eng. Sci.* **8**, 244 (1958).

CHAPTER 10

Mixing of Solids

Curtis W. Clump

Department of Chemical Engineering
Lehigh University, Bethlehem, Pennsylvania

I. Introduction

The mixing of solids, or solids blending, is an operation common to many chemical process industries. Those industries whose products are food, drugs, glass, paints, and others are all users of solids mixing equipment. Not only is the blending of solids frequently encountered industrially but it also has a long history of use behind it. For example, manufacturing of gunpowder is centuries old. Nevertheless, the state of knowledge relative to an understanding of the fundamentals of the operation is not well developed.

Unlike liquid mixing, research on solids mixing has been relatively limited.

One of the reasons for the lack of solids mixing information is that individuals concerned with the mixing of solids have relied upon their accumulated experience to tell how to mix materials and whether or not the final mixture is an adequate one. Such experience is not quantitative nor analytical enough to yield an understanding of the mechanisms of blending operations. Another reason for the lack of solids mixing information is the wide range of flow properties of dry particulate solids which are difficult to describe in quantitative terms. The results obtained for one system of particulate solids may not be applicable to others.

It is the purpose of this chapter to report on the available dry particulate solids mixing information to help in formulating answers to such questions as:

1. What is meant by an adequate mixture?
2. What occurs during a mixing operation?
3. What are the effects of the physical characteristics of the solids being blended upon the time necessary for a satisfactory product?
4. What design features in a mixer are necessary to correct the physical phenomena that oppose good mixing?
5. What is an acceptable standard performance test for a solids mixing operation?

II. Degree of Mixing

A single criterion for an adequate mixture is difficult to achieve because of a diversity of goals for mixing. The researcher is likely to be satisfied with a statistical definition of mixing goodness in terms of composition. On the other hand, the production man may desire to characterize a good mixture by color, size distribution, density, or a variety of other physical ways. Mixing goodness based on composition, and the practical or physical definition of a good mixture based on some physical property may be related, but these relationships are usually empirical. To be useful and of practical value a definition of a satisfactory mixture must (a) be related as closely as possible to specified properties of the final mixture that might serve as a standard, (b) be easily determined, and (c) be adaptable to a wide variety of materials.

As would be expected, any definition that has as its aim the characterization of the degree of mixing will necessitate the study of samples obtained during the mixing operation. Accordingly, any definition becomes no better than the information obtained from the samples.

A. The Ideal Mixture

If a number of samples are taken from a given mixture of materials, and the composition of these samples is identical, then the materials in the mixture are said to be perfectly mixed. The preceding definition applies to a perfect mixture, but its application to solids is limited because the size of the sample is important. Samples containing a single particle will not follow this definition. Even for samples containing as many as 100 particles, sample compositions are not constant, but would vary about some mean value. An ideal mixture may be defined as one in which there will be a statistically random variation in composition of the various samples.

Consider the gross effect of what happens during mixing. If, for example, two solid materials A and B are to be blended in some device and at the start of the blending operation each is segregated in its own pile although some intermingling is possible at the pile interface. So it may be said that at the start the materials are arranged in a definite order. The mixing operation is started and this ordered state is disrupted with more and more intermingling of the two materials as the mixing continues. Eventually the disordered state tends to approach some maximum value with respect to dis-order. This is analogous to the concept of an equilibrium state. This maximum state of disorder is the ideal for this particular mixing operation (L1).

The theoretical equilibrium condition reached after mixing has progressed for some time represents a random distribution of the various materials comprising the mixture. In other words, if the starting materials for mixing are half black and half white, then after mixing has proceeded to the equilibrium state and a sample consisting of one particle is removed there will be an equal chance of the sample being black or white. This statement does not imply that two successive samples will necessarily contain a white and a black particle. It does imply that the probability of finding a particular colored particle at any point is a constant, equal to the proportion of that particular colored particle in the entire mixture. Although the probability of obtaining a sample of a particular color is constant, this does not mean that the proportion of that particular color is constant in every part of the mixer. It means that, if mixing is truly a random process, all particles will distribute themselves independently in such a way that, for every component in the mixture, the probability of finding a particle of this component is the same for all points in the mixture.

B. STATISTICAL CONCEPTS OF DEGREE OF MIXING

Practically, the most significant and meaningful way to characterize a good mixture is how satisfactorily the mixture meets end-use requirements. If the mixture is an acceptable product, then the mixture is a satisfactory one.

However, the most significant way to express theoretically the extent or degree of mixing is by some suitable statistical relation. The ideal mixture definition previously stated embodies statistical concepts. Any property can be substituted for composition in this definition.

Among some of the well-known terms which are involved in statistics are the normal or Gaussian distribution, arithmetic mean, standard deviation, variance, chi-square, confidence interval, *t*-tests, etc.[1] All of these have application in the quantitative determination of the state of mixedness of a batch of solids. The choice of a statistical standard for the degree of mixing is arbitrary, but whatever the statistical criterion, it should be one which best measures the degree of mixing independent of the final use of the blend. Indeed, the most productive use of statistics in a blending operation embraces the skills of the statistician and the practical operator of the equipment. Thus, the statistician must employ a mixing characteristic which is easily understood. A single number, easily computed, which can be examined with regard to its magnitude, would seem desirable as a degree of mixing criterion; that is, a low number might imply a good mix, whereas a high number would imply a poor mixture. Such a number is the standard deviation. The standard deviation as a criterion for the degree of mixing is an easily determined number, and does represent a variation of uniformity within the mixture. Hence, a low standard deviation corresponds to a high uniformity of the composition of the mixture, and therefore a good mixture.

In order to use this particular statistical term it is necessary to know a few basic statistical concepts. As a consequence of sampling and analysis, a group of numbers will be obtained which show the compositions of several sampling spots in the mixed batch. This group of measurements is called a *sample* of *n*-observations. Note that here the term sample means the whole group of measurements. To distinguish between the statistical sample and a sample taken at one particular location, the latter term is called a *spot sample*. Now, the larger the sample, the more reliable the picture that can be drawn about the total batch, or as is known in statistics, the *population*. More will be said about sampling later, but it must be stressed at this time that sampling must be done randomly when obtaining spot samples so as to avoid bias in favor of any one particular location.

1. Frequency Distribution

The data obtained from the group of spot samples may be expressed graphically. An arrangement whereby the observed values are placed in ascending order of magnitude is referred to as an ungrouped frequency distribution. This graph may be obtained by plotting the frequency of samples of identical composition as a function of sample composition. A glance at this graph readily gives such information as the maximum, the minimum, and approximate average of the composition of the sample.

Thus, a graph of a truly representative frequency distribution would give a good picture of the state of mixedness, since it would show how widely the spot samples vary in composition. However, to get a graph representative of

---

[1] An excellent summary of the literature pertinent to the statistical concepts, as they apply to the degree of mixing is found in Ref. (W1). Also, for a detailed discussion of statistics, the reader is referred to (D1, D2).

the total batch would necessitate the acquisition of a very large number of spot samples. Fortunately, statistics provides methods of making objective estimates of some essential characteristics of a frequency distribution. Two of these characteristics of the distribution include: (a) some measure of the value about which the observations tend to center, and (b) the spread or variation of the observations about the central value.

The measure described in (a) above is called the measure of central tendency and may be expressed as the median, mode, or most commonly, the arithmetic mean. The measures which describe the spread or variation of observations are the standard deviation and the variance.

2. The Arithmetic Mean

The arithmetic mean is defined as

$$\bar{x} = \frac{\sum\limits_{i=1}^{n} x_i}{n} \tag{1}$$

where $n =$ the number of spot samples (sample size), and

$x_i =$ the ith value of x which represents the composition, or whatever else is being looked for, in each spot sample.

The arithmetic mean is not a measure of the degree of mixing. When the true mean is known from a knowledge of the batch ingredients, the arithmetic mean can be used to learn if the mean of the measurements on the samples is significantly different from the true mean.

However, the arithmetic mean is useful for two important reasons. Variation of data can be described in terms of the square of the deviations from some central value. The arithmetic mean is that central value for which the sum of the squares of the deviations is a minimum. The arithmetic mean of the sample measurements is also the best estimate of the mean of the population from which the sample is drawn. Perhaps the most important reason for the use of the arithmetic mean is that the means of sample measurements of uniform size tend to be normally distributed regardless of the type of distribution of the population from which the samples were drawn. This characteristic of the sample means permits the use of the normal distribution in making probability statements about the population mean even when the population is not normally distributed.

3. The Mode

The mode of the frequency distribution of n-numbers is the value which occurs most frequently. In a frequency plot against magnitude of the variable, the high point of the curve is at the modal frequency. Should the curve have more than one peak, each peak is referred to as a mode.

The mode is particularly useful in particle size analysis done either by microscopic count or by screen analysis because multimodal distributions are often encountered. In cases of this kind, a single number representing the average particle size is usually not sufficient, and either a graphical or verbal description of the distribution is necessary.

4. The Median

The median is that value of the variable which divides the measured values into two equal portions so that half of the values are higher and half lower than the median value. The median is an important statistic best used when dealing with the order of measurement.

For a symmetrical normal distribution, the arithmetic mean, the mode, and the median coincide at the midpoint of the distribution.

5. Standard Deviation

Even though the mean may be the same for two groups of measurements, a considerable difference may exist in the variation of the measurements of each group. The standard deviation defined below is a measure of this variation or spread in measurements.

The standard deviation, estimated from the samples, is given by

$$S = \sqrt{\sum_{i=1}^{n} (x_i - \bar{x})^2 \bigg/ (n - 1)} \tag{2}$$

This quantity is not the standard deviation of the samples but is the estimate of the population standard deviation calculated from the samples.

The use of Eq. (2) is cumbersome and can be rearranged for easier use.

$$S = \sqrt{\left(\sum_{i=1}^{n} x_i^2 - \bar{x} \sum_{i=1}^{n} x_i \right) \bigg/ (n - 1)} \tag{3}$$

6. Variance

The variance is defined as the square of the standard deviation, S, and is denoted as S^2. If a process has a number of factors contributing to the variance of a measurement, the total variance is equal to the sum of the variances of the individual factors when these are statistically independent. However, the standard deviation of a measurement is not the sum of the standard deviations of the individual factors.

7. Uses of the Standard Deviation

Usually spot samples are removed from a mixer as the mixing operation is carried out and some measurable characteristic (the mean value of the spot samples) is calculated and compared with an acceptable standard. A standard deviation is computed, and if it falls within the control limits the mixing is judged satisfactory. It is apparent from Eq. (2) that the number of spot samples withdrawn from the batch and analyzed influences the standard deviation. In addition, the standard deviation will depend upon the sample size since larger sized samples have a more narrow spread in composition or properties.

To illustrate the use of the standard deviation consider the following process. A right-angled cylinder is being used to mix white and black sand particles. The mixer is horizontal, and filled 30% of capacity with 20% black sand. The mixer rotates at 21 r.p.m. for nine rotations. At the end of the operation, 30 spot samples are taken randomly from various positions within the batch. The samples are counted with respect to total sand particles and the number of black particles. The data and the calculations for the standard deviation are shown in Table I.

Table I
Sample Calculation of Standard Deviation

Sample No.	No. of black particles	Total particles	Proportion black (p_i)	p_i^2
1	36	114	31.58	997.29
2	27	112	24.11	581.29
3	44	111	39.64	1571.33
4	33	114	28.95	838.10
5	22	119	18.49	341.88
6	31	117	26.50	702.25
7	28	119	23.53	553.66
8	34	113	30.09	905.41
9	20	114	17.54	307.65
10	25	111	22.52	507.15
11	21	115	18.26	333.43
12	18	120	15.00	225.00
13	18	111	16.22	263.09
14	22	115	19.13	365.96
15	27	110	24.55	602.71
16	17	104	16.35	267.32
17	25	115	21.74	472.63
18	24	124	19.36	374.81
19	26	108	24.07	579.36
20	10	120	8.33	69.39
21	22	116	18.97	359.86
22	23	115	20.00	400.00
23	13	112	11.61	134.79
24	20	110	18.18	330.51
25	25	115	21.74	472.62
26	17	109	15.60	243.36
27	15	109	13.76	189.34
28	23	113	20.35	414.12
29	27	106	25.47	648.72
30	20	112	17.86	318.98
Total			629.50	14372.01

Thus the mean concentration of black sand for the sample is

$$\bar{p}_i = \frac{\sum\limits_{1}^{30} p_i}{n} = \frac{629.50}{30} = 20.98$$

Using Eq. (3), the standard deviation is found to be

$$S = \sqrt{\left(\sum\limits_{1}^{30} p_i^2 - \bar{p} \sum\limits_{1}^{30} p_i\right)\Big/(n-1)} = \sqrt{\frac{14372.01 - 20.98\,(629.5)}{30 - 1}}$$

$$S = 6.33$$

Now, for illustrative purposes, let us assume that similar data are taken except that this time the mixing operation has been changed so that the cylinder is inclined at some angle. All other conditions and sampling are the same. For this condition, a standard deviation of 7.21 is found. If the standard deviation is used as a measure of mixedness, then the lower the standard deviation, the better the mix. Thus, the use of a horizontal cylinder for mixing gives better mixing than a sloped cylinder.

8. Confidence Limits

Since estimates of a property do vary in precision, it is important that the estimate be accompanied by a statement which describes how close the estimate may be to the quantity which it is desired to estimate. Confidence limits provide a method for stating the precision of an estimate.

For the data in Table I, the 95 confidence limits for the standard deviation of p_i are $0.8S$ and $1.4S$. These limits define the range of variation expected in the value of the standard deviation, S, for 95 out of 100 repetitions of the sampling and analysis of the particle population. The confidence limits for a standard deviation are $(nS^2/\chi_1^2)^{1/2}$ and $(nS^2/\chi_2^2)^{1/2}$, where n is the number of samples or measurements. Chi-square values, χ_1^2 and χ_2^2, are found in tables given in books on statistics (D1, D2).

C. Sampling

When reporting results from a blended batch of solids, complete details of the sampling procedure should be stated. These details should include: the method of taking spot samples, the locations from where the samples were taken, the size and number of samples, the amount of material removed from the mixer during sampling, and the method of analysis.

There are a variety of sampling techniques possible. Sampling can be a very complex statistically planned operation, or as simple as scooping from the mixer a quantity of solids for examination. But, for the sampling technique to yield reproducible results, it must incorporate the following principles:

1. The sampling device that is inserted into the solids, either to remove a small sample or to measure the effects of mixing without removal, must cause as little disturbance to the blended solids as possible.
2. Samples should be taken at random locations throughout the entire batch rather than from one location to avoid bias. Such random locations may be selected by dividing the entire batch into equal parts and assigning a number to each part. Sample locations can be selected by picking numbers at random.
3. To avoid confusing mixer performance with the effects of dumping, sampling should be done within the mixer instead of when the batch is dumped.

Sometimes, the size of the spot sample will depend upon the end use of the mixture. If the batch is to be made into tablets, for example, a convenient sample size is the actual tablet size. However, if samples are removed from the batch and not returned the sum of all samples should not exceed 5% of the total mixture (W1). The ratio of the size of the sample and the amount of material withdrawn from the mixer must be maintained the same in all tests that will be compared. There is no single arbitrary number of spot samples that will be optimum for the evaluation of a mixer. Sampling procedures can become time-consuming and costly when the number of samples becomes large. Collection of large numbers of samples and their analysis is seldom justified when production mixers are tested for performance. However, large sample numbers may be required for meaningful results.

D. SAMPLING DEVICES

Samples may be removed from the interior of a solids mix with little disturbance to the mixture by a "sampling thief." A small thief is illustrated in Fig. 1. This device consists of a solid rod with radial holes of the desired

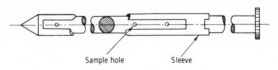

Sample hole   Sleeve

FIG. 1. Sampling thief.

sample size (A1). A sleeve is rotated to uncover these holes when a sample is desired. Further rotation covers the holes again. The thief permits the withdrawal of a sample with little disturbance to the solids within the mixer, provided the thief is less than 3% of the mixer diameter.

E. SAMPLE ANALYSIS

The analysis of spot samples hinges solely on what information is sought. It might be color, conductivity, size distribution, solubility characteristics, or a host of other observable characteristics. The size of the spot sample is quite important when one considers what type of analysis to employ. If particle size distribution is desired, one can count particles of different sizes or one can use a screen analysis. The use of a screen analysis is limited for several reasons. First, the sample size must be large—about 100 gm. This may be no disadvantage if large amounts of materials are in the mixer. Second, screening may be too severe treatment for friable materials. The best precision for particle size determination of a sample is obtained by counting all the particles of each size. This procedure is tedious but accurate when the sample size is small.

Samples can also be analyzed for composition by dissolution in a suitable solvent with subsequent removal and drying of the residue. Chemical reactions such as titrations are sometimes used to indicate the amount of a particular ingredient in a spot sample. Tracers of various kinds also serve as an analytical tool. These tracers may be simple chemical tracers or radioisotopes. For example, Overman and Rohrman (O1) as well as Hoffman (H2) report some interesting uses of radioisotopes to study mixing. Iodine–131, cobalt–60, barium–140, sodium–124, indium–192, and others have all been used. Another interesting use of the tracer technique is reported by Kraftmakher and others (K1) where magnetic iron powders were used. Samples were withdrawn from the mixer and the magnetic susceptibility was measured.

The use of tracers, although convenient for following the course of a mixing operation, can yield results which tend to be misleading. The blending characteristics of the tracers may be quite different from the solids actually being mixed, and thus one may be observing the blending characteristics of the tracer rather than the prime materials.

Fischer (F1) claims that superficial examination with the naked eye is also a powerful analytical tool, for this method can easily discern gross deviations in the blending operation without the necessity for sample removal. This is certainly a good tool whenever the materials being mixed have easily distinguished characteristics. Unfortunately, this is not always the case.

Once a sample has been collected and analyzed it must then be related either to previous or succeeding samples or to some specific yardstick to yield a measure of mixer performance. Comparison between samples can be made as the mixing operation continues. A simple way to compare samples is to compute the variance (S^2) for each sample and to take a ratio of these variances. This ratio, in conjunction with statistical tables and probability, enables one to ascertain whether the samples differ because of significant changes in mixing or whether they differ by pure chance. This analysis of variance is most easily accomplished by use of the F-test of statistics (B2, D1, D2).

III. Mixing Mechanisms

There are three types of particle motion which result in mixing (L2):

1. Convective bulk movement of groups of particles which is analogous to turbulence in fluids.
2. Diffusive movement or scattering of separate particles which is analogous diffusion.
3. Shear movement of particles within a group of relatively slow-moving particles which is analogous to laminar fluid motion.

Commercial mixers combine several of these types of particle motion to obtain mixing.

In theory, if one can set up mathematical expressions that account for these types of particle motions it should be possible to predict the time needed to achieve a desired degree of mixedness. However, no way of predicting such particle motions is available, and thus mixing times can be determined only by a testing program.

IV. Mixing Parameters

Before considering mixers and their evaluation it is important to dwell upon the variables which influence the rate of mixing and the degree of mixing. These variables can be grouped into two categories: those associated with the solids being mixed, and those associated with the mixing equipment.

A. Characteristics of the solids
    1. Particle shape and surface
    2. Particle size distribution
    3. Bulk density and particle density
    4. Moisture content
    5. Friability
    6. Angle of repose
    7. Flowability or lubricity
B. Equipment characteristics
    1. Mixer body dimension and geometry
    2. Agitator dimensions, clearances and geometry (if used)
    3. Size and location of access openings
    4. Construction materials and surface finishes
    5. Loading and emptying equipment details
C. Operating conditions
    1. Weight of each ingredient added and fraction of mixer body volume occupied by ingredients
    2. Method, sequence, and rate of addition of ingredients
    3. Agitator or mixer body rotational speed

There is also an interaction between solids characteristics and the mixer. For example, the relative volume relationship between the mixer and the solids, and the manner, or ease, with which the solids flow over the mixer parts influence mixing.

The physical nature of the solids affect their mixing characteristics. The significance of particle shape, for example, is primarily in its effect on the flow characteristics of a mass of particles. Smooth, round, spherical particles tend to flow more readily than rough irregularly shaped particles. On the other hand, irregular rough particles exhibit greater surface area than smooth regularly shaped particles. A particle having a shape such as a filament flows very poorly. The manner in which the solids flow in the mixing operation certainly is of influence in both the rate of mixing and the degree of mixing. An excellent review of the effects of the physical characteristics of solids on blending is found in Ref. (W1).

Particle size is also of importance. Most often, when solids are being blended, they exist in mixture as a range of sizes rather than uniquely sized. The size distribution of particles within this range greatly influences the behavior of the mass of solids. Segregation of particles according to relative sizes can occur whenever a solid mass comprised of a range of particle sizes is moved. Usually, the smaller particles rise to the top of the mass and the larger particles settle to the bottom. Indeed, the very smallest particles may leave the mass as dust.

Particle density may vary considerably, but by itself it generally has little significance. However, if a mass of particles of widely different densities are being blended, segregation will occur similar to that encountered when blending widely different-sized particles. The bad effects of density differences are not as marked when the particles are small and closely sized.

Moisture content is a factor that influences the flow of solids. Moist, sticky particles retard flow and are a deterrent to the mixing process especially if they stick or adhere to the walls of the mixer or cause agglomeration of particles.

Friability of particles is associated with the ease of breaking a particle. Particles which easily break or dust are called friable. Friability is detrimental in mixing solids because it increases the range of particular sizes. This increase in small particles tends to promote segregation. When friable materials are to be mixed, a mixer should be chosen such as one of the tumbling types to avoid excessive particle breakage.

When one considers the operation of liquid mixing, liquid viscosity plays an important role. The viscosity of a solid mass is not particularly meaningful, but the property of a solid that might be analogous to liquid viscosity is related to angle of repose. This is the angle that must be exceeded to cause one particle to shear against another similar particle. This, in turn, is also related to flowability or lubricity. Solids possessing low angles of repose would exhibit good flowability. Good flowability may not necessarily always be of advantage when mixing solids. Consider the mixing of uniformly sized particles whose physical characteristics are similar except for flowability. The solids which flow easily may become segregated due to their rapid movement toward the bottom of the equipment.

Not only do the particulate solid properties influence their mixing characteristics, but also the equipment geometry. The size and shape of body and stirrer (if used) affect the particle flow patterns and velocities. Even the point of addition of ingredients and the surface finish of equipment parts have some effect on the particle movements within the mixer. The flow patterns existing during filling and emptying provide some mixing action.

The mixer rotational speed has a major effect on particle movement and rate of mixing. The sequence of addition of ingredients can have a major effect. For example, simultaneous addition can provide significant mixing during filling. If a mixer body is filled too full, particle motion and mixing rate may be retarded.

V. Equipment

The basic objective of solids mixing requires primarily intimate intermingling of the materials to be mixed. To meet this requirement, the solids are placed in a vessel of some type which allows the material to be moved in a pattern which achieves mixing. This, however, is not so simple as one might imagine for there is really no one mixer design which universally satisfies all mixing requirements. The complexity and interaction of the many variables already mentioned precludes a universal design. Many shapes have evolved for tumbling solids. Some of these are shown in Fig. 2. The evolution of these

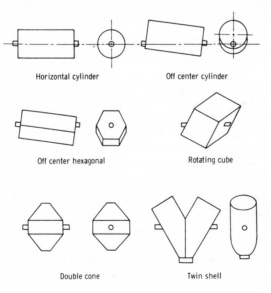

FIG. 2. Schematic outlines of industrial mixers.

shapes, according to Brown (B1), came about as a consequence of manufacturing advantages rather than theoretical consequences.

Mixers may be grouped roughly according to which type of mixing action predominates (W1):

1. Ribbon and spiral mixers; mainly convective and diffusive in action.
2. Drum mixers and tumblers; mainly shearing within the bulk solids with diffusive action at the surface.
3. Ball mills, mullers, colloid mills; mainly shearing but the scrapers used in mullers include convective and diffusive action.
4. Sigma mixers; mainly shearing and convective action.
5. Sifter mixers and high-speed beater mixers; mainly diffusive action.

The choice of which mixer to use is not easy for the best choice depends strongly upon the solids characteristics and upon the desired objective. Free-flowing dry solids may be processed in a variety of mixers. Drum mixers and tumblers are especially suited for dry free-flowing solids since the mixing action is mild. Ribbon mixers are well suited for mixing fibrous materials since the mixing action gives vigorous displacement to the bulk solids. These are but two illustrations of mixer choice. Let us now consider in more detail the performance of certain classes of mixers.

A. Drum Mixers

A drum mixer in its simplest form is a cylinder rotating horizontally about its axis. This type of mixer (Fig. 2) is frequently used. Variations in this simple concept are common. Modifications to this simple mixer include:

1. Inclination of the drum.
2. Placing of flights or baffles within the horizontal drum.
3. Turning the drum end upon end rather than about its horizontal axis.
4. Combinations of all of the above.

A simple horizontal cylindrical mixer is shown in Fig. 3. The cylinder is partially filled with solids and is supported to permit horizontal rotation. The solids tend to be carried round with the cylinder until the angle of repose of the solids is exceeded. At this point the upper layers of material slide toward the base of the pile along preferred shear planes in excess of the angle of repose. Thus there is a constant shuffling of particles from the upper edge of the pile of solids to the bottom. Initially, the mixing action is a shearing one, whereas, as the particles tumble down the surface toward the bottom of the pile a diffusive action occurs. The mixer must be rotated long enough to permit good exchange of particles along the axis as well as at any given cross section. One of the disadvantages of this mixer thus becomes apparent—poor cross flow along the axis.

Fig. 3. Drum mixer (Abbe Engineering Company, New York, N.Y.).

Considerable work aimed at a better understanding of the mixing characteristics of a dum mixer has been carried out by several investigators. These
results are ably discussed in Ref. (O1).

The following is a listing of the significant findings of these works:

1. A correlation relating speed of rotation, cylinder diameter, and volume
per cent loaded has been developed for horizontal rotating cylinders.

$$N = \frac{C}{D^{0.47} X^{0.14}} \tag{4}$$

where N = rotational speed, r.p.m.;
D = inside diameter of cylinder, ft.;
X = volume of cylinder occupied by solids, per cent; and
C = constant.

There are three speeds at which transitions occur in the type of motion of
the solids. First, at low speed a *cascade state* exists in which the particles
simply are carried around the cylinder wall until the angle of repose is exceeded. Then they roll down the inclined surface of the solids as previously described. Second, a *cataract state* develops at higher speeds in which the particles
close to the cylinder wall adhere to it until they reach a certain height; at
which point they drop to the inclined surface and roll down the remainder of
the way. Third, an *equilibrium state* is obtained at still higher speeds in which

the particles do not roll down the inclined surface at all but remain on the cylinder wall until they fall to the bottom of the plane. The best operating conditions for mixing occur between the cataract and equilibrium states. Values for C may be given as: $C = 95$ for the change from cascade motion to cataract motion, $C = 126$ for the lowest speed at which the equilibrium state exists.

2. Segregation of particles is likely to occur in simple rotating drums. This segregation is due to differences in properties between the solids being mixed. Properties such as size, shape, and density differences are important.

3. Specifically no general conclusions can be made relative to power requirements. Oyama (O2) showed that the highest power consumption did not necessarily give the best rate of mixing.

The mixing action of a drum can be improved by addition of internal flights or baffles. Inclination of the drum can also be used in which case baffles are not as effective.

Modifications to the basic cylindrical shape, other than flights or baffles, have included placing the rotating axis at various angles with respect to the principal axis. This tends to improve the horizontal turnover of solids. Rotating cubes and other polyhedron shapes have been employed to improve the horizontal movement of solids.

B. CONICAL (DOUBLE-CONE) MIXERS

A conical mixer is one in which two cones are joined to a small cylindrical section (Fig. 4). This tends to overcome the slow horizontal movement of solids found in a cylindrical-shaped mixer. As a conical vessel rotates, good rolling action of the solids prevail, and because of the varying flow cross section, good crossflow is established. This blender is normally free of flights and baffles so that easy cleaning can be accomplished.

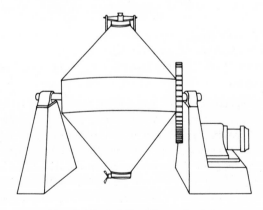

FIG. 4. Conical (double-cone) mixer.

C. V-Shaped Mixers

A relatively recent mixer to be developed which finds considerable industrial use is the V-shaped or twin-shell blender. This consists of two cylinders placed at an angle to form a "V" (Fig. 5). This design provides a nonsymmetrical shape about the rotation axis which provides effective mixing action for a variety of materials. It can also be modified by the incorporation of internal devices to add liquids and break up agglomerates.

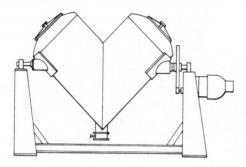

FIG. 5. Twin-shell mixer.

D. Ribbon Mixers

Ribbon mixers have been used for many years in many different applications. This type of equipment (Fig. 6) operates with two spiral ribbons turning inside a fixed shell having a rounded bottom and straight sides. For one arrangement of the ribbons, the outer spiral blade moves the solids axially in one direction and the inner spiral moves the solids in the opposite direction. Axial mixing is less rapid than radial moving, however. To minimize solids being left clinging to the shell wall, the outer ribbon must be located close to the wall to scrape it continually clean. Some attrition of friable particles may occur because of the higher forces acting on the particles between the ribbons and the closely adjacent wall. The ribbon mixer is effective for free-flowing solids and can be adapted for introducing liquids into solids by use of sprays.

Greathead and Simmons (G1) measured mixing patterns in an industrial, ribbon-type mixer with a magnesium sulfate-sodium bicarbonate system. They found that fluctuations in composition lessened with mixing time but did not disappear. The rate of mixing at the surface was much greater than that below the surface. The authors postulated a mixing mechanism of large-scale mixing occurring by movement of solid masses through the material, and small-scale mixing occurring at the boundaries between moving and stationary material. They found also that changes in operating variables produced little effect except when the mixer charge was reduced by half.

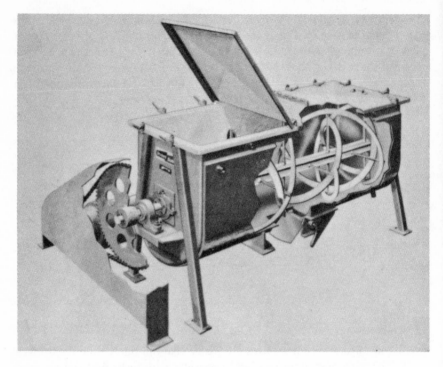

FIG. 6. Ribbon mixer (Strong-Scott Mfg. Co., Minneapolis, Minn.).

E. VERTICAL SPIRAL MIXERS

This type of mixer has a vertical rotating screw which lifts solids from the bottom to the top of a cone-bottomed vessel. This mixer is shown in Fig. 7. When the solids reach the top of the screw, they are flung toward the wall and descend slowly through the bed of particulate solids.

Rathmell (R1) gives the following comparison of ribbon and spiral mixers.

Advantages of spiral mixer:
1. Lower initial cost for same capacity.
2. Less power.
3. Less floor space.

Disadvantages of spiral mixer:
1. Lower output because of longer mixing times.
2. Less uniform product.
3. Wet materials and slurries are more difficult to handle.
4. Greater contamination of one batch by components of a previous batch.

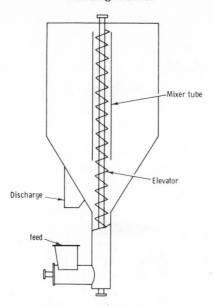

FIG. 7. Vertical mixer.

One of the reasons for a less uniform product is the unequal trajectories of particles at the top of the spiral. The heavier particles are thrown farther by the screw than the lighter particles. However, baffles may be placed in the path of the particles to lessen this segregation.

F. MULLER MIXERS

Muller-type mixers consist of a circular pan with wide, heavy, vertical rollers which ride on the solids in the pan (Fig. 8). In the rotation of the rollers, the inner edge of a roll travels a shorter circle than the outer edge. This provides some slip analogous to the rubbing that occurs in a mortar and pestle. The high local shear forces obtained in this way provide small-scale mixing that breaks up particle agglomerates. Large-scale mixing is provided by scrapers which direct the solids into the path of the rollers. Bullock (B3, B4) and Lofton et al. (L3) discuss thoroughly the use of mullers as mixers. This type of equipment is suitable for both wet and dry materials over a wide range of particle properties. However, mullers are not recommended for simple solids blending because of their high cost, high power requirements, and rough treatment of the solids.

G. SIGMA MIXERS

Sigma mixers get their name primarily from the shape of the blade which rotates in a fixed shell. This type of mixer is not normally used for dry solids

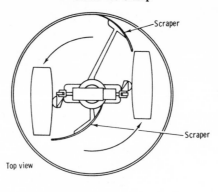

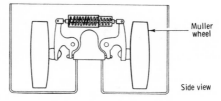

FIG. 8. Muller mixer (B4).

mixing alone. Solids are sometimes mixed in this type of mixer prior to the introduction of liquids. Its prime use is in the mixing of pseudoplastic and plastic materials.

VI. General Comments on Published Equipment—Performance Data

Mixing performance data have been published for a variety of dry particulate solids when mixed in various types of equipment (W1). General conclusions comparing the performance of different types of mixers are of doubtful value because the relative performance of the mixers is strongly influenced by the particulate solids being processed. The mixer best for one pair of dry solids may not be best for others. For this reason, tests using only the particulate solids involved in a specific application are of value in selecting appropriate equipment and operating conditions.

VII. Mixer Selection

Several questions arise in selecting a mixer for dry particulate solids.

1. What are the process requirements and objectives?
    (a) Production rate and batch cycle times (filling, mixing, emptying, cleaning).
    (b) Product properties and uniformity.

2. What mixing performance data are available?
3. What types of equipment meet the process requirements?
4. What size is needed?
5. What rotational speed?
6. What power?
7. What are the investment and operating costs?

Because of a dearth of quantitative information in the literature, equipment choice cannot easily be made without a test program or previous experience with a similar mixing operation. Laboratory equipment tests are frequently used as a basis for mixer selection. Comparative tests of different types of mixers acting on the same material should be made and the performance data examined to indicate which mixer best satisfies the process requirements.

If a rough investment cost comparison is made using as a basis a 30- to 50-cu. ft., 304 stainless steel twin-shell or double-cone blender, then the following can be said (F1):

1. The base unit is the cheapest.
2. Modifications to the base unit such as placing internal devices to aid mixing will increase the price about 50% over the base.
3. Ribbon blenders cost about as much as the modified base unit.
4. Mullers are about five times the cost of the base unit.

VIII. Mixer Testing

It has been pointed out earlier that tests on laboratory and plant-scale mixers are often carried out for use as a basis in selecting mixing procedures and equipment. A standard testing procedure for such evaluations has been established (A1) which is applicable to all batch and continuous mixing equipment. In this procedure, the criteria to be examined in the judging of the performance of solids mixing equipment are set forth. Such items as uniformity and quality of mix, mixing time, power requirements, equipment characteristics, etc., are discussed. The importance of sampling and interpretation of the sampling results is emphasized. The use of standard deviations is recommended as a measure of product uniformity. Statistical procedures are given for determining whether a calculated standard deviation is significantly different from a desired standard deviation and for the limits within which the true standard deviation will be 95% of the time in replicate tests (confidence limits).

Some examples of use of statistical methods for evaluating dry particulate solids mixing are presented in several papers. Stange's description of the use of standard deviation and confidence limits is summarized by Weidenbaum (W1). The use of quality control charts in dry solids mixing is described by

Herdan (H1) and (W1). The use of the chi-square test for determining whether a mixture is random is described by Weidenbaum (W1).

A word of caution seems necessary with respect to the results obtained from a laboratory test. Let us presume that a laboratory test shows a particular type of mixer to be completely satisfactory. In plant operation, the packaged product from the same type of mixer does not meet specifications. Immediately the mixer is suspected. Here, the mixer may be doing exactly what it was expected to do, but in the subsequent handling of the discharge from the mixer, particle segregation and unmixing may occur to give a product which does not meet specifications. Thus the mixer should not be blamed for poor performance, but the subsequent handling of the product.

IX. Scale-Up

Since laboratory pretesting is considered important, some provision for translating the laboratory results to production equipment must be available. One relationship for scale-up may be derived from model theory. Assuming that the ratio of centrifugal to gravitational forces is the only significantly different factor influencing performance of a laboratory sized mixer and a geometrically similar plant-scale unit, scale-up can be made by using the same Froude number. The Froude number expresses the ratio of centrifugal forces to gravitational forces, and is $N_{Fr} = N^2L/g$, where N is the speed, L is a characteristic length, and g is gravitational acceleration.

Thus, by model theory, when diameter, D, is used as the characteristic length,

$$\left(\frac{N^2D}{g}\right)_{\text{lab.}} = \left(\frac{N^2D}{g}\right)_{\text{plant}} \tag{5}$$

At the same value of N^2D/g, the same flow pattern will be obtained in large and small equipment. Equation (4) confirms this relationship for a rotating cylinder.

X. Summary

There are several forces acting on particles in equipment for mixing dry solids. These forces include gravitational and centrifugal forces, frictional forces between particles, and frictional forces between particles and equipment surfaces. Electrostatic forces may be significant in some mixing operations. These forces act on the particles and determine the flow pattern of the mass of particles being mixed within the boundaries imposed by the equipment geometry.

Particle motions occurring in a dry solids mixer cannot be described mathematically. Descriptive terms such as convection, diffusion, and shear are used for particle motions analogous to turbulence, molecular diffusion, and shear in liquids. These movements produce a random dispersion of one kind of particles among the other kinds present. During the mixing of particular solids, the uniformity of composition or properties improves as mixing proceeds. However, some segregation of particles of different size and shape may take place.

The performance of dry solids mixing equipment can be expressed in terms of the uniformity of the product. Methods of describing uniformity of a mixture involve some function of the standard deviation of composition or properties of samples of the product. Equipment performance can be judged by the final uniformity achieved or by the time required to obtain a desired uniformity.

A number of variables affect the performance of dry solids mixing equipment. The most important of these are particle properties and size distribution, the equipment geometry, and the rotational speed of the moving parts.

Rates of mixing in dry solids mixing equipment cannot be predicted. The maximum uniformity of composition of particles which have the same size, shape, and density can be predicted by statistical methods. Data in the literature on rates of mixing are specific for the particular solids and equipment used. These data are of limited value in predicting the performance of other types of equipment for mixing other types of solids. For example, the exponential relationship found for mixing solids in rotated cylinders are not applicable to other types of tumbling equipment.

The day-to-day control of industrial mixing operations to obtain a desired product with properties or composition within specified limits is a frequently occurring problem. Adequate statistical techniques are available for handling these problems.

The type of equipment best suited for mixing particulate solids depends on process and product requirements. Some processes require a product in which the scale of nonuniformity is small. Others which involve large particles do not. Selection of equipment must be based on laboratory tests or experience with similar large-scale mixing operations.

Further work on mixing of dry particulate solids is needed in two major areas:

1. Performance data for typical equipment and solids as illustrations of methods of measurement, calculations of product uniformity, control of product quality, and evaluation of equipment performance. Such data will be helpful in the selection of types of equipment best suited for specified process requirements.
2. Methods of predicting the relationships between mixing performance in small- and large-scale equipment.

List of Symbols

C constant in Eq. (4)
D inside diameter of cylinder, ft.
g gravitational constant, 32.2 ft./min.^{2}
L a characteristic equipment dimension, ft.
N rotational speed, r.p.m.
n number of samples
p fraction of particles that are black
S standard deviation defined by Eq. (2)
X volume % of cylinder occupied by solids
x composition or property of a spot sample
$\bar{x}$ average value of composition or property
χ^2 a statistical parameter called chi-square which can be used to calculate confidence limits (D1, D2)

SUBSCRIPTS

i ith value of a variable

References

(A1) A.I.Ch.E. Test Procedures, "Solids Mixing Equipment." Am. Inst. Chem. Engrs., New York, 1961.
(B1) Brown, C. O., *Ind. Eng. Chem.* **42**, 57A (1950).
(B2) Brownlee, K. A., "Industrial Experimentation." Chemical Publ., New York, 1953.
(B3) Bullock, H. L., *Chem. Eng. Progr.* **51**, 243 (1955).
(B4) Bullock, H. L., *Chem. Eng. Progr.* **53**, 36 (1957).
(D1) Davies, O. L., "Design and Analysis of Industrial Experiments." Hafner, New York, 1952.
(D2) Dixon, W. J., and Massey, F. J., "Introduction to Statistical Analysis." McGraw-Hill, New York, 1951.
(F1) Fischer, J. J., *Chem. Eng.* **67**, 120 (1960).
(G1) Greathead, J. A. A., and Simmons, W. H. C., *Chem. Eng. Progr.* **53**, 194 (1957).
(H1) Herdan, G., "Small Particle Statistics," pp. 342–345. Elsevier, New York, 1953.
(H2) Hoffman, A. M., *Ind. Eng. Chem.* **52**, 781 (1960).
(K1) Kraftmaker, Y. A., Lynlicher, A. N., and Shakhtin, D. M., *Zavodskaya Lab.* **24**, 893 (1958).
(L1) Lacey, P. M. C., *Trans. Inst. Chem. Engrs.* (*London*) **21**, 53 (1953).
(L2) Lacey, P. M. C., *J. Appl. Chem.* **4**, 257 (1954).
(L3) Lofton, W. M., Moore, W. B., Goldsmith, J., and Cooperman, N., "Summary Report to the National Engineering Company: Mixing Studies on the Simpson Intensive Mixer." University of Louisville Institute of Industrial Research, Louisville, Kentucky, 1948.
(O1) Overman, R. T., and Rohrman, F. A., *Chem. Eng.* **68**, 154 (1961).
(O2) Oyama, Y., and Ayaki, K., *Chem. Eng.* (*Tokyo*) **20**, 148 (1956), *Chem. Abstr.* **50**, 8264a (1956).
(R1) Rathmell, C., *Chem. Eng. Progr.* **46**, 116 (1960).
(W1) Weidenbaum, S. S., *Advances in Chem. Eng.* **1**, 240 (1958).

CHAPTER 11

Mechanical Design of Impeller-Type Liquid Mixing Equipment

Richard P. Leedom

E. I. du Pont de Nemours & Company
Wilmington, Delaware

Norman H. Parker*

Croton-on-Hudson, New York

* Consultant.

I. Introduction

This chapter will consider the basic principles of mechanical design of impeller-type mixing equipment. It will include a discussion of each component of the mixer, methods of calculating stresses, deflections, and natural vibration frequencies (critical speeds). Major emphasis will be placed on the practical aspects of equipment design, construction, and installation. Component parts of the mixer to be discussed include: (1) impeller, (2) shaft, (3) shaft coupling, (4) internal steady bearing, (5) stabilizer, (6) shaft seal, (7) drive, and (8) tank elements such as drive supports, baffles, and coils.

Figure 1 is an outline drawing of a typical liquid mixer. It illustrates (1) an impeller, which may be any one of the variety of types discussed previously in Chapter 3; (2) a shaft for transmitting the driving energy to the impeller; (3) one or more rigid shaft couplings which provide a convenient method for installation and disassembly of the unit; (4) a steady bearing which is some-times used to provide support for the shaft and reduce lateral deflections; (5) a stabilizer ring which is used under certain conditions to reduce vibration and deflection of the shaft; (6) a seal, consisting of either a stuffing box, mechanical seal, fluid seal, a throttle bushing or wiper seal to isolate the vessel contents from the external atmosphere; (7) a drive, consisting of the motivating device—electric motor, steam turbine, hydraulic motor, air motor, etc.—flexible coupling, gear reducer, and support stand; (8) baffles which provide resistance to swirl of the liquid; and (9) coils which provide a means for heating and cooling the vessel contents. Further details of coil and jacket construction may be found in Section IV, C of Chapter 5.

Of paramount importance in the design of any mechanical equipment are an analysis and evaluation of the forces acting on the system and an under-standing of how these forces are transmitted and amplified or damped. This applies equally well to fluid agitators. The effects on the equipment of the major forces of concern will be discussed under the applicable heading, but it is appropriate at this time to mention the mechanisms by which these forces are induced.

Shear and bending forces result from the viscous and inertial resistance of the liquid being stirred. Design of equipment to withstand these forces normally is straightforward and requires the application usually of only elementary stress analysis techniques. Less well understood are the intermittent and transient forces, resulting from turbulence in the liquid, which accentuate the effects of the steady viscous and inertial forces, induce bending in the shaft, and excite vibration of the system. Other unbalanced hydraulic forces may be caused by an asymmetrical location of the impeller or asymmetrical design of impeller and baffles. Gravitational forces are associated with a nonvertical shaft. Unbalanced centrifugal forces are produced during shaft and impeller rotation if the center of mass of the rotating elements does not coincide with

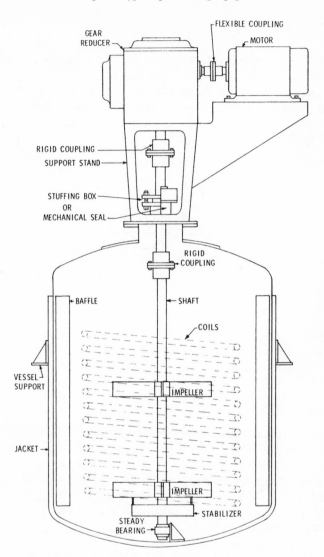

FIG. 1. Typical impeller-type liquid mixer.

the axis of rotation. Frictional forces are encountered in the drive gears, seal, bearings, etc. Periodic forces, characteristic of vibrating systems with a finite stiffness, can cause dangerously high stresses when resonant conditions exist. Except possibly for the gravitational forces, these forces probably cannot be eliminated, but the stresses induced by them in the equipment can be reduced to safe operating limits by judicious design.

II. Impellers

The basic types of impellers and their uses were discussed in Chapter 3. Many versions of these basic types are supplied by mixer manufacturers. Although from the process standpoint one particular design may have an advantage over the others, from the mechanical standpoint the same principles of design are applicable equally to all. For this reason, one type, the flat-blade turbine (Fig. 2), will be selected and discussed in detail.

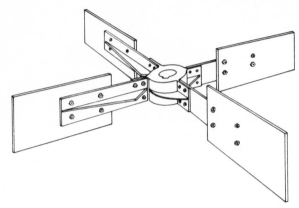

Fig. 2. Flat-blade turbine. (Courtesy of Philadelphia Gear Corp.)

A. TYPES OF CONSTRUCTION

Three basic types of construction are used—one-piece castings, fabricated assemblies, and combinations of both. The type of construction selected usually is determined by convenience or cost of manufacture.

Early mixer applications in open-top tanks led many mixer manufacturers to standardize on low-cost one-piece cast impellers (Fig. 3), but because of the limited size of access openings on pressurized vessels, impeller designs have been modified to the split hub (Fig. 4) and demountable blade types (Figs. 5 and 6) to facilitate installation and removal through those openings. Cast impellers have certain advantages—reasonable uniformity of material of construction and a hard surface caused by rapid cooling of the metal in direct contact with the mold. A basic disadvantage, however, is that heavy blade and hub cross sections often are required and the cost advantage of casting, when corrosion-resistant alloys are used, may be lost. Also heavy impellers increase other costs because impeller weight is a factor affecting shaft size.

Electric and gas welding have made possible simplifications in construction; but, for some alloys, welding alters the properties of the metal and affects its corrosion resistance. When welding is used, it should be confined to areas

FIG. 3. One-piece cast impeller. From Parker (P1).

close to the impeller hub where flow velocities are lowest. The weld projections and rough surfaces are sites for accelerated corrosion especially if the fluid velocity is high. Welded areas near the impeller tips, therefore, should be avoided if possible.

Bolted construction (Figs. 2, 4, 5, and 6) is used to permit assembly and disassembly of impellers. It has the advantages of permitting easy assembly and

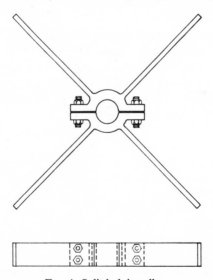

FIG. 4. Split-hub impeller.

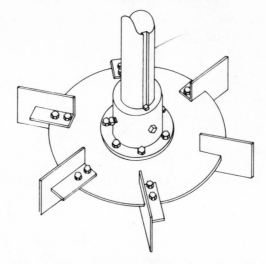

FIG. 5. Demountable-blade disk turbine. (Courtesy of Mixing Equipment Company).

requiring only the simplest tools. However, it has the disadvantages of providing excellent sites for crevice corrosion, of loosening by vibration under certain conditions, and of jamming at times due to stripped threads, corrosion, or foreign material in the threads.

The impeller construction is of interest to the user primarily from the standpoints of convenience in assembly and disassembly, corrosion, strength,

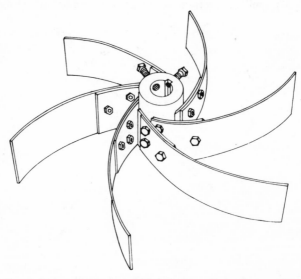

FIG. 6. Demountable curved-blade turbine. From Parker (P1).

weight (insofar as it affects the limiting operating speed), and the existence of crevices as possible harbingers of dirt and contamination from one batch to the next. For specific details on casting or welding as a method of fabrication, the literature of the Steel Founders' Society of America (B3), Alloy Casting Institute (A1), American Society for Metals (L1), or the American Welding Society (P2) can be consulted.

B. IMPELLER BLADE DESIGN

Since the mixing impeller must be designed to resist failure from all the forces acting on it, the first step in designing it is to estimate the magnitude of those forces. The practical way to determine the tangential component of such forces is to divide the shaft torque by the mean effective radius of the impeller. True shaft torque can be obtained only by measurement with a suitable torque-meter; but, for design purposes, it can be approximated satisfactorily from the output of the drive motor adjusted for the gear reducer efficiency and the impeller rotational speed by the formula (R1)

$$T_Q = \frac{63025PE_d}{N} \tag{1}$$

where T_Q = torque, in.-lb.$_F$;
P = horsepower output of the drive motor;
E_d = gear reducer efficiency (generally available from the manu-
facturer);
N = rotational speed, r.p.m.

The effective point of application of this torque to the impeller can be shown to be approximately 75% of the actual impeller radius. Consider the impeller blade to be a flat plate being dragged through the fluid as shown in Fig. 7. The drag force (B2) on a differential length is

$$dF = \frac{C_D \rho V^2 \, dA}{2g_c} \tag{2}$$

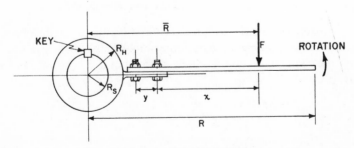

FIG. 7. Turbine blade force diagram.

and the total force, F, in lb.$_F$ is

$$F = \int_{R_h}^{R} \frac{C_D \rho \, (2\pi r N)^2 D_w}{2 \times 12^4 g'_c} \, dr = \frac{2\pi^2 \rho C_D}{3 \times 12^4 g'_c} \, D_w N^2 (R^3 - R_h^3) \tag{3}$$

where D_w = blade width, in.;
R = impeller radius, in.;
R_h = hub radius, in.;
C_D = drag coefficient, dimensionless;
g'_c = conversion factor, lb.ft./lb.$_F$ min.^{2};
ρ = fluid density, lb./ft.^{3}

Similarly, the total torque, T_Q, in in.-lb.$_F$ is

$$T_Q = \frac{2\pi^2 \rho C_D D_w N^2}{4 \times 12^4 g'_c} (R^4 - R_h^4) \tag{4}$$

and the mean effective radius becomes

$$\bar{R} = \frac{T_Q}{F} = \frac{3}{4} \left(\frac{R^4 - R_h^4}{R^3 - R_h^3} \right) \tag{5}$$

If the hub radius is small compared to the impeller radius, then

$$\bar{R} \cong 0.75R \tag{6}$$

Even for a large hub or a disk turbine with blade lengths from $R/2$ to R, $R \cong 0.804R$.

Good design practice dictates that the impeller blades shall be strong enough to withstand the maximum bending moment that the drive system can impart. When Eq. (6) is combined with the equation from elementary mechanics (R1)

$$S = Mc/I, \tag{7}$$

where S = the allowable working stress for the blade metal, p.s.i.;
M = the bending moment, in.-lb.$_F$;
I/c = section modulus for the blade cross section, in.^{3}.

it is possible to calculate the required section modulus along the blade. Since the greatest bending moment occurs at the point where the blade is attached to the hub, the required section modulus at this point can be calculated from Eqs. (1), (6), and (7):

$$\frac{I}{c} = \frac{M}{S} = \frac{F(\bar{R} - R_h)}{S} = \frac{63025 P E_d}{n \, SN} \left(1 - \frac{4}{3} \frac{R_h}{R} \right) \tag{8}$$

where n is the number of blades on the impeller.

This formula applies for any blade cross-sectional configuration and for swept-back or pitched-blade turbines. If the blades are flat plates

$$\frac{I}{c} = \frac{D_w D_t^2}{6} \tag{9}$$

where D_w = blade width, in. (vertical projected height for pitched turbines); D_t = blade thickness, in.

By substitution and rearrangement, the required blade thickness can be calculated as

$$D_t = \left[\frac{378100 P E_d}{nSND_w} \left(1 - \frac{4R_h}{3R} \right) \right]^{1/2} \tag{10}$$

Because the impeller blade is subjected to high frequency oscillations due to turbulence in the liquid being stirred, fatigue failures can become a problem. To avoid this, the design value selected for S should never exceed the endurance limit (D2) or yield point of the blade material, whichever is lower.

Bolted blade construction introduces additional design considerations. The thickness of both the bolted-on-blades and the hub wings must be adjusted for changes in cross section due to the bolt holes or changes in shape such as that shown in Figs. 2 and 5. Tensile stresses in the bolts must be calculated. For the impeller arrangement shown in Fig. 7, the maximum tension, F_b in lb.$_F$, in the bolts can be calculated as

$$F_b = \frac{Fx}{y} \tag{11}$$

The high frequency oscillations of the impeller blades due to turbulence can cause the bolts to loosen. This reduces the stiffness and natural vibration frequency of the impeller assembly and may cause it to fail by overstress or fatigue. For this reason, tight fitting bolts torqued to 80% of their yield point and locked in place with lock washers or wire are preferred for assembling bolted impellers.

C. Impeller Hub Design

When the blades are welded to a central cylindrical hub, the hub is subjected to bending moments by the blades and shear forces by the key, pin, or setscrew which transmits the shaft torque to the impeller. Rigorous analysis of the resulting stress pattern requires the use of thick-walled cylinder calculation methods, but conservative approximations can be made with published formulas (R1) for thin-walled cylinders. These formulas show that the points of greatest stress are apt to be at the key or pin and at the blade-hub junctures. This is especially true if the key is not tight fitting in the keyway or the impeller hub is loose on the shaft. If the key cocks slightly in the keyway, there exists only thin line or point contact between it and the hub. The resulting high stresses can deform the metal.

The bending moments in the hub, due to the blades, are greatest at the juncture of blade and hub, and reach as high as one-half of the bending moment in the blade itself (R1). It is possible, therefore, to calculate one of the limiting values for hub thickness from Eq. (10), by using a stress concentration factor of 1.35 and adding the keyway depth, R_k

$$R_h - R_s = 0.68D_t + R_k \qquad (12)$$

where R_h and R_s are radii of hub and shaft.

The bending moment in a symmetrical impeller hub is a minimum midway between adjacent blades; the keyway, therefore, should be located at that point. The maximum bending moment in the hub at a given torque level will decrease as the number of blades increases.

Torque transmission capability provides a second limiting value for hub thickness. Conservative practice is to use a hub with the same torque transmission capability as the shaft (B1). The minimum hub thickness, $(R_h - R_s)$, for torque transmission can be calculated (R1) from the equation

$$J/c = T_Q/S = \pi(R_h^4 - R_s^4)/2R_h \qquad (13)$$

by combining it with Eq. (1), incorporating a stress concentration factor, f, and rearranging as follows:

$$R_h - R_s \cong \sqrt[3]{\frac{2fT_Q}{\pi S} + R_s^3} - R_s = \sqrt[3]{\frac{63,025 \times 2fPE_d}{\pi NS} + R_s^3} - R_s \qquad (14)$$

Table I lists commonly used hub dimensions for various shaft sizes. Hub sizes shown in the table often will be larger than needed for most agitators because the agitator shaft, designed for stiffness or critical speed, usually has a greater torque transmission capability than that for which it will be used.

Table I

Frequently Used Shaft, Hub, Key, Setscrew, and Taper Pin Dimensions

Shaft diam. (in.)	Hub diam. (in.)	Hub length (in.)	Key size (in.)	Setscrew diam. (in.)	Taper pin No.
$1\frac{13}{16}$–2	$3\frac{3}{4}$	3	$\frac{1}{2} \times \frac{1}{2}$	$\frac{1}{2}$	6
$2\frac{5}{16}$–$2\frac{1}{2}$	$4\frac{5}{8}$	$3\frac{3}{4}$	$\frac{5}{8} \times \frac{5}{8}$	$\frac{5}{8}$	8
$2\frac{13}{16}$–3	$5\frac{1}{2}$	$4\frac{1}{2}$	$\frac{3}{4} \times \frac{3}{4}$	$\frac{5}{8}$	10
$3\frac{5}{16}$–$3\frac{1}{2}$	$6\frac{5}{16}$	$4\frac{11}{16}$	$\frac{7}{8} \times \frac{7}{8}$	$\frac{3}{4}$	10
$3\frac{13}{16}$–4	$7\frac{1}{8}$	$5\frac{1}{4}$	1×1	$\frac{3}{4}$	10
$4\frac{5}{16}$–$4\frac{1}{2}$	8	$5\frac{13}{16}$	$1\frac{1}{4} \times \frac{7}{8}$	$\frac{3}{4}$	10
$4\frac{13}{16}$–5	$8\frac{7}{8}$	$6\frac{5}{16}$	$1\frac{1}{4} \times \frac{7}{8}$	$\frac{7}{8}$	10
$5\frac{9}{16}$–6	$10\frac{1}{2}$	$7\frac{7}{16}$	$1\frac{1}{2} \times 1$	1	11

III. Shafts

A. Mechanical Requirements

The shaft is a vital component of the agitator and frequently limits its mechanical performance. In addition to transmitting torque, the shaft undergoes bending, and if not stiff enough or rigidly supported, it may vibrate badly and cause discomfort to personnel or damage to the equipment. Therefore it must be analyzed for combined torsional and bending stresses, deflection, and critical speed and must be selected to meet the limiting criterion for each. Inadequacy in these respects can result in failure of the shaft by overstress, failure of the seal due to excessive shaft bending, or failure of the bearings due to wear or impact.

B. Torsional and Bending Stresses

The combined shear stress, S_s, due to torsion and bending in the shaft can be calculated from the formula (R1)

$$S_s = (c/J) (M^2 + T_Q^2)^{1/2} \qquad (15)$$

By substituting $\pi D_s^3/16$ for J/c, if the shaft is round, and rearranging, the required minimum shaft diameter can be calculated as

$$D_s = 1.72 \sqrt[6]{\frac{M^2 + T_Q^2}{S_s^2}} \qquad (16)$$

where M = combined total maximum bending moment at the lower support bearing for a cantilever shaft, in.-lb.$_F$;
J/c = shaft section modulus in torsion, in.^{3};
D_s = shaft diameter, in.;
S_s = allowable shear stress, p.s.i.

The torsional shear stress, of course, results from twisting of the shaft by the action of the drive motor and the viscous and inertial resistance of the fluid, but the sources of bending stresses may not be so obvious. Transient unbalanced hydraulic forces, due to turbulence in the liquid or asymmetrical construction of the impeller and baffles, act laterally on the shaft and cause it to bend in a cyclic manner. Gravitational forces exist when a nonvertical shaft is used, centrifugal forces exist while the shaft is rotating if the shaft and impeller are out of balance, and gear or sheave forces can be developed in the agitator drive.

The hydraulic forces due to turbulence exist even when the impeller, vessel, and baffles are symmetrically constructed. Nothing has been published to explain or correlate them, but they can be expected to be functions of the fluid properties, geometry and speed of the impeller, and the drag forces on the impeller. Dimensional analysis shows that appropriate correlating groups for these forces are Reynolds number $(D^2 N\rho/\mu)$ and the force group $Fg_c'/\rho N^2 D^4$. The unbalanced forces, due to lack of symmetry of the equipment, have not been evaluated in the published literature, but it is known that they can be quite large. Preferred design, therefore, is symmetrical arrangement of impeller blades, baffles, and vessel walls.

C. OTHER SHAFT FORCES

Centrifugal forces can be calculated from the formula

$$F_c = 4\pi^2 (W_{si}/g) N^2 \epsilon \tag{17}$$

where W_{si} = combined impeller and shaft weight,

g = acceleration of gravity in ft./min.^{2},

ϵ = radius (in ft.) of the center of mass from the axis of rotation.

It is apparent from this formula that if W, N, or ϵ is large, F_c will be large. Since the process requirements normally determine the agitator speed, and design considerations limit the minimum weight of the equipment, the controllable variable is the eccentricity. The magnitude of ϵ can be minimized by care in the manufacture of the equipment, by insisting upon small tolerances on shaft straightness, by insuring that rigid-coupling faces and impeller bores are centered and square with the axis of rotation, by balancing the assembly, and by designing the equipment with a high enough critical speed to keep stress magnification low [see Eq. (18)].

When the shaft is overhung directly from the gear reducer and the drive gear is mounted on the shaft between the bearings, the stresses due to the gear forces alone are confined to the shaft between the bearings, but the deflection of the cantilever portion of the shaft is increased (Fig. 8). The higher deflection increases the eccentricity of the center of mass and, therefore, increases the centrifugal forces. Because large deflections between the bearings interfere with the proper meshing of the gears, it is not uncommon to have the shaft section between the bearings made short and heavier than the rest of the shaft. As will be shown later, this also is favorable from the critical speed and shaft seal maintenance standpoints.

Misalignment of the stuffing box or bearings, if three bearings are used, will increase shaft stresses and accelerate wear. If the stresses periodically reverse direction, fatigue failure can occur. Although a safety factor normally is incorporated in a well-designed agitator, the safety factor cannot provide for all contingencies. Care, therefore, should be taken to see that all parts are properly aligned as these forces can be quite large.

D. BALANCING

Two types of unbalance must be considered—static and dynamic. Static unbalance, so-called because it can be identified by static means, can be

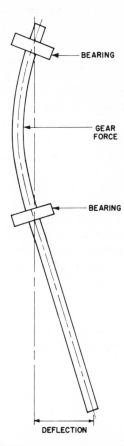

BEARING

GEAR
FORCE

BEARING

DEFLECTION

FIG. 8. Deflection due to gear force.

detected by allowing the rotor to turn freely on its axle, or on a dummy axle, between two parallel knife edges (Fig. 9); it will come to rest with the heavy side on the bottom. Addition of weights on the light side or grinding material off the heavy side until no free rotation is observed will produce balance. Dynamic unbalance, on the other hand, is manifest only during rotation and can exist even when the rotor is statically balanced. It induces a couple in the shaft (Fig. 10) which can be corrected by addition of two weights at opposite ends of the shaft. Size of the weights required must be determined dynamically by test. Thearle (T1) has outlined a procedure for effecting such balance; it requires more skill than static balancing.

When the rotor is "thin" with an axial blade width-to-diameter ratio less than one-sixth, or its rotational speed is low, dynamic balancing is seldom necessary. Most single turbine agitators fit into this category. With multiturbine agitators, however, an unbalanced couple can exist between two impellers. This is illustrated in Fig. 10. Statically, the assembly shown might be in balance, but if the center of mass of the upper impeller is on one side of the axis of rotation and the center of mass of the lower one is on the opposite side, the unbalanced couple will tend to bend the shaft as shown. At high speeds and wide impeller spacing, this couple can be quite large and cause excessive vibration and high bearing wear. Introduction of an equal and opposite couple by addition of two correction weights appropriately placed will balance the assembly dynamically, but this is not always the best practice. A more satisfactory arrangement is to balance statically each "thin" component of the assembly. Even though static balancing must be done to several different pieces of equipment, it is usually less costly than dynamic balancing of the complete assembly. It requires less skill, and can be done in practically any shop. Special equipment is needed for dynamic balancing. Also, if the components are balanced separately, they can be replaced and interchanged at will without upsetting the system. In all but a few cases, component balancing of fluid agitators is preferred to dynamic balancing of the agitator assembly.

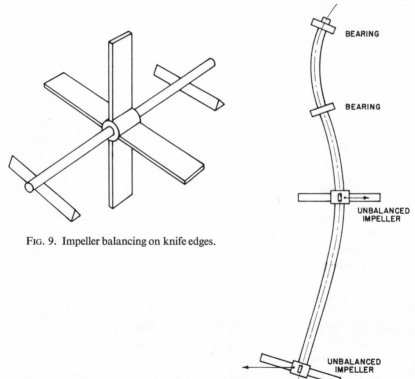

FIG. 9. Impeller balancing on knife edges.

FIG. 10. Twisting couple in multiturbine agitator.

E. Rotational Critical Speeds

As the speed of any rotating shaft, including an agitator shaft, is increased, it may tend to whirl or vibrate violently in a transverse direction. These vibrations increase both the torsional and bending stresses in the equipment and the shaft deflection. They can cause serious damage such as accelerated wear or fatigue failure of the rotating parts and supports. At certain speeds, the vibrations will be in resonance with the natural vibration frequency of the elastic system. At these speeds, called critical speeds, the vibrations can cause complete and rapid destruction of the equipment. The critical speeds correspond with the speeds at which the centrifugal force of the displaced center of mass of the shaft just equals the deflecting forces on it. The shaft vibrates because the centrifugal forces change direction rapidly as the shaft turns.

Shaft vibrations also may be caused by external periodic forces. Such forces may arise from a nearby compressor, centrifuge, steam engine, another agitator, or similar equipment. The vibrations occur at the frequency of the exciting force regardless of the natural frequency of the agitator, but if the two frequencies coincide the resultant forces can become very destructive. Since it is not always possible to anticipate or eliminate external exciting forces, the most practical correction is isolation of the agitator by the use of vibration dampers in the agitator or vessel supports or in the supports of the exciting equipment. Free vibrations, on the other hand, occur when the system is self-exciting; i.e., when periodic forces, such as the unbalanced hydraulic or gear forces discussed previously, are generated within the system.

Since in the present state of knowledge periodic forces generated within the system cannot always be eliminated, their effects must be reduced or controlled by equipment design. Mathematically, it can be shown that in an undamped free vibration system with a superimposed forced vibration resulting from eccentricity of the center of mass of the system, the vibration amplitude will equal the static deflection times $1/[1 - N^2/N_c^2]$ (D1). The quantity $1/[1 - N^2/N_c^2]$ is called the stress magnification factor. If damping is present, the stress magnification factor is (D1)

$$f_{SM} = \{[1 - (N^2/N_c^2)]^2 + [2f_d \cdot (N/N_c)]^2\}^{-1/2} \tag{18}$$

where N_c = critical speed, r.p.m.;
    f_d = damping ratio.

When an agitator is operating immersed in a fluid, some damping is always provided by the fluid. On the other hand, when the agitator operates in air after the liquid has been drained from the vessel, damping will be very small. Under these conditions, if the operating speed is close to the critical speed, the stress magnification factor may be so large that the equipment fails. Good practice dictates that the agitator never be operated within $\pm 30\%$ of the critical speed; preferably, it should not be this close. At an operating speed

70% of the critical speed, the stress magnification factor without damping will be approximately 2. The shaft design stress should not exceed the allowable working stress divided by this factor. If the liquid surface passes through the impeller during draw-off, or if gas is sparged into the liquid beneath the impeller, operating speeds as low as 40% of the critical speed may be required. As a general guide, 1200 ft./min. peripheral speed of the impeller can be taken as the practical limit below which the agitator should be designed to operate at less than 70% of the critical speed and above which the operating speed is 130% of the critical speed.

Theoretically, there are an infinite number of critical speeds for rotating machinery, but in fluid agitator design only the first, and occasionally the second, critical speeds are important. The higher ones are not important because of their remoteness from normal operating speeds.

The first critical speed can be calculated from a relationship that exists between the static deflection and spring constant of an elastic system (the shaft and supports). The relationship can be derived mathematically from the differential equation which equates the undamped elastic restoring force of the shaft to the inertial resistance of the impeller and shaft. For a lumped mass, m, at the outboard end of a cantilever weightless shaft with a spring constant, k, this equation is (D1)

$$m\frac{d^2 Y}{dt^2} + kg_c Y = 0 \tag{19}$$

One solution for this equation is

$$Y = Y_0 \cos\left[(kg_c/m)^{1/2}t\right] + V_0(m/kg_c)^{1/2}\sin\left[(kg_c/m)^{1/2}t\right] \tag{20}$$

where Y and Y_0 = displacements at times t and 0,
$\qquad V_0$ = velocity at $t = 0$,
$\qquad (kg_c/m)^{1/2}$ = the natural frequency of vibration, ω_c.

Each shaft deflection, δ, has a force, F, associated with it according to the relationship

$$F = k\delta. \tag{21}$$

If F is the force of gravity acting on the mass, m,

$$F = mg/g_c \tag{22}$$

and

$$mg/g_c = k\delta \tag{23}$$

Rearrangement of terms yields

$$kg_c/m = g/\delta \tag{24}$$

and then

$$\omega_c^2 = g/\delta \tag{25}$$

or

$$N_c = (60/2\pi)(g/\delta)^{1/2} = 187.7/\sqrt{\delta} \tag{26}$$

where N_c is in cycles/min. and g is in ft./sec.[2]. For a weightless, overhung shaft or cantilever loaded at its extremity (R1),

$$\delta = WL^3/3EI \tag{27}$$

where E = modulus of elasticity of the shaft material,
$\quad\ I$ = moment of inertia,
$\quad W$ = weight of the rotating mass,
$\quad\ L$ = overhung shaft length.

If there are loads at points other than the extremity, their effect can be included by converting them to an equivalent load and adding the equivalent load to W. The equivalent load, W_e, can be calculated from the formula (H2)

$$W_e = [W_1L_1^3 + W_2L_2^3 + W_2L_3^3 \cdots]/L^3 \tag{28}$$

where W_1, W_2, etc., are the separate loads at distances L_1, L_2, etc., from the first point of support or lower bearing.

For a shaft with uniform weight or where a heavy uniformly distributed load is placed along the shaft, the critical speed can be calculated by

$$_sN_c = 234(8EI/W_sL^3)^{1/2} \tag{29}$$

These critical speeds can be combined by means of the Dunkerly formula (D3) as follows to obtain the critical speed for the system

$$\frac{1}{N_c^2} = \frac{1}{_wN_c^2} + \frac{1}{_sN_c^2} \tag{30}$$

These formulas treat the shaft as a rigidly supported cantilever. They apply when the shaft between the bearings is short or has a high moment of inertia compared to the overhung portion. Otherwise, the critical speed should be adjusted downward by the bearing spacing factor

$$f_B = \left[\frac{L}{L + (aI_L/I_a)}\right]^{1/2} \tag{31}$$

where a = length of shaft between the bearings,
$\quad\ L$ = cantilever length,
I_L and I_a = moments of inertia of the cantilever and shaft between the bearings, respectively.

An additional correction which is applied to very high speed shafts is the effect of torsion on the critical speed. Greenhill (G1) showed that the critical speed of a shaft is reduced as the torque increases until at a torque of $\pi EI/L$, the lowest critical speed is reduced to 0 if the shaft is suspended from self-aligning bearings. For most fluid agitators, the torque is sufficiently low that this effect can be ignored.

When the overhung shaft is stepped or constructed from more than one material with different moduli of elasticity, the same relationship between critical speed and deflection applies [Eq. (29)], but a more sophisticated method of calculating deflection must be used. There are at least 10 different methods for so doing (B1, H1)—(1) the graphical funicular-polygon method of K. Culmann and O. Mohr, (2) the area-integration method, (3) the finite difference method, (4) the relaxation method of Sir Richard Southwell, (5) the Laplace transform method, (6) the matrix method, (7) the Q. Castigliano strain-energy method, (8) the conjugate beam method of H. M. Westergaard, (9) the M. Hetenyi trigonometric series method, and (10) the virtual work method.

The method of Culmann and Mohr, according to Bert (B1), is somewhat tedious and time-consuming, but it is probably best when it is necessary to determine deflections along the entire length of the shaft. The area-integration method is rigorous, but is cumbersome to apply. The finite difference method, which involves substituting finite differences for differentials in applicable differential equations, is well suited to computation by a digital computer. The matrix method also is excellent for digital computer solution, particularly if a standard program for inverting higher order matrices is available. Castigliano's strain-energy method is less known than some of the others, but it merits consideration and wider use because it is suitable for rapid calculation by slide rule.

Methods for calculating critical speeds of stepped shafts merit attention because the critical speed of the shaft may be increased as much as 10% to 15% by stepping or making hollow the lower end of the shaft. Too long a step, however, will reduce the critical speed. The following equation was obtained by the conjugate beam (moment area) method described by Hesse (H3) for calculating deflections of shafts:

$$\frac{\delta_S}{\delta_U} = \frac{\begin{aligned}WL^3/3 + w_S L^4/8 + (w_U - w_S)(L_U^3/6)(3L_U/4 + L_S)\\ + (WL_S^3/3)(w_U^2/w_S^2 - 1) + (w_S L_S^4/8)(w_U^2/w_S^2 - 1)\end{aligned}}{(w_U L^4/8 + WL^3/8)} \tag{32}$$

where L = total shaft length, in.;

  L_S = smaller diameter shaft (step) length, in. (see Fig. 11);
  L_U = larger diameter shaft length, in.;
  w_S = smaller diameter shaft weight, lb.$_F$/in.;
  w_U = larger diameter shaft weight, lb.$_F$/in.;
  W = impeller weight, lb.$_F$;
  δ_S = lateral deflection by gravitational force of a two-diam. (stepped) horizontal shaft, in.;
  δ_U = lateral deflection by gravitational force of a single diameter (unstepped) shaft (with w_U, lb.$_F$/in.), in.

The numerator is proportional to the deflection of a stepped or two-diam. shaft, and the denominator is proportional to the deflection of an unstepped or single

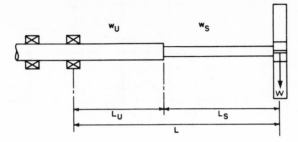

FIG. 11. Stepped agitator shaft.

diameter shaft when gravitational force acts on the masses of the shaft and impeller and the shaft has the same modulus of elasticity throughout its whole length. If the numerator and denominator are divided by $w_U L^4/8$, and the following substitutions are made

$$\delta_S/\delta_U = Y_R$$
$$W/w_U L = F_R$$
$$w_S/w_U = W_R$$
$$L_S/L = L_R$$

Eq. (32) becomes

$$Y_R = \frac{(8/3)F_R + W_R + (1 - W_R)(1 - L_R)^3(1 + L_R/3) + (8/3)L_R^3 F_R(1/W_R^2 - 1) + W_R L_R^4(1/W_R^2 - 1)}{1 + (8/3)F_R}$$

For each of several values of F_R, the values of W_R and L_R for which Y_R is a minimum were calculated. These optimum values of W_R and L_R are plotted in Fig. 12 with F_R as the abscissa. Also plotted in Fig. 12 is the ratio N_R which is the ratio of critical speeds of unstepped (single diameter) and stepped (two-diam.) shafts at the optimum values of W_R and L_R. Equation (26) was used to relate N_R and Y_R as follows:

$$N_R^2 = 1/Y_R$$

The curves in Fig. 12 provide a basis for approximating the optimum length and reduction in weight of step that should be made to maximize the critical speed of a simple cantilever shaft, i.e., a single impeller at the end of a solid overhung shaft. To use Fig. 12, calculate the ratio, F_R, of impeller weight to total shaft weight that would exist if no step were used. From this value, the ratio, W_R, of the stepped shaft weight per foot to the unstepped shaft weight per foot, and the ratio, L_R, of the length of step to total shaft length that will maximize the critical speed can be selected. The ratio, N_R, of the critical speed of the unstepped shaft to the critical speed of the optimum stepped shaft also is shown. Figure 12 applies only for maximizing critical speed by stepping and

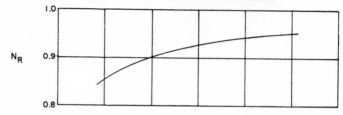

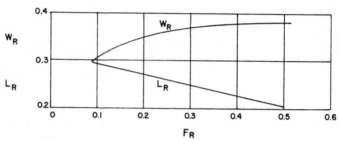

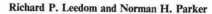

Fig. 12. Optimum proportions of a stepped shaft for minimum critical speed.

cannot be used to calculate actual critical speeds. One of the ten methods listed previously for calculating deflections of stepped shafts must be used to determine the critical speeds of shafts of other proportions.

F. Construction Materials

The material of construction from which a shaft is made will have considerable effect on its performance. The need for a suitable material to resist corrosive chemicals is obvious. From the strength standpoint, there is a large selection of materials with widely different properties, but there are very few materials which have more than a slightly higher value of modulus of elasticity than steel. Beryllium is one of the exceptions, but it is not apt to be used for agitator shafts. Materials are encountered with a lower modulus than steel; a few are listed in Table II. The table also includes the effect of temperature on the modulus and points out the need for designing the agitator on the basis of actual operating conditions. At elevated temperatures, the elastic modulus is lower; consequently, the critical speed and stiffness of the shaft will be lower. This difference should be allowed for in the design of the equipment.

G. Shaft Straightness

Tolerances on shaft straightness would be very desirable to reduce stresses from unbalanced centrifugal forces caused by the center of mass not coinciding with the center of rotation. Unfortunately, no such tolerances or criteria for shaft straightness have been published. Commercial centerless ground shafting

Table II

Modulus of Elasticity

Material	$E \times 10^6$ (lb./in.²)			
	100°C.	300°C.	500°C.	700°C.
Aluminum	10.3	9.1	7.2	2.6
Beryllium	40.0	—	—	—
Hastelloy B	29.5	27.2	28.7	29.3
Inconel	31.0	—	29.5	28
Monel	26	—	25	23.5
Nickel	30	—	29	27.5
Stainless steel	28.5	27.1	26.3	25.5
Steel	29.0	26.0	23.0	20.0
Titanium	15.5	14.4	13.2	11.8

is usually available with a straightness of $\frac{1}{8}$ in. in 5 ft. For high-speed applications, this is not good enough to insure satisfactory service from the agitator, but for low-speed applications, it is sometimes more economical to increase the stiffness of the system by using a slightly larger diameter shaft than to attempt to get the shaft any straighter. In any event, the straightness tolerances of agitator shafting are not cumulative. Even though "nonstraight" shafting can be tolerated in low-speed applications, the impeller should run true so that undesirable vibrations will not be encountered.

H. Couplings

Most agitators with long overhung shafts have rigid couplings to connect the agitator shaft and the gear reducer output shaft. The coupling facilitates shipment, installation, removal, and servicing of the agitator. Although it is an innocuous appearing block of metal, care in its design and fabrication can contribute significantly toward the satisfactory performance of the agitator.

The sleeve (Fig. 13) and the flanged coupling (Figs. 14 and 15) are the two most frequently used types of rigid coupling. The sleeve is used on smaller, less-exacting applications because it is low in cost, but on high torque, large bending moment applications, it is apt to prove less satisfactory than the flanged type. The reasons for this become manifest from the requirements that the rigid coupling must meet.

1. The coupling must be capable of transmitting the agitator torque.
2. The coupling must provide at least as much rigidity as the shaft if the critical speed of the agitator is not to be reduced.
3. The coupling must be strong enough to withstand the bending moments imposed upon it by the unbalanced hydraulic and centrifugal forces.
4. The coupling must provide good alignment between the shafts being connected.

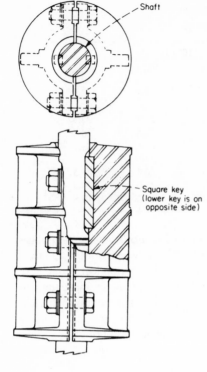

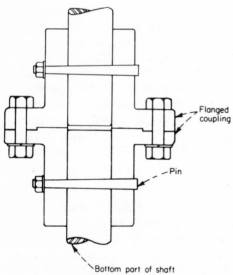

FIG. 13. Clamp-on type of sleeve coupling. From Parker (P1).

FIG. 14. Flanged coupling with taper pins. From Parker (P1).

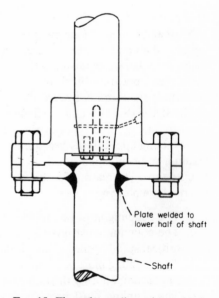

FIG. 15. Flanged coupling with welded joint and tapered joint. From Parker (P1).

5. Because it must be assembled and disassembled in the field, frequently under inconvenient working conditions, the coupling should be relatively easy to take apart and reassemble, and should preferably be self-aligning.
6. The coupling must be capable of taking the thrust due to the weight of the agitator.

Failure of the coupling to fulfill any one of these functions adequately will result in excessive maintenance costs and possibly operating failure. Attention, therefore, should be given to ascertaining that it is well-designed, well-made, and that all parts are match marked or dowelled to insure correct assembly.

When a flanged coupling is used, the greatest rigidity is attained if the coupling halves are mounted on the shaft ends with an interference fit as shown in Fig. 14 or are welded in place as shown in Fig. 15. The interference fit can be obtained by using hydraulic pressure or heat to expand the coupling (Fig. 15) and then letting the coupling shrink in place on the shaft. Since these techniques are not always practical, the wedging action of a taper is sometimes used to insure metal-to-metal contact in a tight fit. A small taper angle permits making the tightest fit, but if it is smaller than the angle of friction between the shaft and coupling, difficulty may be experienced getting the two pieces apart. In practice, therefore, steeper taper angles are used. Clearance fits are not recommended.

Tightness of fit on the shaft is important, because, when loose, the shaft will cock as shown in an exaggerated way in Fig. 16. Hydraulic and centrifugal

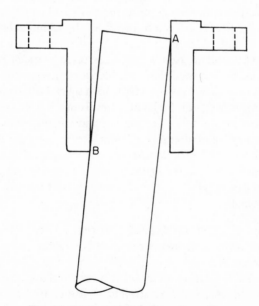

FIG. 16. Loose shaft fit in flanged coupling.

forces at the impeller will be magnified at the coupling in the ratio of total shaft length to length inserted in the coupling. Shaft and coupling will have only line or point contact at *A* and *B*, and the tremendous stresses there will cause "wallowing out" of the coupling with consequent increasing looseness. The looseness causes increased deflection, lower critical speeds, and in turn can be the cause for complete failure. The magnified centrifugal forces increase the stresses in the entire agitator assembly and can provide the exciting force for annoying vibration of the system. Lack of concentricity, angularity of the bores, or lack of squareness of the mating faces of the coupling also increase the impeller mass eccentricity and centrifugal forces to produce similar results.

The coupling must transmit the shaft torque from one coupling half to the other. This can be done by a key between the coupling faces, but more generally it is accomplished by friction between them. To effect this transfer, the coupling bolts should be torqued to 80 % of their yield point. Insufficient tension in the bolts will permit the nuts to work loose while the agitator is in operation, reduce the critical speed, and increase the stresses in the shaft. It too often happens that a shaft fails due to fatigue because a loose coupling caused the shaft stresses to exceed the endurance limit of the shaft material. Even if the loose coupling does not produce such a drastic condition, looseness of the faces will put the flange bolts in shear and bending, a combination which is almost certain to cause failure of the bolts.

I. STEADY BEARINGS

Excessively large shaft deflections at the shaft seal can be reduced, or too low a rotational critical speed can be increased, by adding a bearing inside the vessel at the lower end of the shaft. An intermediate bearing also may be used. Although the two problems usually can be corrected by use of a larger diameter shaft or a stepped or hollowed shaft, the increased investment cost sometimes is not considered justified. This is apt to be so in several situations, for example, when the agitator is used at infrequent intervals in a low activity storage tank, when the agitator is in a noncritical service and can be removed from service for repairs without interrupting the main processing operation, when the agitator is used in a batch process which can be interrupted for repairs without spoiling the batch, or when the agitator must be confined to prevent damage such as operation in a draft tube. Use of the steady bearing, however, introduces many new problems.

1. Alignment of the steady bearing and gear reducer support bearings is difficult or impossible unless the shaft is straight beyond normal commercial tolerances.
2. Use of a steady bearing requires extra care in the alignment of the shaft coupling and in procuring straight shafting, If, because of coupling misalignment or inadequate shaft straightness, the shaft must be forced

into the steady bearing, a high load is imposed immediately upon the bearing and a bending moment is set up in the shaft. When the shaft rotates, the steady bearing and the reducer support bearings wear excessively, and the shaft undergoes reverse bending so that fatigue failure becomes a distinct possibility.

3. Lubrication of the steady bearing is difficult unless the process fluid has lubricating properties. Although there are some materials available today which can be operated without lubricants, any of these materials will give better service if lubricated.

4. If there are abrasive solids present in the liquid, wear and consequent replacements can be a large burden on the operation.

5. Corrosion problems frequently are associated with the steady bearing or its supports.

6. Replacement of the steady bearing can be quite costly if it fails during use and the batch is spoiled. Just the manpower and material costs for bearing replacement can be expensive.

Although improvements in shafting manufacture and bearing materials are making the application of steady bearings more tolerable, most of the problems associated with them always will exist. Whenever possible, therefore, the use of a steady bearing should be avoided. Where one must be used, a design with replaceable sleeves for shaft and stationary bearing surfaces is advantageous. Figure 17 shows a commercial design.

J. STABILIZERS

Even though it is not considered good practice to operate an agitator in the speed range between 70 and 130 % of the first critical speed, there are times, such

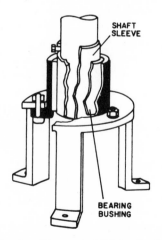

FIG. 17. Steady bearing. (Courtesy of Mixing Equipment Company.)

as when a variable speed drive is used, when it is impractical to avoid this range. Under such operating conditions, periodic shaft deflections may be very large; they will be especially large if simultaneously, the fluid is being drained out of the vessel and the liquid surface passes through the impeller. The unbalanced hydraulic forces associated with these deflections are very large and may cause the impeller and shaft to vibrate violently. To damp such vibrations, a stabilizer sometimes is used.

A stabilizer is a device installed on the bottom end of a cantilever shaft below the bottom impeller to provide damping by fluid drag when it is moved through the liquid (Fig. 18). It is frequently a circular cylinder with its longitudinal axis coincident with the shaft axis, but other shapes, such as spheres or symmetrically shaped geometric configurations, may be used. When a transient unbalanced hydraulic force and the reinforcing centrifugal action push the impeller away from the axis of rotation, the movement creates a counterbalancing drag force (F_D) on the stabilizer equal to

$$F_D = C_D\rho(AV^2/2g_c)$$

where ρ = fluid density,
  A = projected area of the stabilizer in the direction of movement,
  V = velocity of motion (B2).

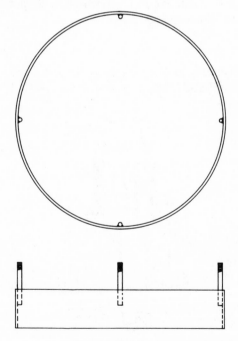

FIG. 18. Stabilizer.

This drag force provides damping which, as shown by Eq. (18), reduces both the amplitude of vibration and the stress magnification factor. The stabilizer, however, increases the weight and effective length of the shaft and thus reduces the agitator critical speed; it should be sized, therefore, so that it makes the denominator in Eq. (18) greater than 1.

IV. Gear Reducers

As a component of the fluid agitator, the gear reducer performs several functions. Its primary purpose, of course, is to provide a means for reducing the speed between the prime mover (electric motor, etc.) and the agitator shaft. A second function is to provide support for the cantilever agitator shaft when other supports are not provided. To perform these functions most effectively, special requirements are imposed upon the reducer which are not always met by the ordinary gear reduction units selected from most commercial catalogs.

1. The agitator reducer output shaft usually is vertically downward, but sometimes a bottom-entering or side-entering drive is used which has the shaft vertically upward or horizontal.
2. The overhung load on the agitator reducer output shaft is frequently higher than the commercial unit counterpart is designed to take. The unbalanced forces due to turbulence in the liquid are leveraged by the long overhung shaft, and the centrifugal forces acting on the shaft are amplified by the stress magnification factor discussed in Section III,E. Heavier shaft bearings, therefore, must be used to withstand the higher loads.
3. Because of critical speed considerations, combined torque and reverse bending of the shaft due to transient hydraulic forces, and limited allowable shaft deflections at the seal, the agitator reducer output shaft often must be extra large to provide the required shaft stiffness. When an ordinary commercial gear reducer is applied to agitator service, a unit with unusually large torque rating may have to be used just to secure the required size output shaft and required overhung load rating of the bearings.
4. The bearings supporting the output shaft of a reducer used in agitator service often will seem to be oversized but the transient impact type forces to which the bearings are subjected and the need for stiffness in the bearings themselves make the heavy-duty bearings necessary. The stiffest assembly will be obtained if all the "end play" has been eliminated from the output shaft bearings by preloading them slightly. The bearings must have sufficient capacity to take the operating load and any load due to expansion of the shaft against the bearings during operation.
5. A large bearing span lowers the shaft critical speed [Eq. (31)], increases deflection of the cantilever shaft, and causes poor meshing of the reducer

gears due to increased deflections where the gears are mounted on the output shaft. On the other hand, a small bearing span or increased shaft diameter between the bearings is sometimes used to increase the shaft stiffness. The shorter bearing span puts a higher load on the bearings and further manifests the need for heavy support bearings. The effect of shaft deflection on gear meshing and the effect of gear forces on shaft deflection can be isolated by separate support of the shaft—either by an internal quill or by external bearings—and torque transmission to the output shaft by a flexible coupling.

6. A service factor normally is applied to the required rating of the gear reducer to allow for the effects of load variability. An example is a factor of 1.4 for the ratio of gear reducer power rating to agitator power transmitted by the gear reducer. Such a ratio is typical of gear reducers used 24 hours per day with moderate shock loading like that produced by stirring liquid in a vessel under turbulent flow conditions. Gear reducer manufacturers provide recommended service factors for their equipment.

Because the drives for turbine-type agitators have output shafts extending vertically downward, the possibility of gear and bearing lubricant leakage is

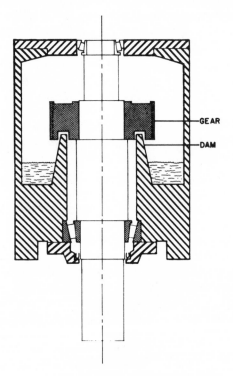

FIG. 19. Agitator-drive dry well.

increased especially when the oil seals start to wear. To obviate this objectionable condition, it is good practice to install a dam, or "dry well," which
surrounds the shaft and extends above the static oil level (see Fig. 19).

Other features of the agitator drive which may affect its selection are the
ease with which the reduction ratio can be changed if the need arises, the
vertical or lateral space available, type of prime mover required—viz., a
standard foot-mounted or flanged-type electric motor, input shaft overhung
load if a belt or chain drive is used, ease of mounting and demounting the
reducer, and method of mounting so that alignment can be maintained. Since
all of these considerations assume more or less importance, depending upon
the specific application requirements, little can be said beyond mentioning
them as possible considerations affecting drive selection for special cases.

V. Shaft Seals

When toxic or flammable materials are to be agitated, when the process
materials must be protected against contamination, when evaporation of
volatile materials will cause excessive losses, or when the process pressure is
different from atmospheric pressure, a shaft seal must be provided to isolate
the vessel contents from the atmospheric surroundings and prevent outflow of
fluid from the vessel or inflow of foreign material to the vessel. A number of
methods are available for effecting this seal, but the two most commonly used
methods are stuffing boxes and mechanical seals.

A. STUFFING BOXES

The stuffing box (see Fig. 20) effects a seal by radial expansion of a soft
material—such as rubber, plastic, or fibrous packing—against the shaft and
the carefully machined cylindrical housing surrounding it. Radial expansion

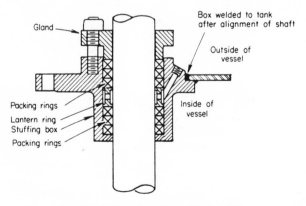

FIG. 20. Stuffing box. From Parker (P1).

of the packing is accomplished by axial compression of the packing against a metal lip at the bottom of the stuffing box. A gland follower exerts the compressive force when it is tightened against the packing by two or more accessible external bolts. The follower sometimes is constructed in more than one piece to facilitate repacking the stuffing box, but at higher pressures split construction may not be practical because it is a weaker assembly. A lantern ring frequently is inserted inside the stuffing box between the rings of packing. This ring may be split to facilitate replacement. The lantern ring provides a channel for circulating fluid through the stuffig box. The fluid lubricates and cools the contacting parts, reduces wear and deterioration of the packing, and improves the effectiveness of the seal by filling up pores and void spaces with fluid. When a suitable fluid is used, it also dissolves any leaking vapors and carries them away to a convenient disposal point.

To be effective, packing must have the following properties:

1. It must be compressible and easily deformed.
2. It must be impervious to air and the gases or liquids being sealed.
3. It must be corrosion-resistant to all the materials with which it will come in contact.
4. It must be resistant to service temperatures.
5. It must have a low coefficient of friction against the agitator shaft or be easily lubricated to prevent excessive wear of the packing and shaft.
6. It must be resilient so that it will not be permanently deformed by normal operating deflections of the agitator shaft.
7. It must have a high compressive strength and be nonextrudable in order to resist the large forces exerted by the agitator shaft against it and to transmit the axial compressive forces from one packing ring to the next.

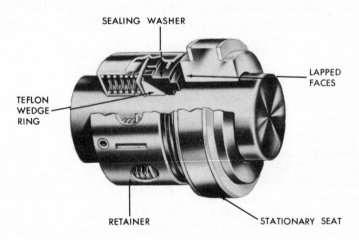

FIG. 21. Mechanical seal. (Courtesy of Crane Packing Co.)

B. Mechanical Seals

A mechanical seal (Fig. 21), consists of two optically flat ring-shaped surfaces—one fixed and the other mounted on the shaft—which rotate against each other in such close proximity that little or nothing can pass between them. To prevent rapid wear, the mating surfaces must have a high lubricity or be lubricated by the cooling fluid. Sufficient force must be maintained on the two surfaces to insure keeping them together, but the force should not be so large that excessive wear will result. It is important, therefore, that the applied force be controlled and that it be applied consistently without reversing its direction. An insufficient applied force or a reversed force permits the mating faces to open and leak.

The closing force on the seal faces is the resultant of several forces.

1. The pressure drop across the seal faces always tends to open them. This pressure usually is assumed to vary linearly across the faces so that the force tending to open the faces is nearly one-half the product of the pressure drop and seal face area. However, this force may be considerably less according to Mayer (M1) who maintains that no fluid pressure exists between the seal faces at face pressures above a certain minimum.
2. Forces from one or more springs which counterbalance or exceed the forces tending to open the faces.
3. Other fluid forces which are a function of the geometrical design of the seal and the pressure drop across it. When the pressure on one side of the seal face is much higher than the hydraulic pressure on the other side, extra strong springs are required in an unbalanced seal (Fig. 22) to keep the seal faces closed. Since the springs would exert an excessive force if the pressure difference were reduced, the net fluid force on the seal head can be balanced (reduced to zero if desired) by stepping the shaft under the seal so that the area C between the step and outside diameter of the face is less than the area of the seal face, B, as shown in Fig. 22b. The ratio of the two areas (C/B) is called the balance per cent.

A pressure difference of 100 p.s.i. across the seal face is the practical limit above which a balanced seal should be considered, but the actual limiting pressure difference is determined by the PV rating of the seal face materials and the seal power consumption. The PV rating (product of load in pounds per square inch and surface speed in feet per minute) is an index used to describe the limiting wear characteristics of bearing materials. The power consumed by friction in the mechanical seal is converted into heat. If an excessive temperature is developed, this will cause degradation of the seal lubricant, weakening of the seal structural members, and acceleration of seal face wear.

Methods for calculating seal power consumption are as follows:

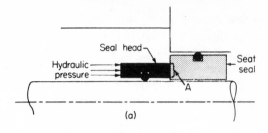

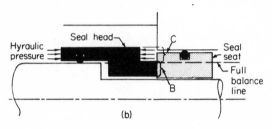

Fig. 22. Mechanical seal balancing: (a) unbalanced and (b) fully balanced. From Tankus (T2).

1. When a lubricating film exists between the faces and no surface rubbing occurs (S1)

$$P = \pi^3 D_{sf}^3 N^2 r_{fw} \mu / 33{,}000 z_{sf} g_c'$$ (33)

2. When the faces are not lubricated and rub against each other (M1)

$$P = 144\pi^2 D_{sf}^2 N p \mu_f r_{fw} / 33{,}000$$ (34)

where P = horsepower;

D_{sf} = mean diameter of seal face, ft.;

r_{fw} = width of face, ft.;

p = pressure between the faces, p.s.i.;

μ_f = coefficient of friction between the seal faces;

μ = viscosity of the lubricant, lb./ft. min.;

z_s = face separation, ft.

Advantages claimed for the mechanical seal are economy, zero or controlled leakage over a long service life, reduced operating power requirements, elimination of shaft or shaft sleeve wear, freedom from periodic maintenance, and relative insensitivity to shaft deflection or end play. Disadvantages are the higher first cost and the need for careful handling and installation because it is a precision piece of equipment.

The life of the seal depends upon the reliability of the materials of construction. Under commercial conditions, a 2-year, 24-hour/day seal life is common, and under unusual conditions 15 to 20 years is possible. The limits on service conditions are determined by the effects of temperature, pressure, speed, and lubricity on the PV rating of the seal. (Limiting face speeds of 18,000 ft./min. regardless of the materials in contact is not often a consideration because fluid agitators usually operate at comparatively low speeds.) The normal pressure limit is 2000 p.s.i. for mechanical seals, but improvements have been made which in some cases permit handling pressures as high as 3000 p.s.i.

C. Clearance Seals

Clearance seals are used when the pressure differentials are beyond the design limitations of the face-type mechanical seal, and a moderate amount of leakage can be tolerated. They usually are designed so that the annular clearance between the rotating shaft and stationary housing is a minimum consistent with the radial motion of the shaft. If the clearance is too small, of course, the shaft and housing will rub and cause wear, galling, or scoring of the shaft unless materials with good bearing properties are used.

There are two types of clearance seals: the labyrinth and the bushing or ring seal (K1). The former is usually less expensive, but allows a higher leakage rate. It is often used on air compressors where leakage has only a minor effect on efficiency, but it seldom is used for agitators because process vapor leakage is often objectionable. The bushing, a close fitting, stationary sleeve within which the shaft rotates, is more frequently used for agitator shafts because of its lower leakage rate and the possibilities it offers for designing it to provide support for the shaft. Volumetric leakage rate in laminar flow of an incompressible fluid through a fine annulus is given by (K1)

$$Q = \frac{720 g_c \pi D_s r_{sc}^3 \Delta p}{\mu L_b} (1 + \tfrac{3}{2} e^2) \tag{35}$$

where Q = volumetric flow rate, in.^{3}/sec.;
  r_{sc} = radial clearance, in.;
  Δp = pressure drop, p.s.i.;
  μ = absolute viscosity, lb./ft. min.;
  L_b = length of bushing, in.;
  e = eccentricity ratio which for perfect centering of the shaft is 0, and 1 when the shaft rubs on the sleeve.

If a seal operates at a high temperature, differential expansion may become a problem when the shaft and bushing are made from dissimilar materials, or when there is a substantial temperature gradient between them. To prevent seizing, it is best to select a bushing material which has a slightly higher coefficient of expansion than the shaft material. If rubbing contact is made, the heat generated will cause the bushing to expand and free itself.

Some of the problems that might be encountered in the use of bushings are flashing of the sealed liquid as the pressure on it falls through the seal, excessive power consumption by the seal, and changes in leakage rate due to expansion of the seal annulus or change in viscosity of the seal fluid. Flashing of liquid in the seal due to pressure "letdown" or temperature increase may cause dissolved solids to be deposited in the annulus. If the solids build up sufficiently to score the shaft or to cause friction forces to become excessive, the seal will become inoperative.

Power consumption by the seal bushing due to viscous shear can be calculated from the formula (K1)

$$P = \pi^3 \mu D_s^3 L_b N^2 / 12^3 \times 550 \times 60 g_c r_{sc} \qquad (36)$$

in which P is in horsepower and N is in r.p.m. This energy is dissipated as frictional heat in the bushing and causes the temperature to rise. The increased temperature not only reduces the viscosity of the fluid, but causes the clearance to increase due to differential expansion of the bushing and shaft. To some extent, the effect is self-regulating because both the lower viscosity and wider clearance permit increased leakage which in turn provides some cooling effect. The lower viscosity also causes power consumption to drop and heat generation to diminish.

D. Centrifugal, Wiper, and Lip Seals

Other types of dynamic seals less frequently used for agitator shafts are fluid centrifugal seals, wiper or lip seals, and split ring seals (K1). These types are limited to use in low pressure service where less exacting applications are encountered and some leakage can be tolerated.

E. Seal Selection

The choice of a seal depends upon the requirements of the process. The mechanical seal is preferred for applications requiring minimum leakage and infrequent maintenance. The leakage rate is usually less than 1% of the rate encountered with a stuffing box. A stuffing box, on the other hand, has a lower first cost and can readily be tightened or repacked often without removing the unit from service. Frequent adjustments are needed to keep the packing tight, and complete replacement of the packing is required periodically. The bushing is used for very high pressure seals because other types do not work satisfactorily. The fluid centrifugal seal is cheap, but slight pressure surges will blow the fluid out. The wiper and split ring seals are usually limited to cases where dirt and foreign material must be kept out of the vessel. They are not designed to take much shaft runout and, therefore, cannot hold a large pressure differential.

VI. Baffles and Auxiliaries

A common method for preventing rotation of the fluid in the vessel is the use of four or more radial wall baffles one-tenth to one-twelfth the vessel diameter in width equally spaced around the vessel wall and running the full length of the wall. A rational basis for designing the baffles is to assume that the full torque developed by the agitator is resisted by a force applied at the midpoint of the baffle and concentrated at a point directly opposite the impeller. This is conservative. Actually, the resisting force will be distributed over a length of the baffle, but data are not available to show how. With these assumptions, the resisting force may be calculated as follows:

$$F_B = \frac{T_Q}{n_b[(T/2) - (B/2) - B_c]} = \frac{63,025PE_d}{(n_b/2)(T - B - 2B_c)N} \tag{37}$$

where F_B = force acting on one baffle,
$\quad n_b$ = number of baffles,
$\quad B$ = radial baffle width,
$\quad B_c$ = clearance between the baffles and the vessel wall.
If flat plates are used for baffles and F_B is assumed to be applied midway between two baffle supports, the required thickness of the baffles can be calculated from

$$\frac{I}{c} = \frac{BB_t^2}{6} = \frac{M}{S} = \frac{F_B L_s}{4S} \tag{38}$$

and

$$B_t = \left[\frac{3F_B L_s}{2BS}\right]^{1/2} \tag{39}$$

where B_t = baffle thickness, in.;
$\quad L_s$ = spacing between baffle supports, in.;
$\quad S$ = allowable bending stress, p.s.i.

The allowable stress should be less than the fatigue limit for the baffle material because F_B is a periodic force, the frequency of which is related to the agitator speed and number of impeller blades.

Fixed formulas cannot be written down for analyzing all auxiliaries, such as dip pipes and heating coils, but a general approach is to calculate the drag force on such auxiliaries, assuming the liquid velocity is a function of impeller speed (see Chapter 4). The bending moments induced in the equipment by this force can then be used as a basis for determining the required strength. The natural vibration frequency of such auxiliaries also should be calculated from their static deflections, assuming application of a force equal to their weight [see Eqs. (26) and (29)].

VII. Summary

The basic principles of mechanical design of impeller-type mixing equipment are well developed. If the forces imposed upon each of the components of the equipment are known, it is possible with existing known methods to establish the dimensions of the components required to withstand the stresses developed and to be within desired tolerances on deflections due to these stresses. Methods have been established for calculating natural vibration frequencies also. It is possible, therefore, with currently available procedures to specify the mixer design so that mechanical difficulties due to excessive stresses, deflections, or vibrations can be avoided.

Data have not been published on the intermittent and transient forces on the impeller which result from turbulence in the liquid. Such forces induce bending of the agitator shaft and may be the exciting force for excessive shaft vibration. Data also are lacking on the forces due to asymmetrical design and location of the impeller and baffles. Allowable tolerances on shaft straightness, shaft deflections, and static or dynamic unbalance of shaft and impeller also need to be developed.

List of Symbols

a distance between bearings, in.

A area, ft.^{2}

B baffle width, in.

B_c baffle clearance, in.

B_t baffle thickness, in.

c distance of outermost part of a beam to the neutral axis for bending, in.

C_D drag coefficient, dimensionless

D impeller diameter, ft.

D_s shaft diameter, in.

D_{sf} mean seal face diameter, ft.

D_t impeller blade thickness, in.

D_w projected vertical height of impeller blade, in.

e eccentricity ratio (distance between centers of inner and outer circular boundaries of annulus divided by average clearance)

E modulus of elasticity, lb.$_F$/in.^{2}

E_d gear reducer efficiency, dimensionless

f stress concentration factor

f_B bearing spacing factor, dimensionless

f_d damping ratio

f_{SM} stress magnification factor

F force, lb.$_F$

F_b tensile force in impeller blade bolts, lb.$_F$

F_B force applied to baffles, lb.$_F$

F_e centrifugal force due to an eccentric mass, lb.$_F$

F_D drag force, lb.$_F$

F_R ratio of impeller weight to unstepped shaft weight

g gravitational acceleration, ft./min.^{2}

g_c gravitational conversion factor in equation, $f = ma/g_c$, lb. ft./lb.$_F$ sec.^{2}

g_c' gravitational conversion factor, lb. ft./lb.$_F$ min.^{2}

I bending moment of inertia, in.^{4}

I/c bending section modulus, in.^{3}

I_a moment of inertia of shaft between bearings, in.^{4}

I_L moment of inertia of cantilevered shaft, in.^{4}

J polar moment of inertia, in.^{4}

k spring constant, lb.$_F$/ft.

L shaft length from outboard drive bearing to cantilevered end, in.

$L_1, L_2, L_3, \ldots,$ distances from impellers, 1, 2, 3, . . ., to first point of support of shaft, in.

L_b bushing length, in.

L_R ratio of step length to total shaft length

L_s spacing between baffle supports, in.

L_S smaller diameter shaft (step) length, in.

L_U larger diameter shaft length, in.

m mass, lb.

M bending moment, in. lb.$_F$

n number of impeller blades

n_b number of baffles

N rotational speed, r.p.m.

N_c rotational critical speed, r.p.m.

N_R ratio of rotational critical speeds of unstepped and stepped shafts

$_sN_c$ rotational critical speed of a shaft with an equally distributed load, r.p.m.

$_wN_c$ rotational critical speed of a weightless shaft with a concentrated load, r.p.m.

p pressure, lb.$_F$/in.^{2}

P power, hp.

Q flow rate, in.^{3}/sec.

r radial dimension, in.

r_{fw} seal face width, ft.

r_{sc} radial clearance of annular seal, in.

$\bar{R}$ mean effective radius, in.

R impeller radius, in.

R_h hub radius, in.

R_k radial key depth, in.

R_s shaft radius, in.

S allowable working stress, lb.$_F$/in.^{2}

t time, sec.

T vessel diameter, in.

T_Q torque, in.-lb.$_F$

V velocity, ft./sec.

w_S smaller diameter shaft weight, lb.$_F$/in.

w_U larger diameter shaft weight, lb.$_F$/in.

W weight, lb.$_F$

$W_1, W_2, W_3, \ldots,$ weights of impellers 1, 2, 3, . . ., lb.$_F$

W_e equivalent weight of shaft and impellers, lb.$_F$

W_R ratio of stepped shaft weight per foot to unstepped shaft weight per foot

W_s shaft weight, lb.$_F$

W_{si} combined shaft and impeller weight, lb.$_F$

x distance, in.
y distance, in.
Y displacement, ft.
Y_R ratio of stepped shaft deflection to unstepped shaft deflection
z_s seal face axial separation, ft.
δ displacement, in.
ϵ eccentricity of center of mass from axis of rotation, ft.
ω_c natural frequency of vibration, c.p.s.
μ fluid viscosity, lb./ft. min.
μ_f coefficient of friction, dimensionless
ρ fluid density, lb./ft.^{3}

References

(A1) Alloy Casting Institute, "Corrosion Resistant Alloys," A.C.I. Data Sheets. Alloy Casting Institute, New York, 1954, 1957.

(B1) Bert, C. W., *Machine Design* 32, 128–133 (Nov. 24, 1960).

(B2) Binder, R. C., "Advanced Fluid Mechanics." Prentice-Hall, Englewood Cliffs, New Jersey, 1958.

(B3) Briggs, C. W., "Steel Castings Handbook." 3rd ed. Steel Founders' Soc. of America, Cleveland, 1960.

(D1) Den Hartog, J.P., "Mechanical Vibrations." McGraw-Hill, New York, 1956.

(D2) Dolan, T. J., *in* "Mechanical Properties in Metals Engineering Design" (O. J. Horger, ed.), 2nd ed. McGraw-Hill, New York, 1965.

(D3) Dunkerly, S., *Phil. Trans. Roy. Soc. (London)*, **185A**, Pt. 1, 279–360 (1895).

(G1) Greenhill, A. G., *Proc. Inst. Mech. Engrs. (London)*, pp. 182–225 (1883).

(H1) Heckenkamp, F. W., *Design News* **18**, 134–137 (Dec. 11, 1963).

(H2) Hesse, H. C., *Prod. Eng.* **21**, 90–93 (Dec. 1950).

(H3) Hesse, H. C., *Prod. Eng.*, **23**, 175–179 (March 1952).

(K1) Kuchler, T. C., Clearance seals. *Machine Design—The Seals Book* **36**, 16–19 (June 11, 1964).

(L1) Lyman, T., ed., "Metals Handbook." Vol. I, 8th ed., pp. 122–145. Am. Soc. Metals, Novelty, Ohio, 1961.

(M1) Mayer, E., *Machine Design* **32**, 106 (March 3, 1960).

(P1) Parker, N. H., *Chemical Eng.* **71**, 165 (June, 1964).

(P2) Phillips, A. L., ed., "Welding Handbook." 5th ed. Am. Welding Society, New York, 1962–1964.

(R1) Roark, R. J., "Formulas for Stress and Strain." McGraw-Hill, New York, 1954.

(S1) Schmidt, G. W., "Packing and Mechanical Seals." Crane Packing Co., Morton Grove, Illinois, 1959

(T1) Thearle, E. L., *Trans. ASME, APM* 56–19, **56**, 745–753 (1934).

(T2) Tankus, H., *Machine Design* **32**, 27 (March, 1967).

Author Index

Numbers in parentheses are reference numbers. Numbers in italic show the page on which the complete reference is listed.

Subject Index

A

Absorption with chemical reaction, 152
Absorption of oxygen in sodium sulphite solutions, 61
Agglomeration, 238
Agitator drive, 313–315
Agitator shafts, mechanical design, 297–305
Anchor agitators for solids suspension, 229, 234
Asphalt mixing, 209
Axial dispersion in piston flow, 136, 138
Axial mixing, *see also* Back-mixing
Axial mixing in reactors, 136, 138, 159

B

Back-mixing, *see also* Axial mixing
Back-mixing in gas phase of aerated agitated vessels, 98–99
Baffles, angled, 244
  circulation pattern from, 230
  design, 321
  forces on, 321
  in solids suspension, 256
  vortex elimination, 252
Banbury continuous mixers, 197, 199–200
  capacities, 200
  power, 216, 218
  scale-up, 216, 218
Banbury mixers, 197–199
  capacity data, 198
  sizes, 198
  viscosity ratio effect, 193
Batch polymerization, 141, 142
Batch stirred-tank reactors, 140–141
Beken blade, 211–212
  cleaning, 212
Bench scale tests, *see* Pilot tests
Bingham plastic materials, 239
Boundary layer theory, mass transfer rates from, 56
Bubble-liquid mass transfer coefficients, 64–67

C

Carbon dioxide dissolution in water, 71
Catalyst suspension, 228
Cavitation at impellers, 250
Cellulose acetate suspension, 233, 254
Chemical reaction and mass transfer, 59–61, 91–92, 157
Circulation pattern, with anchor agitator, 234
  with cone impeller, 233
  in flocculator, 234
  with propeller, 233
  slurry viscosity effect on, 241
  slurry density effect on, 241
  with turbines, 230, 232
  universally directed, 230
  upwardly directed, 230
Clay mixing, 209, 238
Clearance seals, 319–320
  power consumption, 320
Coalescence, bubbles in liquids, 40–44
  of liquid-liquid dispersions, 29, 30, 32, 38–45
  effect on reaction yield, 159
  surface solute effect on, 45
Complete suspension, definition, 227
Compounding pigment concentrates, *see* Pigment concentrate compounding
Cone impeller, 229, 232, 233
Conical mixers, 278
Continuous-flow stirred-tank reactors, 116–132
Continuous mixers, countercurrent slurry flow, 251
  roll mills, 200–203
Continuous mixers, Banbury type, *see* Banbury continuous mixers
Couplings, 307–310
  flanged, 307, 308
  sleeve, 307, 308
Critical shaft speed, 301–306
  bearing spacing factor, 303